ASTÉRISC

RELATIVE p-ADIC HODGE THEORY: FOUNDATIONS

Kiran S. Kedlaya
Ruochuan Liu

Société Mathématique de France 2015
Publié avec le concours du Centre National de la Recherche Scientifique

K. S. Kedlaya

Department of Mathematics, Room 7202, University of California, San Diego, 9500 Gilman Drive, La Jolla, CA 92093-0112, USA.

E-mail : kedlaya@ucsd.edu

R. Liu

Department of Mathematics, University of Michigan, 1844 East Hall, 530 Church Street, Ann Arbor, MI 48109-1043, USA.

E-mail : ruochuan@umich.edu

RELATIVE p-ADIC HODGE THEORY: FOUNDATIONS

Kiran S. Kedlaya, Ruochuan Liu

Abstract. — We describe a new approach to relative p-adic Hodge theory based on systematic use of Witt vector constructions and nonarchimedean analytic geometry in the style of both Berkovich and Huber. We give a thorough development of φ-modules over a relative Robba ring associated to a perfect Banach ring of characteristic p, including the relationship between these objects and étale $\mathbb{Z}_p$-local systems and $\mathbb{Q}_p$-local systems on the algebraic and analytic spaces associated to the base ring, and the relationship between (pro-)étale cohomology and φ-cohomology. We also make a critical link to mixed characteristic by exhibiting an equivalence of tensor categories between the finite étale algebras over an arbitrary perfect Banach algebra over a nontrivially normed complete field of characteristic p and the finite étale algebras over a corresponding Banach $\mathbb{Q}_p$-algebra. This recovers the homeomorphism between the absolute Galois groups of $\mathbb{F}_p((\pi))$ and $\mathbb{Q}_p(\mu_{p^\infty})$ given by the field of norms construction of Fontaine and Wintenberger, as well as generalizations considered by Andreatta, Brinon, Faltings, Gabber, Ramero, Scholl, and most recently Scholze. Using Huber's formalism of adic spaces and Scholze's formalism of perfectoid spaces, we globalize the constructions to give several descriptions of the étale local systems on analytic spaces over p-adic fields. One of these descriptions uses a relative version of the Fargues-Fontaine curve.

***Résumé* (Théorie de Hodge p-adique dans le cas relatif : fondations)**

Nous décrivons une approche nouvelle de la théorie de Hodge p-adique dans le cas relatif, fondée sur l'utilisation systématique des constructions des vecteurs de Witt et de la géométrie analytique non archimédienne à la manière de Berkovich et de Huber. Nous donnons un traitement approfondi des φ-modules sur un anneau de Robba relatif associé à un anneau de Banach parfait de caractéristique p, qui inclut le lien entre ces objets et les $\mathbb{Z}_p$-systèmes locaux étales et les $\mathbb{Q}_p$-systèmes locaux sur les espaces algébriques et analytiques associés à l'anneau de base, ainsi que le lien entre la cohomologie (pro-)étale et la φ-cohomologie. Nous établissons aussi un lien critique avec la caractéristique mixte en exhibant une équivalence de catégories tensorielles entre les algèbres étales finies sur une algèbre de Banach parfaite arbitraire sur un corps non trivialement normé complet de caractéristique p, et les algèbres étales finies sur une $\mathbb{Q}_p$-algèbre de Banach correspondante. Ceci redonne l'homéomorphisme entre les groupes de Galois absolus de $\mathbb{F}_p((\pi))$ et de $\mathbb{Q}_p(\mu_{p^\infty})$ donné par le corps des normes de Fontaine et Wintenberger, ainsi que des généralisations considérées par Andreatta, Brinon, Faltings, Gabber, Ramero, Scholl, et plus récemment Scholze. En utilisant le formalisme des espaces adiques de Huber et le formalisme des espaces perfectoïdes de Scholze, nous globalisons ces constructions afin de donner plusieurs descriptions des systèmes locaux étales sur des espaces analytiques sur des corps p-adiques. L'une de ces descriptions utilise une version relative de la courbe de Fargues-Fontaine.

CONTENTS

CHAPTER 0

INTRODUCTION

After its formalization by Deligne [**35**], the subject of *Hodge theory* may be viewed as the study of the interrelationship among different cohomology theories associated to algebraic varieties over $\mathbb{C}$, most notably singular (Betti) cohomology and the cohomology of differential forms (de Rham cohomology). From work of Fontaine and others, there emerged a parallel subject of *p-adic Hodge theory* concerning the interrelationship among different cohomology theories associated to algebraic varieties over a finite extension K of $\mathbb{Q}_p$, most notably étale cohomology with coefficients in $\mathbb{Q}_p$ and algebraic de Rham cohomology.

In ordinary Hodge theory, the relationship between Betti and de Rham cohomologies is forged using the *Riemann-Hilbert correspondence*, which relates topological data (local systems) to analytic data (integrable connections). In p-adic Hodge theory, one needs a similar correspondence relating de Rham data to étale $\mathbb{Q}_p$-local systems, which arise from the étale cohomology functor on schemes of finite type over K. However, in this case the local systems turn out to be far more plentiful, so it is helpful to first build a correspondence relating them to some sort of intermediate algebraic objects. This is achieved by the theory of (φ, Γ)*-modules*, which gives some Morita-type dualities relating étale $\mathbb{Q}_p$-local systems over K (*i.e.*, continuous representations of the absolute Galois group G_K on finite-dimensional $\mathbb{Q}_p$-vector spaces) with modules over certain mildly noncommutative topological algebras. The latter appear as topological monoid algebras over certain commutative *period rings* for certain continuous operators (the eponymous φ and Γ).

One of the key features of Hodge theory is that it provides information not just about individual varieties, but also about families of varieties through the mechanism of *variations of Hodge structures*. Only recently has much progress been made in developing any analogous constructions in p-adic Hodge theory; part of the difficulty is that there are two very different directions in which relative p-adic Hodge theory can be developed. In the remainder of this introduction, we first give a bit more background about (φ, Γ)-modules, and contrast the *arithmetic* and *geometric* forms of relative p-adic Hodge theory. We then describe the results of this paper in detail, indicate some points of contact with recent work of Scholze [**114, 115**], and describe some future goals.

0.1. Artin-Schreier theory and (φ, Γ)-modules

For K a perfect field of characteristic p, the discrete representations of the absolute Galois group G_K of K on finite dimensional $\mathbb{F}_p$-vector spaces form a category equivalent to the category of *φ-modules* over K, *i.e.*, finite-dimensional K-vector spaces equipped with isomorphisms with their φ-pullbacks. This amounts to a nonabelian generalization of the Artin-Schreier description of $(\mathbb{Z}/p\mathbb{Z})$-extensions of fields of characteristic p [**37**, Exposé XXII, Proposition 1.1].

A related result is that the continuous representations of G_K on finite free $\mathbb{Z}_p$-modules form a category equivalent to the category of finite free $W(K)$-modules equipped with isomorphisms with their φ-pullbacks. Here $W(K)$ denotes the ring of Witt vectors over K, and φ denotes the unique lift to $W(K)$ of the absolute Frobenius on K. One can further globalize this result to arbitrary smooth schemes over K, in which the corresponding category becomes a category of *unit-root F-crystals*; see [**28**, Theorem 2.2] or [**81**, Proposition 4.1.1].

Fontaine's theory of (φ, Γ)-modules [**47**] provides a way to extend such results to mixed-characteristic local fields. The observation underpinning the theory is that a sufficiently wildly ramified extension of a mixed-characteristic local field behaves Galois-theoretically just like a local field of positive characteristic. For example, for K_0 a finite unramified extension of $\mathbb{Q}_p$ (or the completion of an infinite algebraic unramified extension of $\mathbb{Q}_p$) with residue field k_0, the fields $K_0(\mu_{p^\infty})$ and $k_0((T))$ have homeomorphic Galois groups. One can describe representations of the absolute Galois group of a local field by restricting to a suitably deeply ramified extension, applying an Artin-Schreier construction, then adding appropriate descent data to get back to the original group. This assertion is formalized in the theory of *fields of norms* introduced by Fontaine and Wintenberger [**49, 127**]; some of the analysis depends on Sen's calculation of ramification numbers in p-adic Lie extensions [**117**]. See [**22**, Part 4] for a detailed but readable exposition.

0.2. Arithmetic vs. geometric

As noted earlier, there are two different directions in which one can develop relative forms of p-adic Hodge theory. We distinguish these as *arithmetic* and *geometric*.

In arithmetic relative p-adic Hodge theory, one still treats continuous representations of the absolute Galois group of a finite extension of $\mathbb{Q}_p$. However, instead of taking representations simply on vector spaces, one allows finite locally free modules over affinoid algebras over $\mathbb{Q}_p$. Interest in the arithmetic theory arose originally from the consideration of p-adic analytic families of automorphic forms and their associated families of Galois representations; such families include Hida's p-adic interpolation of ordinary cusp forms [**74**], the eigencurve of Coleman–Mazur [**25**], and further generalizations. Additional interest has come from the prospect of a p-adic local Langlands correspondence which would be compatible with formation of analytic families on both the Galois and automorphic sides. For the group $\mathrm{GL}_2(\mathbb{Q}_p)$, such a correspondence has recently emerged from the work of Breuil, Colmez, Emerton, Paškūnas *et al.* (see for instance [**27**]) and has led to important advances concerning modularity of Galois representations, in the direction of the Fontaine-Mazur conjecture [**92**], [**41**].

In the arithmetic setting, there is a functor from Galois representations to families of (φ, Γ)-modules, constructed by Berger and Colmez [**13**]. However, this functor is not an equivalence of categories; rather, it can only be inverted locally [**34**], [**90**]. It seems that in this setting, one is forced to study families of Galois representations in the context of the larger category of families of (φ, Γ)-modules. For instance, one sees this distinction in the relative study of Colmez's *trianguline* Galois representations [**26**], as in the work of Bellaïche [**9**] and Pottharst [**105**].

By contrast, in geometric relative p-adic Hodge theory, one continues to consider continuous representations acting on finite-dimensional $\mathbb{Q}_p$-vector spaces. However, instead of the absolute Galois group of a finite extension of $\mathbb{Q}_p$, one allows étale fundamental groups of affinoid spaces over finite extensions of $\mathbb{Q}_p$. The possibility of developing an analogue of (φ, Γ)-module theory in this setting emerged from the work of Faltings, particularly his *almost purity theorem* [**42, 43**], and prior to this paper had been carried out most thoroughly by Andreatta and Brinon [**1, 2**]. A similar construction was described by Scholl [**112**]. (See also the exposition by Olsson [**103**].)

0.3. Analytic spaces associated to Banach algebras

We now turn to the topics addressed by this particular paper. The first substantial chunk of the paper concerns some geometric spaces associated to nonarchimedean commutative Banach rings, including the *Gel'fand spectrum* in the sense of Berkovich and the *adic spectrum* considered by Huber. These constructions are somewhat more exotic than the spaces of maximal ideals occurring in Tate's theory of rigid analytic spaces, but Tate's construction starts to behave poorly when one considers Banach rings other than affinoid algebras, and breaks down completely if one considers non-noetherian rings. Since such rings play a crucial role in our work, we are forced to use the alternate constructions.

One important feature of our work is the dialogue between the Gel'fand and adic spectra. The latter is topos-theoretically complete, whereas the Gel'fand spectrum is the maximal Hausdorff quotient of the adic spectrum. While the Gel'fand spectrum is a natural dwelling place for most of our local arguments, we transfer back to adic spectra in order to globalize.

When we globalize (by glueing together adic spectra), we do not end up with the most general sort of spaces considered by Huber: we only encounter spaces which are *analytic*, meaning that the residue fields associated to all points carry nontrivial valuations. By contrast, Huber's theory also includes spaces more closely related to ordinary schemes and formal schemes. For another, we are mostly interested only in spaces which have a certain finiteness property (that of being *taut*) which roughly means they are approximated well by their maximal Hausdorff quotients. (For instance, Berkovich's *strictly analytic spaces* can be promoted to taut adic spaces in such a way that the original spaces occur as the maximal Hausdorff quotients.) However, one cannot hope to banish nontaut spaces entirely from p-adic Hodge theory: for instance, they appear in the work of Hellmann [**70, 69**] on the moduli spaces for Breuil-Kisin modules described by Pappas and Rapoport [**104**]. The study of these spaces seems to include features of both arithmetic and geometric relative p-adic Hodge theory, which appears to render both Tate and Berkovich spaces insufficient.

0.4. Perfectoid fields and algebras

As noted earlier, one of the main techniques of p-adic Hodge theory is the relationship between the absolute Galois groups of certain fields of mixed and positive characteristic, such as $\mathbb{Q}_p(\mu_{p^\infty})$ and $\mathbb{F}_p((T))$. For relative p-adic Hodge theory, it is necessary to extend this correspondence somewhat further. However, instead of an approach dependent on ramification theory, we use a construction based on analysis of Witt vectors.

Suppose first that L is a perfect field of characteristic p complete for a multiplicative norm, with valuation ring $\mathfrak{o}_L$. Let $W(\mathfrak{o}_L)$ be the ring of p-typical Witt vectors over $\mathfrak{o}_L$. One can generate certain fields of characteristic 0 by quotienting $W(\mathfrak{o}_L)$ by certain principal ideals and then inverting p. For instance, for L the completed perfect closure of $\mathbb{F}_p((\overline{\pi}))$ and

$$z = \sum_{i=0}^{p-1} [\overline{\pi} + 1]^{i/p} \in W(\mathfrak{o}_L),$$

we may identify $W(\mathfrak{o}_L)/(z)$ with the ring of integers of the completion of $\mathbb{Q}_p(\mu_{p^\infty})$. The relevant condition (that of being *primitive of degree* 1 in the sense of Fargues and Fontaine [**45**]) is a Witt vector analogue of the property of an element of $\mathbb{Z}_p[\![T]\!]$ being associated to a monic linear polynomial whose constant term is not invertible in $\mathbb{Z}_p$ (which allows use of the division algorithm to identify the quotient by this element with $\mathbb{Z}_p$). See Definition 3.3.4.

Using this construction, we obtain a correspondence between certain complete fields of mixed characteristic (which we call *perfectoid fields*, following [**114**]) and perfect fields of characteristic p together with appropriate principal ideals in the ring of Witt vectors over the valuation ring (see Theorem 3.5.3). It is not immediate from the construction that the former category is closed under formation of finite extensions, but this turns out to be true and not too difficult to check (see Theorem 3.5.6). In particular, we recover the field of norms correspondence.

To extend this correspondence to more general Banach algebras, we exploit a relationship developed in [**87**] between the Berkovich space of a perfect uniform Banach ring R of characteristic p and the Berkovich space of the ring $W(R)$. (For instance, any subspace of $\mathcal{M}(R)$ has the same homotopy type as its inverse image under μ [**87**, Corollary 7.9].) This leads to a correspondence between certain Banach algebras[(1)] in characteristic 0 (which we call *perfectoid algebras*, again following [**114**]) and perfect Banach algebras in characteristic p, which is compatible with formation of both rational subspaces and finite étale covers (see Theorem 3.6.5 and Theorem 3.6.21). We also obtain a result in the style of Faltings's *almost purity theorem* [**42, 43**] (see also [**52, 53**]), which underlies the aforementioned generalization of (φ, Γ)-modules introduced by Andreatta and Brinon [**1, 2**].

0.5. Robba rings and slope theory

Another important technical device in usual p-adic Hodge theory is the classification of Frobenius-semilinear transformation on modules over certain power series by *slopes*, in rough analogy with the classification of vector bundles on curves. This originated in work of the first author [**82**]; we extend this work here to the relative setting. (The relevance of such results to p-adic Hodge theory largely factors through the work of Berger [**10, 12**], to which we will return in a later paper.)

In [**82**], one starts with the *Robba ring* of germs of analytic functions on open annuli of outer radius 1 over a p-adic field, and then passes to a certain "algebraic closure" thereof. The latter can be constructed from the ring of Witt vectors over the completed algebraic closure of a power series field. One is thus led naturally to consider similar constructions starting from the ring of Witt vectors over a general analytic field; the analogues of the results of [**82**] were worked out by the first author in [**83**].

Using the previously described work largely as a black box, we are able to introduce analogues of Robba rings starting from the ring of Witt vectors of a perfect uniform $\mathbb{F}_p$-algebra (and obtain some weak analogues of the theorems of Tate and Kiehl), and to study slopes of Frobenius modules thereof. We obtain semicontinuity of the slope polygon as a function on the spectrum of the base ring (Theorem 7.4.5) as well as a slope filtration theorem when this polygon is constant (Theorem 7.4.9). (Similar results in the arithmetic relative setting have recently been obtained by the second author [**97**].)

We also obtain a description of Frobenius modules over relative Robba rings in the style of Fargues and Fontaine, using vector bundles over a certain scheme (Theorem 6.3.12). When the base ring is an analytic field, the scheme in question is connected, regular, separated, and noetherian of dimension 1; it might thus be considered to be a *complete absolute curve*. (The adjective *absolute* means that the curve cannot be seen as having relative dimension 1 over a point; it is a scheme over $\mathbb{Q}_p$, but

1. The definition of perfectoid algebras can be made at several levels of generality; we work in more generality than in [**114**], where perfectoid algebras must be defined over a perfectoid field, but less than in [**48**] or [**53**], where perfectoid algebras need not contain $1/p$.

not of finite type.) However, for more general base rings, the resulting scheme is not even noetherian. In any case, one obtains a p-adic picture with strong resemblance to the correspondence between stable vector bundles on compact Riemann surfaces and irreducible unitary fundamental group representations, as constructed by Narasimhan and Seshadri [**102**].

0.6. φ-modules and local systems

By combining the preceding results, we obtain a link between étale local systems and φ-modules (and cohomology thereof), in what amounts to a broad nonabelian generalization of Artin-Schreier theory as well as a generalization of the field of norms correspondence. This link is most naturally described on the category of *perfectoid spaces*, obtained by glueing the adic spectra of perfectoid Banach algebras.

To describe étale local systems and their cohomology on more general spaces, including rigid and Berkovich analytic spaces, one must combine the previous theory with a descent construction from some local perfectoid covers of the space. For foundational purposes, an especially convenient mechanism for this is provided by the *pro-étale topology* proposed by Scholze [**115**]. In this framework, the rings of p-adic periods (such as the extended Robba ring) become sheaves for the pro-étale topology, and the analogue of a (φ, Γ)-module is a sheaf of φ-modules over a ring of period sheaves. Note that there is no explicit analogue of Γ; its role is instead played by the sheaf axiom for the pro-étale topology.

In this language, we obtain φ-module-theoretic descriptions of étale $\mathbb{Z}_p$-local systems (Theorem 9.3.7) and their pro-étale cohomology (Theorem 9.4.2), as well as $\mathbb{Q}_p$-local systems (Theorem 9.3.13) and their pro-étale cohomology (Theorem 9.4.5). We also see that when the base space is reduced to a point, our categories of φ-modules are equivalent to the corresponding categories of (φ, Γ)-modules arising in classical p-adic Hodge theory (§9.5).

0.7. Contact with the work of Scholze

After preparing the initial version of this paper, we discovered that some closely related work had been carried out by Peter Scholze, which ultimately has appeared in the papers [**114, 115**]. The ensuing rapid dissemination of Scholze's work has had the benefit of providing an additional entry point into the circle of ideas underlying our work. However, it also necessitates a discussion of the extent to which the two bodies of work interact and overlap, which we now provide (amplifying the brief discussion appearing in the introduction to [**114**]).

We begin with some historical remarks. The genesis of our work lies in the first author's paper [**87**], the first version of which appeared on arXiv in April 2010. There one first finds the homeomorphism of topological spaces which now underlies the perfectoid correspondence. This homeomorphism is again described in the first author's 2010 ICM lecture [**86**], together with some preliminary discussion of how it could be used to construct tautological local systems on Rapoport-Zink period spaces. In late 2010, we began to prepare the present paper so as to provide foundations for carrying out the program advanced in [**86**]. In early 2011, we learned that Scholze had used similar ideas for a totally different (and rather spectacular) purpose: to resolve some new cases of the weight-monodromy conjecture in ℓ-adic étale cohomology (which subsequently appeared as [**114**]). Draft versions of papers were exchanged in both directions, which in Scholze's case included his work on the de Rham-étale comparison isomorphism (which subsequently appeared as [**115**]).

Based on this exchange, we elected to adopt some key formal ideas from Scholze's work but to retain independent derivations of all of our results, even in cases of overlap. This decision was dictated in part by some minor but nonnegligible foundational differences between the two works. We now describe some points of agreement and disagreement with [**114, 115**].

- As noted previously, we make heavy use of the Gel'fand spectrum associated to a Banach ring, translating into the language of adic spectra for glueing constructions; by contrast, Scholze works exclusively with adic spectra.
- The term *perfectoid* is adopted from Scholze; we had not initially assigned a word to this concept.
- The *perfectoid correspondence* for fields (Theorem 3.5.3) is described in [**114**] without reference to Witt vectors, as is the compatibility with finite extensions (Theorem 3.5.6); this necessitates the use of some almost ring theory in the form of [**52**, Theorem 5.3.27]. Besides being necessary for the construction of period sheaves, we find the arguments using Witt vectors somewhat more transparent; see [**89**] for a demonstration of this point in the form of an exposition of the classical theory of (φ, Γ)-modules.
- Scholze considers only perfectoid algebras over perfectoid fields, whereas our definition of a perfectoid algebra (Definition 3.6.1) does not include this restriction. This is partly because our description of the perfectoid correspondence for algebras (Theorem 3.6.5) includes enough extra data (in terms of Witt vectors) to enable lifting from characteristic p back to characteristic 0 without reference to an underlying field. This more general approach has also been adopted by Gabber and Ramero [**53**].
- The compatibility of the perfectoid correspondence with finite étale covers (Theorem 3.6.21) is established by reduction to the field case, much as in [**114**], but again the use of Witt vectors takes the place of almost ring theory. As a result, instead of proving almost purity in the course of proving Theorem 3.6.21, we deduce it as a corollary (Theorem 5.5.9).
- Our study of perfectoid algebras includes some results with no analogues in [**114**], including compatibility with strict morphisms (Proposition 3.6.9) and with morphisms of dense image (Theorem 3.6.17). The latter implies that the uniform completed tensor product of two perfectoid algebras (over a not necessarily perfectoid base) is again perfectoid (Corollary 3.6.18).

- Our definition of a *perfectoid space* is essentially that of Scholze, except that we do not insist on working over a perfectoid base field. Our derivations of the basic properties are as in [**114**] except that we make internal references in place of the equivalent cross-references to [**114**].
- For passing from φ-modules to (φ, Γ)-modules, we adopt Scholze's *pro-étale topology* essentially unchanged from [**115**]. Note that this is not simply the pro-category associated to the étale topology, but requires an extra Mittag-Leffler condition; by contrast, in the category of schemes, Bhatt and Scholze [**17**] have shown that the pro-category associated to the étale topology can be used with similar effect.

0.8. Further goals

To conclude, we indicate some questions we intend to address in subsequent work.

- The usual Robba ring in p-adic Hodge theory is *imperfect*, that is, its Frobenius endomorphism is not surjective. By contrast, our analogue of the Robba ring corresponding to a perfectoid algebra has bijective Frobenius. For certain purposes (*e.g.*, approximations of the p-adic Langlands correspondence), it is desirable to descend the theory of (φ, Γ)-modules from the perfect Robba ring to an imperfect version when possible. This occurs in classical p-adic Hodge theory *via* the theorem of Cherbonnier and Colmez [**24**]; some cases in relative p-adic Hodge theory are covered by the generalization of Cherbonnier-Colmez given by Andreatta and Brinon [**2**]. However, it should be possible to embed these constructions into a more general framework; some first steps in this direction are taken in [**88**].
- Another goal is to construct certain "tautological" local systems on period spaces of p-adic Hodge structures (filtered (φ, N)-modules). Such period spaces arise, for instance, in the work of Rapoport and Zink on period mappings on deformation spaces of p-divisible groups [**107**]. The construction we have in mind is outlined in [**86**]; it is likely to be greatly assisted by the work of Scholze and Weinstein on the (perfectoid) moduli space of p-divisible groups [**116**]. The non-minuscule case is particularly intriguing, as there the natural parameter space is no longer a period domain (as suggested somewhat cavalierly in [**86**]); rather, it is a presently hypothetical space analogous to Hartl's moduli space of Hodge-Pink structures in the equal characteristic case [**65**].
- Yet another goal is the integration of Scholze's approach to the de Rham-étale comparison isomorphism into the framework of relative (φ, Γ)-modules, and its extension to the crystalline and semistable cases. For instance, if $X \to Y$ is a proper morphism of adic spaces, there should be higher direct image functors from de Rham relative (φ, Γ)-modules on X to the corresponding objects on Y. A closely related issue is to study the analogue of Fontaine's de Rham, crystalline, and semistable conditions and in particular to generalize the "de Rham

implies potentially semistable" theorem (originally established by Berger using the André-Kedlaya-Mebkhout monodromy theorem for p-adic differential equations).

Acknowledgments

Thanks to Fabrizio Andreatta, Max Bender, Vladimir Berkovich, Brian Conrad, Chris Davis, Laurent Fargues, Jean-Marc Fontaine, Ofer Gabber, Arthur Ogus, Peter Scholze, Michael Temkin, and Liang Xiao for helpful discussions. Thanks also to Scholze for providing early versions of his papers [**114, 115**] and to the anonymous referees for highly instructive feedback. Kedlaya was supported by NSF CAREER grant DMS-0545904, NSF grant DMS-1101343, DARPA grant HR0011-09-1-0048, MIT (NEC Fund, Cecil and Ida Green professorship), and UC San Diego (Stefan E. Warschawski professorship). Liu was partially supported by IAS under NSF grant DMS-0635607 and the Recruitment Program of Global Experts of China. Additionally, both authors were supported by NSF grant DMS-0932078 while in residence at MSRI during fall 2014.

CHAPTER 1

ALGEBRO-GEOMETRIC PRELIMINARIES

Before proceeding to analytic geometry, we start with some background facts from algebraic geometry.

Hypothesis 1.0.1. — Throughout this paper, fix a prime number p.

Convention 1.0.2. — When we refer to a *tensor category*, we will always assume it is equipped not just with the usual monoidal category structure, but also with a *rank function* into some abelian group. Equivalences of tensor categories (or for short *tensor equivalences*) will be assumed to respect rank.

1.1. Finite, flat, and projective modules

Convention 1.1.1. — Throughout this paper, all rings are assumed to be commutative and unital unless otherwise indicated.

Definition 1.1.2. — Let M be a module over a ring R. We say that M is *pointwise free* if $M \otimes_R R_\mathfrak{p}$ is a free module over $R_\mathfrak{p}$ for each maximal ideal $\mathfrak{p}$ of R (and hence for each prime ideal). The term *locally free* is sometimes used for this, but it is better to make this term match its usual meaning in sheaf theory by saying that M is *locally free* if there exist $f_1, \dots, f_n \in R$ generating the unit ideal such that $M \otimes_R R[f_i^{-1}]$ is a free module over $R[f_i^{-1}]$ for $i = 1, \dots, n$.

We say that M is *projective* if it is a direct summand of a free module. The following conditions are equivalent [**20**, §II.5.2, Théorème 1].

(a) M is finitely generated and projective.

(b) M is a direct summand of a finite free module.

(c) M is finitely presented and pointwise free.

(d) M is finitely generated and pointwise free of locally constant rank. (The *rank* of M at $\mathfrak{p} \in \mathrm{Spec}(R)$ is defined as $\dim_{R_\mathfrak{p}/\mathfrak{p}R_\mathfrak{p}}(M_\mathfrak{p}/\mathfrak{p}M_\mathfrak{p})$.)

(e) M is finitely generated and locally free.

For R reduced, it is enough to check that M is finitely generated of locally constant rank; see [**39**, Exercise 20.13]. For an analogous argument for Banach rings, see Proposition 2.8.4.

For M finitely presented, we may define the *Fitting ideals* $\mathrm{Fitt}_i(M)$ as in [**39**, §20.2]; these are finitely generated ideals of R satisfying $\mathrm{Fitt}_0(M) \subseteq \mathrm{Fitt}_1(M) \subseteq \cdots$ and $\mathrm{Fitt}_i(M) = R$ for i sufficiently large. The construction commutes with base change: for any ring homomorphism $R \to S$, we have $\mathrm{Fitt}_i(M \otimes_R S) = \mathrm{Fitt}_i(M)S$ [**39**, Corollary 20.5]. The module M is finite projective of constant rank n if and only if $\mathrm{Fitt}_i(M) = 0$ for $i = 0, \dots, n-1$ and $\mathrm{Fitt}_n(M) = R$ [**39**, Proposition 20.8].

Definition 1.1.3. — Let M be a module over a ring R. We say M is *faithfully flat* if M is flat and $M \otimes_R N \neq 0$ for every nonzero R-module N. Since tensor products commute with direct limits, M is faithfully flat if and only if it satisfies the following conditions.

(a) For any injective homomorphism $N \to P$ of finite R-modules, $M \otimes_R N \to M \otimes_R P$ is injective.

(b) For any nonzero finite R-module N, $M \otimes_R N$ is nonzero.

For other characterizations, see [**20**, §I.3.1, Proposition 1].

Lemma 1.1.4. — *A flat ring homomorphism $R \to S$ is faithfully flat if and only if every maximal ideal of R is the contraction of a maximal ideal of S.*

Proof. — See [**20**, §I.3.5, Proposition 9]. □

Remark 1.1.5. — Recall that for any ring R, quasicoherent sheaves on $\mathrm{Spec}(R)$ correspond to R-modules *via* the global sections functor. Under this correspondence, the property of a sheaf being finitely generated, finitely presented, or finite projective implies the corresponding property for its module of global sections [**122**, Tags 01PB, 01PC, 05JM].

1.2. Comparing étale algebras

We will expend a great deal of effort comparing finite étale algebras over different rings. A key case is given by base change from a ring to a quotient ring, in which the henselian property plays a key role. (A rather good explanation of this material can be obtained from [**52**] by specializing from almost ring theory to ordinary ring theory.)

Definition 1.2.1. — As in [**60**, Définition 17.3.1], we say a morphism of schemes is *étale* if it is locally of finite presentation and formally étale. (The latter condition is essentially a unique infinitesimal lifting property; see [**60**, Définition 17.1.1].) A morphism of rings is étale if the corresponding morphism of affine schemes is étale.

For any ring R, let $\mathbf{FÉt}(R)$ denote the tensor category of finite étale algebras over the ring R, with morphisms being arbitrary morphisms of R-algebras. Such morphisms are themselves finite and étale (*e.g.*, by [**60**, Proposition 17.3.4]). Any $S \in \mathbf{F\acute{E}t}(R)$ is finite as an R-module and finitely presented as an R-algebra, and hence finitely presented as an R-module by [**59**, Proposition 1.4.7]. Also, S is flat over R by [**60**, Théorème 17.6.1]. By the criteria described in Definition 1.1.2, S is finite projective as an R-module. Conversely, an R-algebra S which is finite projective as an R-module is finite étale if and only if the R-module homomorphism $S \to \operatorname{Hom}_R(S,R)$ taking x to $y \mapsto \operatorname{Trace}_{S/R}(xy)$ is an isomorphism. (Namely, since étaleness is an open condition [**60**, Remarques 17.3.2 (iii)], this reduces to the case where R is a field, which is straightforward.)

For short, we will describe a finite étale R-algebra which is faithfully flat over R (or equivalently of positive rank everywhere over R, by Lemma 1.1.4 and the going-up theorem) as a *faithfully finite étale* R-algebra.

Definition 1.2.2. — Let R be a ring and let U be an element of $\mathbf{F\acute{E}t}(R)$. Then there exists an idempotent element $e_{U/R} \in U \otimes_R U$ mapping to 1 *via* the multiplication map $\mu : U \otimes_R U \to U$ and killing the kernel of μ; see for instance [**52**, Proposition 3.1.4]. For V another R-algebra and $e \in U \otimes_R V$ an idempotent element, let $\Gamma(e) : V \to e(U \otimes_R V)$ be the morphism of R-algebras sending $x \in V$ to $e(1 \otimes x)$. In case $\Gamma(e)$ is an isomorphism, we obtain a morphism $\psi_e : U \to V$ of R-algebras by applying the natural map $\Delta(e) : U \to e(U \otimes_R V)$ sending x to $e(x \otimes 1)$ followed by $\Gamma(e)^{-1}$. Conversely, for $\psi : U \to V$ a morphism of R-algebras, the idempotent $e_\psi = (1 \otimes \psi)(e_{U/R}) \in U \otimes_R V$ has the property that $\Gamma(e_\psi)$ is an isomorphism (see [**52**, Proposition 5.2.19]).

Lemma 1.2.3. — *For R a ring, $U \in \mathbf{F\acute{E}t}(R)$, and V an R-algebra, the function $\psi \mapsto e_\psi$ defines a bijection from the set of R-algebra morphisms from U to V to the set of idempotent elements $e \in U \otimes_R V$ for which $\Gamma(e)$ is an isomorphism. The inverse map is $e \mapsto \psi_e$.*

Proof. — See [**52**, Lemma 5.2.20]. □

Lemma 1.2.4. — *Let R be a ring and let I be an ideal contained in the Jacobson radical of R.*

(a) *No two distinct idempotents of R are congruent modulo I.*

(b) *For any integral extension S of R, the ideal IS is contained in the Jacobson radical of S.*

Proof. — For (a), let $e, e' \in R$ are idempotents with $e - e' \in I$. Then $(e + e' - 1)^2 = 1 - (e - e')^2 \in 1 + I$, so $e + e' - 1$ is a unit in R.

For (b), note that the set of $y \in S$ which are roots of monic polynomials over R whose nonleading coefficients belong to I is an ideal. Consequently, if $x \in 1 + IS$, it is a root of a monic polynomial $P \in R[T]$ for which $P(T-1)$ has all nonleading coefficients in I. It follows that the constant coefficient P_0 of P belongs to $\pm 1 + I$ and so is a unit; since $P_0 = P_0 - P(x)$ is divisible by x (being the evaluation at x of the polynomial $P_0 - P$), x is also a unit. □

Proposition 1.2.5. — *The following statements hold.*

(a) *Let $R \to R'$ be a homomorphism of rings such that for any invertible module M over R, every element of M which generates $M \otimes_R R'$ also generates M. (For example, this holds if* $\mathrm{Pic}(R) \to \mathrm{Pic}(R')$ *is injective and every element of R which maps to a unit in R' is itself a unit.) Suppose that for each $S \in$* $\mathbf{F\acute{E}t}(R)$*, every idempotent element of $S \otimes_R R'$ is the image of a unique idempotent element of S. Then the base change functor* $\mathbf{F\acute{E}t}(R) \to \mathbf{F\acute{E}t}(R')$ *is rank-preserving and fully faithful.*

(b) *Let R be a ring, and let I be an ideal contained in the Jacobson radical of R. Suppose that for each $S \in$* $\mathbf{F\acute{E}t}(R)$*, every idempotent element of S/IS lifts to S. Then the base change functor* $\mathbf{F\acute{E}t}(R) \to \mathbf{F\acute{E}t}(R/I)$ *is rank-preserving and fully faithful.*

Note that in case (a), if $R \to R'$ is injective, then S injects into $S \otimes_R R'$ because S is locally free as an R-module (see Definition 1.2.1), so the last condition simply becomes that every idempotent element of $S \otimes_R R'$ belongs to S.

Proof. — The rank-preserving property is evident in both cases. To check full faithfulness in case (a), note that the hypothesis on the homomorphism $R \to R'$ implies that a map between finite projective R-modules is an isomorphism if and only if its base extension to R' is an isomorphism (because the isomorphism condition amounts to invertibility of the determinant). Consequently, for $U, V \in \mathbf{F\acute{E}t}(R)$ and $e \in U \otimes_R V$ an idempotent element, the map $\Gamma(e) : V \to e(U \otimes_R V)$ is an isomorphism if and only if its base extension to R' is an isomorphism. Lemma 1.2.3 then implies that every morphism $U \otimes_R R' \to V \otimes_R R'$ of R'-algebras descends to a morphism $U \to V$ of R-algebras, as desired.

To check full faithfulness in case (b), note that for $U, V \in \mathbf{F\acute{E}t}(R)$ and $\overline{e} \in (U/IU) \otimes_{R/IR} (V/IV)$ an idempotent element, by Lemma 1.2.4 there is at most one idempotent $e \in U \otimes_R V$ lifting $\overline{e}$. Moreover, because I is contained in the Jacobson radical of R, for any invertible module M over R, every element of M which generates M/I also generates M by Nakayama's lemma. We may thus apply (a) to deduce the claim. □

Definition 1.2.6. — A pair (R, I) consisting of a ring R and an ideal $I \subseteq R$ is said to be *henselian* if the following conditions hold.

(a) The ideal I is contained in the Jacobson radical of R.

(b) For any monic $f \in R[T]$, any factorization $\overline{f} = \overline{g}\overline{h}$ in $(R/I)[T]$ with $\overline{g}, \overline{h}$ monic and coprime lifts to a factorization $f = gh$ in $R[T]$.

For example, if R is I-adically complete, then (R, I) is henselian by the usual proof of Hensel's lemma. A local ring R with maximal ideal $\mathfrak{m}$ is *henselian* if the pair $(R, \mathfrak{m})$ is henselian.

Remark 1.2.7. — There are a number of equivalent formulations of the definition of a henselian pair. For instance, let (R, I) be a pair consisting of a ring R and an ideal I contained in the Jacobson radical of R. By [**54**, Theorem 5.11], (R, I) is henselian if and only if every monic polynomial $f = \sum_i f_i T^i \in R[T]$ with $f_0 \in I, f_1 \in R^\times$ has a root in I. (In other words, it suffices to check the lifting condition for $\overline{g} = T$.) See [**108**, Exposé XI, §2] for some other formulations.

Theorem 1.2.8. — *Let (R, I) be a henselian pair. Then the base change functor* $\mathbf{F\acute{E}t}(R) \to \mathbf{F\acute{E}t}(R/I)$ *is a tensor equivalence.*

Proof. — See [**93**, Satz 4.4.7], [**94**, Satz 4.5.1], or [**63**]. See also [**52**, Theorem 5.5.7] for a more general assertion in the context of almost ring theory. □

Remark 1.2.9. — Let $\{R_i\}_{i \in I}$ be a direct system in the category of rings. We may then define the direct 2-limit $\varinjlim_i \mathbf{F\acute{E}t}(R_i)$. For $R = \varinjlim_i R_i$, there is a natural functor $\varinjlim_i \mathbf{F\acute{E}t}(R_i) \to \mathbf{F\acute{E}t}(R)$ given by base extension to R. This functor is fully faithful by [**60**, Proposition 17.7.8 (ii)] (since affine schemes are quasicompact). To see that it is essentially surjective, start with $S \in \mathbf{F\acute{E}t}(R)$. Since S is finitely presented as an R-algebra, by [**59**, Lemme 1.8.4.2] it has the form $S_i \otimes_{R_i} R$ for some $i \in I$ and some finitely presented R_i-algebra S_i. By [**60**, Proposition 17.7.8 (ii)] again, there exists $j \geqslant i$ such that $S_j = S_i \otimes_{R_i} R_j$ is finite étale over R_j. We conclude that $\varinjlim_i \mathbf{F\acute{E}t}(R_i) \to \mathbf{F\acute{E}t}(R)$ is a tensor equivalence.

1.3. Descent formalism

We will make frequent use of faithfully flat descent for modules, as well as variations thereof (*e.g.*, for Banach rings). It is convenient to frame this sort of argument in standard abstract descent formalism, since this language can also be used to discuss glueing of modules (see Example 1.3.3). We set up in terms of cofibred categories rather than fibred categories, as appropriate for studying modules over rings rather than sheaves over schemes; in the latter context we will use sheaf-theoretic language instead.

Definition 1.3.1. — Let $F : \mathcal{F} \to \mathcal{C}$ be a covariant functor between categories. For X an object (resp. f a morphism) in $\mathcal{C}$, let $F^{-1}(X)$ (resp. $F^{-1}(f)$) denote the class of objects (resp. morphisms) in $\mathcal{C}$ carried to X (resp. f) *via* F.

For $f : X \to Y$ a morphism in $\mathcal{C}$ and $E \in F^{-1}(X)$, a *pushforward* of E along f is a morphism $\tilde{f} : E \to f_*E \in F^{-1}(f)$ such that any $g \in F^{-1}(f)$ with source E factors uniquely through $\tilde{f}$. (We sometimes call the target f_*E a pushforward of E as well, understanding that it comes equipped with a fixed morphism from E.) We say $\mathcal{F}$ is a *cofibred category* over $\mathcal{C}$, or that $F : \mathcal{F} \to \mathcal{C}$ defines a cofibred category, if pushforwards always exist and the composition of two pushforwards (when defined) is always a pushforward.

Definition 1.3.2. — Let $\mathcal{C}$ be a category in which pushouts exist. Let $F : \mathcal{F} \to \mathcal{C}$ be a functor defining a cofibred category. Let $f : X \to Y$ be a morphism in $\mathcal{C}$. Let $\pi_1, \pi_2 : Y \to Y \sqcup_X Y$ be the coprojection maps. Let $\pi_{12}, \pi_{13}, \pi_{23} : Y \sqcup_X Y \to Y \sqcup_X Y \sqcup_X Y$ be the coprojections such that π_{ij} carries the first and second factors of the source into the i-th and j-th factors in the triple coproduct (in that order). A *descent datum* in $\mathcal{F}$ along f consists of an object $M \in F^{-1}(Y)$ and an isomorphism $\iota : \pi_{1*}M \to \pi_{2*}M$ between some choices of pushforwards, satisfying the following *cocycle condition.* Let M_1, M_2, M_3 be some pushforwards of M along the three coprojections $Y \to Y \sqcup_X Y \sqcup_X Y$. Then ι induces a map $\iota_{ij} : M_i \to M_j$ *via* π_{ij}; the condition is that $\iota_{23} \circ \iota_{12} = \iota_{13}$.

For example, any object $N \in F^{-1}(X)$ induces a descent datum by taking M to be a pushforward of N along f and taking ι to be the map identifying $\pi_{1*}M$ and $\pi_{2*}M$ with a single pushforward of M along $X \to Y \sqcup_X Y$. Any such descent datum is said to be *effective.* We say that f is an *effective descent morphism* for $\mathcal{F}$ if the following conditions hold.

(a) Every descent datum along f is effective.

(b) For any $M, N \in \mathcal{F}$ with $F(M) = F(N)$, the morphisms $M \to N$ in $\mathcal{F}$ lifting the identity morphism are in bijection with morphisms between the corresponding descent data. (We leave the definition of the latter to the reader.)

Example 1.3.3. — Let $\mathcal{C}$ be the category of rings. Let $\mathcal{F}$ be the category of modules over rings, with morphisms defined as follows: for $R_1, R_2 \in \mathcal{C}$ and $M_i \in \mathcal{F}$ a module over R_i, morphisms $M_1 \to M_2$ consist of pairs (f, g) with $f : R_1 \to R_2$ a morphism in $\mathcal{C}$ and $g : f_*M_1 \to M_2$ a morphism of modules over R_2. Let $F : \mathcal{F} \to \mathcal{C}$ be the functor taking each module to its underlying ring; this functor defines a cofibred category with pushforwards defined as expected.

Let $R \to R_1, \dots, R \to R_n$ be ring homomorphisms corresponding to open immersions of schemes which cover $\operatorname{Spec} R$, and put $S = R_1 \oplus \cdots \oplus R_n$. Then $R \to S$ is an effective descent morphism for $\mathcal{F}$; this is another way of stating the standard fact that any quasicoherent sheaf on an affine scheme is represented uniquely by a module over the ring of global sections [**56**, Théorème 1.4.1]. This fact is generalized by Theorem 1.3.4.

Theorem 1.3.4. — *Any faithfully flat morphism of rings is an effective descent morphism for the category of modules over rings (Example 1.3.3).*

Proof. — See [**62**, Exposé VIII, Théorème 1.1]. □

Theorem 1.3.5. — *For $f : R \to S$ a faithfully flat morphism of rings, an R-module U is finite (resp. finite projective) if and only if $f^*U = U \otimes_R S$ is a finite (resp. finite projective) S-module. An R-algebra U is finite étale if and only if f^*U is a finite étale S-algebra.*

Proof. — For the first assertion, see [**62**, Exposé VIII, Proposition 1.10]. For the second assertion, see [**62**, Exposé IX, Proposition 4.1]. □

For a morphism of rings which is faithful but not flat (e.g., a typical adic completion of a nonnoetherian ring), it is difficult to carry out descent except for modules which are themselves flat. Here is a useful example due to Beauville and Laszlo [**8**]. (Note that even the noetherian case of this result is not an immediate corollary of faithfully flat descent, because we do not specify a descent datum on $\widehat{R}$ itself; see [**4**, §2].)

Proposition 1.3.6. — *Let R be a ring. Suppose that $t \in R$ is not a zero divisor and that R is t-adically separated. Let $\widehat{R}$ be the t-adic completion of R.*

(a) *For any flat R-module M, the sequence*

$$0 \longrightarrow M \longrightarrow (M \otimes_R R[t^{-1}]) \oplus (M \otimes_R \widehat{R}) \longrightarrow M \otimes_R \widehat{R}[t^{-1}] \longrightarrow 0,$$

in which the last nontrivial arrow is the difference between the two base extension maps, is exact.

(b) *Let M_1 be a finite projective module over $R[t^{-1}]$, let M_2 be a finite projective module over $\widehat{R}$, and let $\psi_{12} : M_1 \otimes_{R[t^{-1}]} \widehat{R}[t^{-1}] \cong M_2 \otimes_{\widehat{R}} \widehat{R}[t^{-1}]$ be an isomorphism of $\widehat{R}[t^{-1}]$-modules. Then there exist a finite projective R-module M, an isomorphism $\psi_1 : M \otimes_R R[t^{-1}] \cong M_1$, and an isomorphism $\psi_2 : M \otimes_R \widehat{R} \cong M_2$ such that $\psi_{12} \circ \psi_1 = \psi_2$; moreover, these data are unique up to unique isomorphism. In particular, the morphism $R \to R[t^{-1}] \oplus \widehat{R}$ is an effective descent morphism for the category of finite projective modules over rings.*

In order to carry out analogous arguments in other contexts (as in Proposition 2.7.5), it is helpful to introduce some formalism. We will see later how to recover the Beauville-Laszlo theorem in this framework (Remark 2.7.9).

Definition 1.3.7. — Let

$$\begin{array}{ccc} R & \longrightarrow & R_1 \\ \downarrow & & \downarrow \\ R_2 & \longrightarrow & R_{12} \end{array}$$

be a commuting diagram of ring homomorphisms such that the sequence

$$0 \longrightarrow R \longrightarrow R_1 \oplus R_2 \longrightarrow R_{12} \longrightarrow 0 \tag{1.3.7.1}$$

of R-modules, in which the last nontrivial arrow is the difference between the given homomorphisms, is exact. By a *glueing datum* over this diagram, we will mean a datum consisting of modules M_1, M_2, M_{12} over R_1, R_2, R_{12}, respectively, equipped with isomorphisms $\psi_1 : M_1 \otimes_{R_1} R_{12} \cong M_{12}$, $\psi_2 : M_2 \otimes_{R_2} R_{12} \cong M_{12}$. We say such a glueing datum is *finite* or *finite projective* if the modules are finite or finite projective over their corresponding rings.

When considering a glueing datum, it is natural to consider the kernel M of the map $\psi_1 - \psi_2 : M_1 \oplus M_2 \to M_{12}$. There are natural maps $M \to M_1$, $M \to M_2$ of R-modules, which by adjunction correspond to maps $M \otimes_R R_1 \to M_1$, $M \otimes_R R_2 \to M_2$.

Lemma 1.3.8. — *Consider a finite glueing datum for which $M \otimes_R R_1 \to M_1$ is surjective. Then we have the following.*

(a) *The map $\psi_1 - \psi_2 : M_1 \oplus M_2 \to M_{12}$ is surjective.*

(b) *The map $M \otimes_R R_2 \to M_2$ is also surjective.*

(c) *There exists a finitely generated R-submodule M_0 of M such that for $i = 1, 2$, $M_0 \otimes_R R_i \to M_i$ is surjective.*

Proof. — The surjection $M \otimes_R R_1 \to M_1$ induces a surjection $M \otimes_R R_{12} \to M_{12}$, and hence a surjection $M \otimes_R (R_1 \oplus R_2) \to M_{12}$. Since this map factors through $\psi_1 - \psi_2$, the latter is surjective. This yields (a).

For each $\mathbf{v} \in M_2$, $\psi_2(\mathbf{v})$ lifts to $M \otimes_R (R_1 \oplus R_2)$; we can thus find $\mathbf{w}_i$ in the image of $M \otimes_R R_i \to M_i$ such that $\psi_1(\mathbf{w}_1) - \psi_2(\mathbf{w}_2) = \psi_2(\mathbf{v})$. Put $\mathbf{v}' = (\mathbf{w}_1, \mathbf{v} + \mathbf{w}_2) \in M_1 \oplus M_2$; note that $\mathbf{v}' \in M$ by construction. Consequently, the image of $M \otimes_R R_2 \to M_2$ contains both $\mathbf{w}_2$ and $\mathbf{v} + \mathbf{w}_2$, and hence also $\mathbf{v}$. This yields (b), from which (c) is immediate since M_i is a finite R_i-module. □

Lemma 1.3.9. — *Suppose that for every finite projective glueing datum, the map $M \otimes_R R_1 \to M_1$ is surjective.*

(a) *For any finite projective glueing datum, M is a finitely presented R-module and $M \otimes_R R_1 \to M_1$, $M \otimes_R R_2 \to M_2$ are bijective.*

(b) *Suppose in addition that the image of* $\mathrm{Spec}(R_1 \oplus R_2) \to \mathrm{Spec}(R)$ *contains* $\mathrm{Maxspec}(R)$. *Then with notation as in (a), M is a finite projective R-module.*

Proof. — Choose M_0 as in Lemma 1.3.8 (c). Choose a surjection $F \to M_0$ of R-modules with F finite free, and put $F_1 = F \otimes_R R_1$, $F_2 = F \otimes_R R_2$, $F_{12} = F \otimes_R R_{12}$, $N = \ker(F \to M)$, $N_1 = \ker(F_1 \to M_1)$, $N_2 = \ker(F_2 \to M_2)$, $N_{12} = \ker(F_{12} \to M_{12})$. From Lemma 1.3.8, we have a commutative diagram

(1.3.9.1)
$$\begin{array}{ccccccccc}
 & & 0 & & 0 & & 0 & & \\
 & & \downarrow & & \downarrow & & \downarrow & & \\
0 & \longrightarrow & N & \longrightarrow & N_1 \oplus N_2 & \longrightarrow & N_{12} & \dashrightarrow & 0 \\
 & & \downarrow & & \downarrow & & \downarrow & & \\
0 & \longrightarrow & F & \longrightarrow & F_1 \oplus F_2 & \longrightarrow & F_{12} & \longrightarrow & 0 \\
 & & \downarrow & & \downarrow & & \downarrow & & \\
0 & \longrightarrow & M & \longrightarrow & M_1 \oplus M_2 & \longrightarrow & M_{12} & \longrightarrow & 0 \\
 & & \vdots & & \downarrow & & \downarrow & & \\
 & & 0 & & 0 & & 0 & &
\end{array}$$

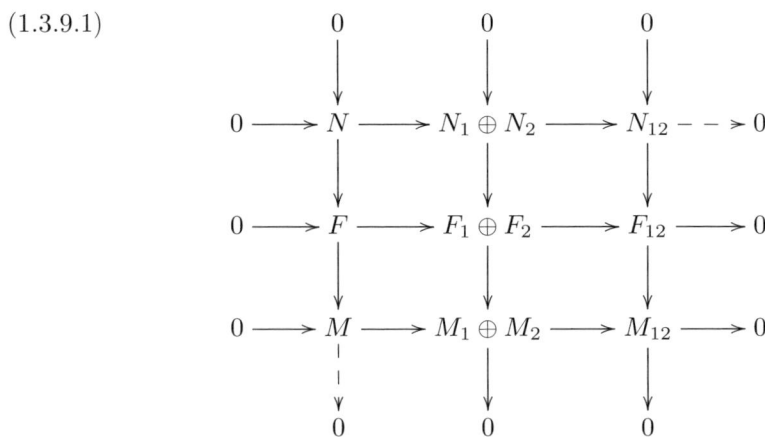

with exact rows and columns, excluding the dashed arrows. Since M_i is projective, the exact sequence

$$0 \longrightarrow N_i \longrightarrow F_i \longrightarrow M_i \longrightarrow 0$$

splits, so

$$0 \longrightarrow N_i \otimes_{R_i} R_{12} \longrightarrow F_{12} \longrightarrow M_{12} \longrightarrow 0$$

is again exact. Thus N_i is finite projective over R_i and admits an isomorphism $N_i \otimes_{R_i} R_{12} \cong N_{12}$. By Lemma 1.3.8 again, the dashed horizontal arrow in (1.3.9.1) is surjective. By diagram chasing, the dashed vertical arrow in (1.3.9.1) is also surjective; that is, we may add the dashed arrows to (1.3.9.1) while preserving exactness of the rows and columns. In particular, M is a finitely generated R-module; we may repeat the argument with M replaced by N to deduce that M is finitely presented.

For $i = 1, 2$, we obtain a commutative diagram

$$\begin{array}{ccccccccc}
 & & N \otimes_R R_i & \longrightarrow & F_i & \longrightarrow & M \otimes_R R_i & \longrightarrow & 0 \\
 & & \downarrow & & \| & & \downarrow & & \\
0 & \longrightarrow & N_i & \longrightarrow & F_i & \longrightarrow & M_i & \longrightarrow & 0
\end{array}$$

with exact rows: the first row is derived from the left column of (1.3.9.1) by tensoring over R with R_i, while the second row is derived from the middle column of (1.3.9.1), Since the left vertical arrow is surjective (Lemma 1.3.8 once more), by the five lemma, the right vertical arrow is injective. We thus conclude that the map $M \otimes_R R_i \to M_i$, which was previously shown (Lemma 1.3.8) to be surjective, is in fact a bijection. This yields (a).

For n a nonnegative integer and $i \in \{1, 2, 12\}$, let $U_{n,i}$ be the closed-open subset of $\mathrm{Spec}(R_i)$ on which M_i has rank n. This set is the nonzero locus of some idempotent $e_{n,i} \in R_{n,i}$. Since $M_{12} \cong M_1 \otimes_{R_1} R_{12} \cong M_2 \otimes_{R_2} R_{12}$, $U_{n,12}$ can be characterized as the inverse image of either $U_{n,1}$ or $U_{n,2}$; this means that the images of e_1 and e_2 in R_{12} are both equal to e_{12}. It follows that $e = e_1 \oplus e_2$ is an idempotent in R mapping to e_i in R_i; its nonzero locus is an open subset U_n of $\mathrm{Spec}(R)$ whose inverse image in $\mathrm{Spec}(R_i)$ is $U_{n,i}$. This means that to prove that M is projective, we may reduce to the case where M_1 and M_2 are finite projective of some constant rank n.

Since M is finitely presented, we may define its Fitting ideals $\mathrm{Fitt}_i(M)$ as in Definition 1.1.2. Since $M_1 \oplus M_2$ is finite projective over $R_1 \oplus R_2$ of constant rank n, $\mathrm{Fitt}_i(M)(R_1 \oplus R_2) = \mathrm{Fitt}_i(M_1 \oplus M_2)$ equals 0 for $i = 0, \dots, n-1$ and $R_1 \oplus R_2$ for $i = n$. Since $R \to R_1 \oplus R_2$ is injective, this immediately implies that $\mathrm{Fitt}_i(M) = 0$ for $i = 0, \dots, n-1$.

Now assume that the image of $\mathrm{Spec}(R_1 \oplus R_2) \to \mathrm{Spec}(R)$ contains $\mathrm{Maxspec}(R)$. Then for each $\mathfrak{p} \in \mathrm{Maxspec}(R)$, M must have rank n at $\mathfrak{p}$ by comparison with some point in $\mathrm{Spec}(R_1 \oplus R_2)$, so $\mathrm{Fitt}_n(M)_{\mathfrak{p}} = \mathrm{Fitt}_n(M_{\mathfrak{p}}) = R_{\mathfrak{p}}$. It follows that the inclusion $\mathrm{Fitt}_n(M) \subseteq R$ is an equality, yielding (b). □

Corollary 1.3.10. — *Suppose that the hypotheses of Lemma 1.3.9 (b) are satisfied. Then the natural functor*

$$\mathbf{F\acute{E}t}(R) \longrightarrow \mathbf{F\acute{E}t}(R_1) \times_{\mathbf{F\acute{E}t}(R_{12})} \mathbf{F\acute{E}t}(R_2)$$

is an equivalence of categories.

Proof. — Choose $A_1 \in \mathbf{F\acute{E}t}(R_1), A_2 \in \mathbf{F\acute{E}t}(R_2), A_{12} \in \mathbf{F\acute{E}t}(R_{12})$ equipped with isomorphisms $A_1 \otimes_{R_1} R_{12} \cong A_2 \otimes_{R_2} R_{12} \cong A_{12}$, and view this package as a finite projective glueing datum. By Lemma 1.3.9 plus our extra assumptions, the kernel A of $A_1 \oplus A_2 \to A_{12}$ is a finite projective R-module and the natural maps $A \otimes_R R_1 \to A_1$, $A \otimes_R R_2 \to A_2$ are isomorphisms. Using the exact sequence

$$(1.3.10.1) \qquad 0 \longrightarrow A \longrightarrow A_1 \oplus A_2 \longrightarrow A_{12} \longrightarrow 0,$$

the multiplication maps on A_1, A_2, A_{12} define a multiplication map on A, making it a flat R-algebra. By Lemma 1.3.9 again, we also have an exact sequence

$$(1.3.10.2) \quad 0 \longrightarrow \mathrm{Hom}_R(A, R) \longrightarrow \mathrm{Hom}_{R_1}(A_1, R_1) \oplus \mathrm{Hom}_{R_2}(A_2, R_2) \longrightarrow \\ \longrightarrow \mathrm{Hom}_{R_{12}}(A_{12}, R_{12}) \longrightarrow 0.$$

Using (1.3.10.1), (1.3.10.2), and the snake lemma, we see that the the trace pairing on A defines an isomorphism $A \to \mathrm{Hom}_R(A, R)$. This proves the claim. □

1.4. Étale local systems

The étale topology on schemes only gives rise to a useful notion of locally constant sheaves if one restricts attention to torsion coefficients. In order to consider étale local systems of $\mathbb{Z}_p$-modules or $\mathbb{Q}_p$-vector spaces, the traditional approach is to keep track of $\mathbb{Z}_p$-modules as inverse systems, then arrive at $\mathbb{Q}_p$-vector spaces by formally inverting p and performing étale descent. One complicating feature of this approach is that the inversion of p only happens locally; that is, an étale $\mathbb{Q}_p$-local system does not generally admit a $\mathbb{Z}_p$-lattice. Another inconvenience is that the objects in these categories are not literally defined as sheaves. After first introducing the usual definitions, we describe the alternate point of view introduced recently by Bhatt and Scholze [**17**], in which étale local systems are reinterpreted as genuine locally constant sheaves for a modified topology called the *pro-étale topology*. The construction is inspired by (but somewhat simpler than) the similarly named construction for adic spaces introduced by Scholze in [**115**], which we will also make use of in §9.

Definition 1.4.1. — For X a scheme, the *small étale site* $X_{\acute{e}t}$ of X is the category of étale X-schemes and étale morphisms, equipped with the Grothendieck topology generated by set-theoretically surjective families of morphisms.

For n a positive integer, a *lisse sheaf* of $\mathbb{Z}/p^n\mathbb{Z}$-modules on $X_{\acute{e}t}$ is a sheaf of flat $\mathbb{Z}/p^n\mathbb{Z}$-modules which is represented by a finite étale X-scheme. A *lisse $\mathbb{Z}_p$-sheaf* on $X_{\acute{e}t}$ is an inverse system $T = \{\cdots \to T_1 \to T_0\}$ in which each T_n is a lisse sheaf of $\mathbb{Z}/p^n\mathbb{Z}$-modules and each arrow $T_{n+1} \to T_n$ identifies T_n with the cokernel of multiplication by p^n on T_{n+1}. A lisse $\mathbb{Z}_p$-sheaf on $X_{\text{ét}}$ is also called an *(étale) $\mathbb{Z}_p$-local system* on X. Such objects may be constructed using faithfully flat descent (Theorem 1.3.4 and Theorem 1.3.5). Let $\mathbb{Z}_p$-$\mathbf{Loc}(X)$ denote the category of $\mathbb{Z}_p$-local systems on X.

An *isogeny $\mathbb{Z}_p$-local system* on X is an element of the isogeny category of $\mathbb{Z}_p$-local systems on X. Let $\mathbb{Z}_p$-$\mathbf{ILoc}(X)$ denote the category of isogeny $\mathbb{Z}_p$-local systems on X.

A *$\mathbb{Q}_p$-lisse sheaf* on $X_{\text{ét}}$, also called an *(étale) $\mathbb{Q}_p$-local system* on X, is an element of the stack associated to the fibred category of isogeny $\mathbb{Z}_p$-local systems, *i.e.*, a descent datum in isogeny $\mathbb{Z}_p$-local systems for the étale topology (compare [**32**, Definition 4.1]). Let $\mathbb{Q}_p$-$\mathbf{Loc}(X)$ denote the category of $\mathbb{Q}_p$-local systems on X.

Remark 1.4.2. — The *small finite étale site* $X_{\text{fét}}$ of X is the subcategory of $X_{\text{ét}}$ in which all internal morphisms and all structure morphisms are finite étale, with the induced topology. One may similarly define categories of $\mathbb{Z}_p$-local systems, isogeny $\mathbb{Z}_p$-local systems, and $\mathbb{Q}_p$-local systems with respect to $X_{\text{fét}}$. To distinguish the two sets of definitions, let us temporarily write $\mathbb{Z}_p$-$\mathbf{Loc}(X_{\text{fét}})$ and $\mathbb{Z}_p$-$\mathbf{Loc}(X_{\text{ét}})$ for the two resulting categories of étale $\mathbb{Z}_p$-local systems, and similarly for $\mathbb{Z}_p$-$\mathbf{ILoc}$ and $\mathbb{Q}_p$-$\mathbf{Loc}$.

The fact that lisse sheaves of $\mathbb{Z}/p^n\mathbb{Z}$-modules are represented by finite étale schemes means that the restriction functor $\mathbb{Z}_p$-$\mathbf{Loc}(X_{\text{fét}}) \to \mathbb{Z}_p$-$\mathbf{Loc}(X_{\text{ét}})$ is a tensor equivalence. The same then holds for $\mathbb{Z}_p$-$\mathbf{ILoc}(X_{\text{fét}}) \to \mathbb{Z}_p$-$\mathbf{ILoc}(X_{\text{ét}})$ but not in general for $\mathbb{Q}_p$-$\mathbf{Loc}(X_{\text{fét}}) \to \mathbb{Q}_p$-$\mathbf{Loc}(X_{\text{ét}})$ unless $X = \operatorname{Spec}(K)$ for K a field (in which case $X_{\text{fét}} = X_{\text{ét}}$).

Remark 1.4.3. — By Remark 1.4.2, for any rings A, B, any tensor equivalence $\mathbf{F\acute{E}t}(A) \cong \mathbf{F\acute{E}t}(B)$ induces tensor equivalences $\mathbb{Z}_p$-$\mathbf{Loc}(\operatorname{Spec}(A)) \cong \mathbb{Z}_p$-$\mathbf{Loc}(\operatorname{Spec}(B))$, $\mathbb{Z}_p$-$\mathbf{ILoc}(\operatorname{Spec}(A)) \cong \mathbb{Z}_p$-$\mathbf{ILoc}(\operatorname{Spec}(B))$. However, this is not sufficient to produce an equivalence $\mathbb{Q}_p$-$\mathbf{Loc}(\operatorname{Spec}(A)) \cong \mathbb{Q}_p$-$\mathbf{Loc}(\operatorname{Spec}(B))$; for this, it would suffice to have an isomorphism of the étale topoi associated to $\operatorname{Spec}(A)$ and $\operatorname{Spec}(B)$.

Remark 1.4.4. — Let X be a connected scheme. Choose a geometric point $\overline{x}$ of X and use it as the base point to define the étale fundamental group $\pi_1^{\text{ét}}(X, \overline{x})$ in the sense of [**62**, Exposé V, §7]. Then the category of étale $\mathbb{Z}_p$-local systems on X is equivalent to the category of continuous representations of $\pi_1^{\text{ét}}(X, \overline{x})$ on finite free $\mathbb{Z}_p$-modules. (See also [**61**, Exposé VI, §1.2.4].)

Similarly, the category of isogeny $\mathbb{Z}_p$-local systems on X is equivalent to the category of continuous representations of $\pi_1^{\text{ét}}(X, \overline{x})$ on finite-dimensional $\mathbb{Q}_p$-vector spaces. Underlying this statement is the fact that $\pi_1^{\text{ét}}(X, \overline{x})$ is by definition profinite, since it is defined in terms of the category of finite étale covering spaces. Consequently, any continuous map from $\pi_1^{\text{ét}}(X, \overline{x})$ into $\operatorname{GL}_n(\mathbb{Q}_p)$ has compact image, and so factors through some conjugate of $\operatorname{GL}_n(\mathbb{Z}_p)$. That is, there is a lattice in $\mathbb{Q}_p^n$ stable under the action of $\pi_1^{\text{ét}}(X, \overline{x})$, which can even be constructed explicitly as follows. Define a sequence of lattices $T_0, T_1, \dots$ in $\mathbb{Q}_p^n$ by taking T_0 to be arbitrary and T_{m+1} to be the lattice generated by the image of T_m under $\pi_1^{\text{ét}}(X, \overline{x})$. Then $T_1 = T_2 = \cdots$

The obvious functor from isogeny $\mathbb{Z}_p$-local systems to $\mathbb{Q}_p$-local systems is fully faithful but not always essentially surjective, as in the following examples suggested by the referee.

Example 1.4.5. — Let k be a field. Let X be the union of two copies of $\mathbb{P}^1_k$ (for k an arbitrary field) glued along $\{0, \infty\}$. Form an étale $\mathbb{Q}_p$-local system V of rank 1 by glueing the trivial rank 1 local systems on $X - \{0\}$ and $X - \{\infty\}$ *via* the morphism which is the identity on one copy of $\mathbb{P}^1_k - \{0, \infty\}$ and multiplication by p on the other copy. Using Remark 1.4.4, we may see that V is not an isogeny $\mathbb{Z}_p$-local system: if it were, the corresponding representation into $\mathbb{Q}_p^\times$ would have noncompact image (because the image would have to contain $p^{\mathbb{Z}}$).

While Example 1.4.5 involves a scheme which is not irreducible, by using étale descent rather than just Zariski descent we can produce an example involving an irreducible (but not normal) scheme.

Example 1.4.6. — With notation as in Example 1.4.5, let Y be a copy of $\mathbb{P}^1_k$ with the points 0 and ∞ glued together. The scheme X then arises as a finite étale cover of Y of degree 2, induced by the map $\mathbb{P}^1_k \cup \mathbb{P}^1_k \to \mathbb{P}^1_k$ acting as the identity on the first factor and the map $x \mapsto 1/x$ on the second factor. Let $\tau : X \to X$ be the nontrivial involution of X over Y. Then $V \oplus \tau^* V$ descends to an étale $\mathbb{Q}_p$-local system of rank 2 on Y.

Remark 1.4.7. — For $T \in \mathbb{Z}_p\text{-}\mathbf{Loc}(X)$, let $T \otimes \mathbb{Q}_p$ denote the corresponding object in $\mathbb{Z}_p\text{-}\mathbf{ILoc}(X)$. For Y an X-scheme, let T_Y be the pullback of T to Y.

For m a nonnegative integer, let F_m be the functor taking each X-scheme Y to the pairs (T', ι) in which $T' \in \mathbb{Z}_p\text{-}\mathbf{Loc}(Y)$ and $\iota : T_Y \otimes \mathbb{Q}_p \to T' \otimes \mathbb{Q}_p$ is an isomorphism such that $p^m \iota \in \operatorname{Mor}(T_Y, T')$, $p^m \iota^{-1} \in \operatorname{Mor}(T', T_Y)$. Then F_m is representable by a finite étale X-scheme $L_m(T)$; we identify $L_m(T_Y)$ with $L_m(T) \times_X Y$. In the context of Remark 1.4.4, this construction corresponds to identifying lattices in $\mathbb{Q}_p^n$ between $p^{-m}\mathbb{Z}_p^n$ and $p^m \mathbb{Z}_p^n$.

The value of this construction is that it allows for certain statements about lattices in isogeny $\mathbb{Z}_p$-local systems to be translated into assertions about finite étale X-schemes, despite the fact that a $\mathbb{Z}_p$-local system over X is defined in terms of an infinite sequence of finite étale covers. Here are some examples pertinent to the proof of Lemma 1.4.8.

(a) There is a closed subscheme $I_m(T)$ of $L_m(T) \times_X L_m(T)$ which is finite étale over X with the following property: for every X-scheme Y, $I_m(T)(Y)$ is the subset of $(L_m(T) \times_X L_m(T))(Y)$ corresponding to pairs $((T'_1, \iota_1), (T'_2, \iota_2))$ for which $\iota_2 \circ \iota_1^{-1} \in \operatorname{Mor}(T'_1, T'_2)$. In the context of Remark 1.4.4, this construction corresponds to the inclusion relation on lattices in $\mathbb{Q}_p^n$ between $p^{-m}\mathbb{Z}_p^n$ and $p^m \mathbb{Z}_p^n$. (To construct $I_m(T)$, by faithfully flat descent we may reduce to the case where T is constant modulo p^{4m}, in which case the argument is straightforward.)

(b) Let $s_1, \dots, s_k : X \to L_m(T)$ be sections of the map $L_m(T) \to X$. Then there exists a unique section $s : X \to L_m(T)$ with the following property: for every X-scheme Y and every section $s' : Y \to L_m(T_Y)$, $(s \times_X Y) \times_Y s' : Y \to L_m(T_Y) \times_Y L_m(T_Y)$ factors through $I_m(T_Y)$ if and only if $(s_i \times_X Y) \times_Y s' : Y \to L_m(T_Y) \times_Y L_m(T_Y)$ factors through $I_m(T_Y)$ for $i = 1, \dots, k$. In the context of Remark 1.4.4, this construction corresponds to forming the lattice of $\mathbb{Q}_p^n$ generated by a finite number of other lattices. (Again, the construction proceeds by faithful flat descent to reduce to the case where T is constant modulo p^{2m}.)

(c) Let $f : Y \to L_m(T)$ be a morphism of faithfully finite étale X-schemes. Then there exists a faithfully finite étale X-scheme Y' such that $Y \times_X Y'$ splits over Y' as a finite disjoint union of copies of Y' (by induction on the degree of $Y \to X$). The pullback of f then gives rise to a collection of sections of $L_m(T_{Y'}) \to Y'$, to which we may apply the construction of (b); since the latter is canonical, it acquires a descent datum back to X, so we end up with a section $s : X \to L_m(T)$.

Whereas $\mathbb{Z}_p$-local systems and $\mathbb{Q}_p$-local systems descend along surjective étale morphisms of schemes, isogeny $\mathbb{Z}_p$-local systems do not in general. However, one does obtain descent for finite étale morphisms.

Lemma 1.4.8. — *Any faithfully finite étale morphism $R \to R'$ of rings is an effective descent morphism for isogeny $\mathbb{Z}_p$-local systems.*

Proof. — Suppose that $V' \in \mathbb{Z}_p$-$\mathbf{ILoc}(\mathrm{Spec}(R'))$ carries a descent datum relative to $\mathrm{Spec}(R)$. Choose $T' \in \mathbb{Z}_p$-$\mathbf{Loc}(\mathrm{Spec}(R'))$ giving rise to V'. We construct a sequence of morphisms $T' = T'_0 \to T'_1 \to \cdots$ in $\mathbb{Z}_p$-$\mathbf{Loc}(\mathrm{Spec}(R'))$ which are isomorphisms in $\mathbb{Z}_p$-$\mathbf{ILoc}(\mathrm{Spec}(R'))$, as follows.

Put $R'' = R' \otimes_R R'$ and let $\pi_1, \pi_2 : \mathrm{Spec}(R'') \to \mathrm{Spec}(R')$ be the two projections. Given T'_i, put $T''_i = \pi_1^*(T'_i)$; for each m, we identify $L_m(T''_i)$ with $L_m(T'_i) \times_{\mathrm{Spec}(R'),\pi_1} \mathrm{Spec}(R'')$. We may fix a sufficiently large m at the beginning; the descent datum on V' defines a section s of the projection $L_m(T''_i) \to \mathrm{Spec}(R''_i)$. By applying Remark 1.4.7 (c) to the composition $\pi_1 \circ s : \mathrm{Spec}(R'') \to L_m(T''_i) \to L_m(T'_i)$, we obtain a new object $T'_{i+1} \in \mathbb{Z}_p$-$\mathbf{Loc}(\mathrm{Spec}(R'))$ and an isomorphism $T'_i \otimes \mathbb{Q}_p \to T'_{i+1} \otimes \mathbb{Q}_p$. Moreover, this morphism descends to a morphism $T'_i \to T'_{i+1}$ because the pullback of s along the diagonal morphism $\mathrm{Spec}(R') \to \mathrm{Spec}(R'')$ corresponds to the identity morphism on T'.

We now see that $T'_1 = T'_2 = \cdots$ by reducing to the case where R and R' are local (and hence connected) and applying Remark 1.4.4. This means that T'_1 acquires a descent datum from V and thus descends to an object in $\mathbb{Z}_p$-$\mathbf{Loc}(\mathrm{Spec}(R))$. □

Remark 1.4.9. — Suppose that X is normal and noetherian. Then X is the disjoint union of finitely many irreducible components. On each component, étale $\mathbb{Q}_p$-local systems correspond precisely to continuous representations of the étale fundamental group (*e.g.*, see [**17**, Lemma 7.4.7]). Consequently, the natural functor $\mathbb{Z}_p$-**ILoc**$(X) \to \mathbb{Q}_p$-**Loc**(X) is an equivalence of categories.

We now describe the alternate approach to local systems given in [**17**], in which local systems become genuine sheaves for an alternate topology.

Definition 1.4.10. — A morphism $f : Y \to X$ of schemes is *weakly étale* if both f and $\Delta_f : Y \to Y \times_X Y$ are flat. For example, if $X = \mathrm{Spec}(A)$, $Y = \mathrm{Spec}(B)$ and B is a direct limit of étale A-algebras, then f is weakly étale.

The *pro-étale site* of a scheme X, denoted by $X_{\text{proét}}$, is the site consisting of weakly étale X-schemes and fpqc coverings. For any topological space T and any scheme X, the presheaf

$$\mathcal{F}_T : U \longmapsto \mathrm{Map}_{\mathrm{cont}}(U, T)$$

on $X_{\text{proét}}$ is a sheaf [**17**, Lemma 4.2.12], called the *constant sheaf* with values in T. A sheaf which is locally of this form is said to be *locally constant.* (In [**17**] one finds also a discussion of *constructible* sheaves, which we do not consider here.)

For any n, given a lisse sheaf of $\mathbb{Z}/p^n\mathbb{Z}$-modules on $X_{\text{ét}}$, we may pull back from $X_{\text{ét}}$ to $X_{\text{proét}}$ to obtain a $\mathcal{F}_{\mathbb{Z}/p^n\mathbb{Z}}$-module on $X_{\text{proét}}$ which is locally free of finite rank. By taking inverse limits, we obtain a functor from $\mathbb{Z}_p$-**Loc**(X) to sheaves of $\mathcal{F}_{\mathbb{Z}_p}$-modules on $X_{\text{proét}}$ which are locally free of finite rank. We also obtain a functor from $\mathbb{Q}_p$-**Loc**(X) to sheaves of $\mathcal{F}_{\mathbb{Q}_p}$-modules on $X_{\text{proét}}$ which are locally free of finite rank.

Theorem 1.4.11. — *For any scheme X, the category $\mathbb{Z}_p$-**Loc**(X) (resp. $\mathbb{Q}_p$-**Loc**(X)) is naturally equivalent to the category of sheaves of $\mathcal{F}_{\mathbb{Z}_p}$-modules (resp. $\mathcal{F}_{\mathbb{Q}_p}$-modules) on $X_{\text{proét}}$ which are locally free of finite rank (and in particular locally constant).*

Proof. — See [**17**, §6.8]. □

Remark 1.4.12. — For F a finite extension of $\mathbb{Q}_p$, one can define étale $\mathfrak{o}_F$-local systems, isogeny étale $\mathfrak{o}_F$-local systems, and étale F-local systems on a scheme X by analogy with the case $F = \mathbb{Q}_p$. In fact, these can be interpreted as objects of the corresponding category over $\mathbb{Q}_p$ plus the extra structure of an action of $\mathfrak{o}_F$. We will only need this observation in the case where $F = \mathbb{Q}_{p^d}$ is a finite unramified extension of $\mathbb{Q}_p$ of degree d, in which case we label the resulting categories $\mathbb{Z}_{p^d}$-**Loc**(X), $\mathbb{Z}_{p^d}$-**ILoc**(X), and $\mathbb{Q}_{p^d}$-**Loc**(X).

1.5. Semilinear actions

Convention 1.5.1. — For S a ring equipped with an endomorphism φ and M an S-module, a *semilinear φ-action* on M will always mean an *isomorphism* (not just an endomorphism) $\varphi^* M \to M$ of S-modules. We will commonly interpret such an action as a φ-semilinear map $M \to M$.

Although we have not found a precise reference, we believe that the following is a standard lemma in algebraic K-theory, specifically from the study of polynomial extensions (as in [**7**, Chapter XII]).

Lemma 1.5.2. — *Let S be a ring equipped with an endomorphism φ. Let M be a finitely generated S-module equipped with a semilinear φ-action. Then there exists a finite free S-module F equipped with a semilinear φ-action and a φ-equivariant surjection $F \to M$.*

Proof. — Choose generators $\mathbf{v}_1, \dots, \mathbf{v}_n$ of M, and use them to define a surjection $E = S^n \to M$ of S-modules. Let $T : \varphi^* M \to M$ be the given isomorphism. Choose $A_{ij}, B_{ij} \in S$ so that $T(\mathbf{v}_j \otimes 1) = \sum_i A_{ij}\mathbf{v}_i$, $T^{-1}(\mathbf{v}_j) = \sum_i B_{ij}(\mathbf{v}_i \otimes 1)$; by writing

$$\mathbf{v}_k = T(T^{-1}(\mathbf{v}_k)) = T\left(\sum_j B_{jk}(\mathbf{v}_j \otimes 1)\right) = \sum_{i,j} A_{ij}B_{jk}\mathbf{v}_i,$$

we see that the columns of the matrix $C = AB - 1$ are elements of $N = \ker(E \to M)$.

Let D be the block matrix $\left(\begin{smallmatrix} A & C \\ 1 & B \end{smallmatrix}\right)$. By using row operations to clear the bottom left block, we find that $\det(D) = \det(AB - C) = 1$. Consequently, D is invertible over S, so we may use it to define an isomorphism $\varphi^* F \to F$ for $F = E \oplus E$. This isomorphism carries $\varphi^*(N \oplus E)$ into $N \oplus E$, so we obtain a φ-equivariant surjection $F \to M$ as desired. □

Corollary 1.5.3. — *Let S be a ring equipped with an endomorphism φ. Let M be a finite projective S-module equipped with a semilinear φ-action. Then there exists another finite projective S-module N admitting a semilinear φ-action such that $M \oplus N$ is a free S-module.*

Proof. — Apply Lemma 1.5.2 to construct a finite free S-module F equipped with a semilinear φ-action and a φ-equivariant S-linear surjection $F \to M$, then put $N = \ker(F \to M)$. □

Definition 1.5.4. — Let S be a ring equipped with an endomorphism φ. Let M be a module over S equipped with a semilinear φ-action. We then write

$$H^0_\varphi(M) = \ker(\varphi - 1, M), \qquad H^1_\varphi(M) = \operatorname{coker}(\varphi - 1, M),$$

and $H^i_\varphi(M) = 0$ for $i \geqslant 2$. The groups $H^i_\varphi(M)$ may be interpreted as the Yoneda extension groups $\operatorname{Ext}^i(S, M)$ in the category of left modules over the twisted polynomial ring $S\{\varphi\}$, by tensoring M over S with the free resolution

$$0 \longrightarrow S\{\varphi\} \stackrel{\varphi-1}{\longrightarrow} S\{\varphi\} \longrightarrow S \longrightarrow 0$$

of S. (For a detailed development of Yoneda extension groups, see for instance [**75**, §IV.9].)

Remark 1.5.5. — In Definition 1.5.4, if M is a module over S equipped with a semilinear φ^d-action for some positive integer d, we may identify $H^i_{\varphi^d}(M)$ with $H^i_\varphi(N)$ for $N = M \oplus \varphi^* M \oplus \cdots \oplus (\varphi^{d-1})^* M$.

Remark 1.5.6. — Let S be a ring equipped with an endomorphism φ. Let M be a finite projective module over S equipped with a semilinear φ-action. Then there is a natural way to equip the dual module $M^\vee = \mathrm{Hom}_R(M, R)$ with a φ-module structure so that the pairing map $M \otimes_R M^\vee \to R$ is φ-equivariant.

CHAPTER 2

SPECTRA OF NONARCHIMEDEAN BANACH RINGS

We set notation and terminology concerning spectra of nonarchimedean (commutative) Banach rings. We will consider two separate but related notions of spectrum, the *Gel'fand spectrum* of Berkovich [**14, 15**] and the *adic spectrum* of Huber [**77, 78, 79**].

Convention 2.0.1. — For M a matrix over a ring equipped with a submultiplicative seminorm α, we write $\alpha(M)$ for $\sup_{i,j}\{\alpha(M_{ij})\}$.

2.1. Seminorms on groups and rings

We begin by setting notation regarding seminorms. We will later have to consider also *semivaluations*, which take values not in the real numbers but in more general ordered abelian groups; see §2.4.

Definition 2.1.1. — Consider the following conditions on an abelian group G and a function $\alpha : G \to [0, +\infty)$.

(a) For all $g, h \in G$, we have $\alpha(g - h) \leqslant \max\{\alpha(g), \alpha(h)\}$.

(b) We have $\alpha(0) = 0$.

(b′) For all $g \in G$, we have $\alpha(g) = 0$ if and only if $g = 0$.

We say α is a *(nonarchimedean) seminorm* if it satisfies (a) and (b), and a *(nonarchimedean) norm* if it satisfies (a) and (b′). Any seminorm α induces a norm on $G/\ker(\alpha)$.

If α, α' are two seminorms on the same abelian group G, we say α *dominates* α', and write $\alpha \geqslant \alpha'$ or $\alpha' \leqslant \alpha$, if there exists $c > 0$ for which $\alpha'(g) \leqslant c\alpha(g)$ for all $g \in G$. If α and α' dominate each other, we say they are *equivalent*; in this case, α is a norm if and only if α' is.

Definition 2.1.2. — Let G, H be two abelian groups equipped with nonarchimedean seminorms α, β, and let $\varphi : G \to H$ be a homomorphism. We say φ is *bounded* if $\alpha \geqslant \beta \circ \varphi$, and *isometric* if $\alpha = \beta \circ \varphi$. (An intermediate condition is that φ is *submetric*, meaning that $\alpha(g) \geqslant \beta(\varphi(g))$ for all $g \in G$.)

The *quotient seminorm* induced by α is the seminorm $\overline{\alpha}$ on image(φ) defined by

$$\overline{\alpha}(h) = \inf\{\alpha(g) : g \in G, \varphi(g) = h\}.$$

If H is also equipped with a seminorm β, we say φ is *strict* if the two seminorms $\overline{\alpha}$ and β on image(φ) are equivalent; this implies in particular that φ is bounded. We say φ is *almost optimal* if $\overline{\alpha}$ and β coincide. We say φ is *optimal* if every $h \subset$ image(φ) admits a lift $g \in G$ with $\alpha(g) = \overline{\alpha}(h)$. Any optimal homomorphism is almost optimal, and any optimal homomorphism is strict, but not conversely. Also beware that a composition $g \circ f$ of strict morphisms is not guaranteed to be strict unless f is surjective or g is injective.

Remark 2.1.3. — Berkovich uses the term *admissible* in place of *strict*, but the latter is well-established in the context of topological groups, as in [**19**, §III.2.8]. However, there is no perfect choice of terminology; our convention will create some uncomfortable linguistic proximity during the discussion of *strict p-rings*.

Definition 2.1.4. — For G an abelian group equipped with a nonarchimedean seminorm α, equip the group of Cauchy sequences in G with the seminorm whose value on the sequence $g_0, g_1, \dots$ is $\lim_{i\to\infty} \alpha(g_i)$. The quotient by the kernel of this seminorm is the *separated completion* $\widehat{G}$ of G under α. For the unique continuous extension of α to $\widehat{G}$, the homomorphism $G \to \widehat{G}$ is isometric, and injective if and only if α is a norm (in which case we call $\widehat{G}$ simply the *completion* of G).

Definition 2.1.5. — Let A be a ring. Consider the following conditions on a (semi)norm α on the additive group of A.

(c) For all $g, h \in A$, we have $\alpha(gh) \leqslant \alpha(g)\alpha(h)$.

(c′) We have (c), and for all $g \in A$, we have $\alpha(g^2) = \alpha(g)^2$. (Equivalently, $\alpha(g^n) = \alpha(g)^n$ for all $g \in A$ and all positive integers n. In particular, $\alpha(1) \in \{0, 1\}$.)

(c″) We have (c′), $\alpha(1) = 1$, and for all $g, h \in A$, we have $\alpha(gh) = \alpha(g)\alpha(h)$.

We say α is *submultiplicative* if it satisfies (c), *power-multiplicative* if it satisfies (c′), and *multiplicative* if it satisfies (c″). Note that if α is a submultiplicative seminorm and α' is a power-multiplicative seminorm, then α dominates α' if and only if $\alpha(a) \geqslant \alpha'(a)$ for all $a \in A$.

Example 2.1.6. — For any abelian group G, the *trivial norm* on G sends 0 to 0 and any nonzero $g \in G$ to 1. For any ring A, the trivial norm on A is submultiplicative in all cases, power-multiplicative if and only if A is reduced, and multiplicative if and only if A is an integral domain. (As usual, the zero ring is not considered to be a domain.)

Definition 2.1.7. — For A a ring equipped with a submultiplicative seminorm α, define

$$\mathfrak{o}_A = \{a \in A : \alpha(a) \leqslant 1\}$$
$$\mathfrak{m}_A = \{a \in A : \alpha(a) < 1\}$$
$$\kappa_A = \mathfrak{o}_A/\mathfrak{m}_A.$$

If $\alpha(1) \leqslant 1$, then $\mathfrak{o}_A$ is a ring and $\mathfrak{m}_A$ is an ideal of $\mathfrak{o}_A$. If A is a field equipped with a multiplicative norm, then $\mathfrak{o}_A$ is a valuation ring with maximal ideal $\mathfrak{m}_A$ and residue field κ_A.

Example 2.1.8. — Let I be an arbitrary index set. For each $i \in I$, specify a ring A_i and a power-multiplicative seminorm α_i on A_i. Put $A = \prod_{i \in I} A_i$, and define the function $\alpha : A \to [0, +\infty]$ by setting $\alpha((a_i)_{i \in I}) = \sup_i\{\alpha_i(a_i)\}$. Then the subset A_0 of A on which α takes finite values is a subring on which α restricts to a power-multiplicative seminorm. This example is in some sense universal; see Theorem 2.3.10.

Definition 2.1.9. — Let A be a ring equipped with a submultiplicative seminorm α. The *spectral seminorm* on A is the power-multiplicative seminorm α_{sp} defined by $\alpha_{\mathrm{sp}}(a) = \lim_{s\to\infty} \alpha(a^s)^{1/s}$. (The existence of the limit is an exercise in real analysis known as *Fekete's lemma.*) Note that equivalent choices of α define the same spectral seminorm.

Definition 2.1.10. — Let A, B, C be rings equipped with submultiplicative seminorms α, β, γ, and let $A \to B, A \to C$ be bounded homomorphisms. The *product seminorm* on the ring $B \otimes_A C$ is defined by taking $f \in B \otimes_A C$ to the infimum of $\max_i\{\beta(b_i)\gamma(c_i)\}$ over all presentations $f = \sum_i b_i \otimes c_i$. The separated completion of $B \otimes_A C$ for the product seminorm is denoted $B \widehat{\otimes}_A C$ and called the *completed tensor product* of B and C over A.

Remark 2.1.11. — The definition of a submultiplicative seminorm α on A does not include the condition that $\alpha(1) \leqslant 1$. However, if we define the *operator seminorm* α' by the formula

$$\alpha'(a) = \inf\{c \geqslant 0 : \alpha(ab) \leqslant c\alpha(b) \text{ for all } b \in A\}, \tag{2.1.11.1}$$

then α' is a submultiplicative norm, $\alpha'(1) \leqslant 1$, and $\alpha'(a) \leqslant \alpha(a)$ for all $a \in A$. Moreover, if $\alpha(1) > 0$, then we may take $b = 1$ in (2.1.11.1) to deduce that $\alpha'(a) \geqslant \alpha(1)^{-1}\alpha(a)$. Consequently, in all cases α' is equivalent to α (this being trivially true if $\alpha(1) = 0$).

2.2. Banach rings and modules

Definition 2.2.1. — Throughout this paper, an *analytic field* is a field equipped with a nontrivial multiplicative nonarchimedean norm under which it is complete. For K an analytic field, any finite extension of K admits a unique structure of an analytic field extending K [**18**, Theorem 3.2.3/2]. The inclusion of the nontriviality condition is a convention which is not universal: it is notably absent in Berkovich's work.

However, this condition will be needed in order to work with adic spectra; it also shows up as a hypothesis in some other key results, such as the open mapping theorem (Theorem 2.2.8).

A *Banach ring* is a commutative ring R equipped with a submultiplicative norm under which it is complete. We allow the zero ring as a Banach ring, so that the completed tensor product is defined on the category of Banach rings. (What we call Banach rings would more commonly be called *commutative Banach rings*, but we will not use noncommutative Banach rings in this paper.)

A *Banach algebra* over a Banach ring R is a Banach ring S equipped with the structure of an R-algebra in such a way that the map $R \to S$ is bounded.

From §2.4 on, we will only consider Banach rings containing a topologically nilpotent unit. For more discussion of this condition, see Remark 2.3.9.

Lemma 2.2.2. — *Let I be a nontrivial ideal in a Banach ring A. Then the closure of I is also a nontrivial ideal. In particular, any maximal ideal in A is closed.*

Proof. — If the closure were trivial, then I would contain an element x for which $1 - x \in \mathfrak{m}_A$. But then the series $\sum_{i=0}^{\infty}(1-x)^i$ would converge in A to an inverse of x, contradicting the assumption that I is a nontrivial ideal. □

For A a Banach ring, it is easy to check (using Remark 1.2.7) that the pair $(\mathfrak{o}_A, \mathfrak{m}_A)$ is henselian. The following refinement of this observation will also prove to be useful. See also Proposition 2.6.8.

Lemma 2.2.3. — *Let $\{(A_i, \alpha_i)\}_{i \in I}$ be a direct system in the category of Banach rings and submetric homomorphisms. Equip the direct limit A of the A_i in the category of rings with the infimum α of the quotient seminorms induced by the α_i.*

(a) *The pair $(A, \ker(\alpha))$ is henselian.*

(b) *The pair $(\mathfrak{o}_A, \mathfrak{m}_A)$ is henselian.*

Proof. — In both cases, we check the criterion of Remark 1.2.7. To check (a), note first that $\ker(\alpha)$ is contained in the Jacobson radical of A: if $a - 1 \in \ker(\alpha)$, then there exists some index i for which $a - 1$ is an element of A_i of norm less than 1. Since A_i is complete, this forces a to be invertible. With that in mind, let $f = \sum_i f_i T^i \in A[T]$ be a monic polynomial with $f_0 \in \ker(\alpha), f_1 \in A^\times$. We construct a root of f using the Newton-Raphson iteration as follows. Put $x_0 = 0$. Given $x_l \in A$ for some nonnegative integer l such that $x_l \in \ker(\alpha)$, $f'(x_l)$ is invertible modulo $\ker(\alpha)$ and hence is a unit. We may thus define $x_{l+1} = x_l - f(x_l)/f'(x_l)$ and note that $x_{l+1} \in \ker(\alpha)$. For any sufficiently large $i \in I$, the sequence $\{x_l\}$ is Cauchy in A_i, and so has a limit which is a root of f.

To check (b), let $f = \sum_i f_i T^i \in \mathfrak{o}_A[T]$ be a monic polynomial with $f_0 \in \mathfrak{m}_A, f_1 \in \mathfrak{o}_A^\times$; then f admits a root r in $\mathfrak{m}_{\widehat{A}}$. Choose any $s \in A$ with $\alpha(r - s) < 1$, and put $x_0 = s$, $x_{l+1} = x_l - f(x_l)/f'(x_l)$. For any sufficiently large $i \in I$, the sequence $\{x_l\}$ is Cauchy in A_i, and so has a limit which is a root of f. □

Lemma 2.2.4. — *Retain notation as in Lemma 2.2.3.*

(a) *The base change functor* $\mathbf{FÉt}(A) \to \mathbf{FÉt}(\widehat{A})$ *is rank-preserving and fully faithful.*

(b) *Suppose that* α *is a multiplicative seminorm and* $K = A/\ker(\alpha)$ *is a field. Then the base change functor* $\mathbf{FÉt}(A) \to \mathbf{FÉt}(\widehat{K})$ *is an equivalence of categories.*

Proof. — To check (a), we first observe that by Lemma 2.2.3 (a), $\ker(\alpha)$ is contained in the Jacobson radical of A. Next, for any $x \in A$ which becomes a unit in $\widehat{A}$, we can find $y \in A$ for which $\alpha(xy - 1) < 1$, so xy is a unit in some A_i and so x is a unit in A. Next, any invertible module M over A is the base extension of some invertible module M_i over some A_i; if $M_i \otimes_{A_i} \widehat{A}$ admits a generator, then so does $M_i \otimes_{A_i} A_j$ for any sufficiently large j by Lemma 2.2.13 below. Finally, for $S \in \mathbf{FÉt}(A)$, note that any idempotent $S \otimes_A \widehat{A}$ can have at most one preimage in $S \otimes_A A/\ker(\alpha)$ (since $A/\ker(\alpha)$ injects into $\widehat{A}$ and S is a projective A-module) and hence at most one preimage in S (by Lemma 1.2.4). We conclude by Proposition 1.2.5 that to check (a), it suffices to verify that for each $S \in \mathbf{FÉt}(A)$, every idempotent of $S \otimes_A \widehat{A}$ arises from some idempotent of S.

Since $S \in \mathbf{FÉt}(A)$, by Remark 1.2.9 we can choose an index $i \in I$ for which $S = S_i \otimes_{A_i} A$ for some $S_i \in \mathbf{FÉt}(A_i)$. Since S_i is a finite locally free A_i-module (see Definition 1.2.1), we can choose a finite free A_i-module F_i admitting a direct sum decomposition $F_i \cong S_i \oplus T_i$. Choose a basis $x_1, \dots, x_n$ of F_i and let $y_1, \dots, y_n$ be the projections of $x_1, \dots, x_n$ onto S_i. For $h, k \in \{1, \dots, n\}$, write $y_h y_k$ in F_i as $\sum_l c_{hkl} x_l$ with $c_{hkl} \in A_i$, so that in S_i we have $y_h y_k = \sum_l c_{hkl} y_l$. Put $c = \max\{1, \sup_{h,k,l}\{\alpha_i(c_{hkl})\}\}$.

For each $j \in I$ with $i \leqslant j$, let β_j be the restriction to $S_j = S_i \otimes_{A_i} A_j$ of the supremum norm on $F_j = F_i \otimes_{A_i} A_j$ defined by the basis $x_1, \dots, x_n$. Note that $\beta_j(xy) \leqslant c\beta_j(x)\beta_j(y)$ for all j and all $x, y \in S_j$. Similarly, let β be the supremum seminorm on $S \otimes_A \widehat{A}$ defined by the basis $x_1, \dots, x_n$, so that $\beta(xy) \leqslant c\beta(x)\beta(y)$ for all $x, y \in S \otimes_A \widehat{A}$. In particular, any nonzero idempotent element $e \in S \otimes_A \widehat{A}$ satisfies $\beta(e) \geqslant c^{-1}$.

Let $e \in S \otimes_A \widehat{A}$ be an idempotent element. Choose $\epsilon > 0$ with $\epsilon \max\{\beta(e), 1\} < 1$. Since $e^2 = e$ in $S \otimes_A \widehat{A}$, we can choose $j \in I$ and $x \in S_j$ with $\beta(x - e) < c^{-1}$ and $\beta_j(x^2 - x) \leqslant c^{-1}\epsilon$. Define the sequence $x_0, x_1, \dots$ by $x_0 = x$ and $x_{l+1} = 3x_l^2 - 2x_l^3$. We then have $\beta_j(x_l^2 - x_l) \leqslant c^{-1}\epsilon^{l+1}$ by induction on l, by writing

$$x_{l+1}^2 - x_{l+1} = 4(x_l^2 - x_l)^3 - 3(x_l^2 - x_l)^2.$$

Also, $x_{l+1} - x_l = (x_l^2 - x_l)(1 - 2x_l)$, so by induction on l, $\beta_j(x_l) \leqslant \max\{\beta_j(x), 1\}$ and $\beta(x_l) \leqslant \max\{\beta(x), 1\}$. Using the equation $x_{l+1} - x_l = (x_l^2 - x_l)(1 - 2x_l)$ again, we see that the x_l form a Cauchy sequence, whose limit y in S_j must satisfy $y^2 = y$. In addition, $\beta(y - x) \leqslant c^{-1}\epsilon \max\{\beta(x), 1\}$, so $\beta(y - e) < c^{-1}$. Since $(y - e)^2$ is an idempotent element of $S \otimes_A \widehat{A}$, this is only possible if $(y - e)^2 = 0$; since $y(y - e)^2 = y - ey$ and $e(y - e)^2 = e - ey$, this yields $y = e$. This completes the proof of (a).

To check (b), note that the hypotheses ensure that the completion $\widehat{K}$ of K is an analytic field. It suffices to show that an arbitrary finite separable field extension $\widehat{L}$ of $\widehat{K}$ occurs in the essential image of the base change functor. By the primitive element theorem, we can write $\widehat{L} \cong \widehat{K}[T]/(P)$ for some monic separable polynomial $P \in \widehat{K}[T]$. By Hensel's lemma (or more precisely Krasner's lemma), we also have $\widehat{L} \cong \widehat{K}[T]/(Q)$ for any monic polynomial $Q \in \widehat{K}[T]$ whose coefficients are sufficiently close to those of P. In particular, we may choose $Q \in K[T]$, in which case we may write $\widehat{L} = L \otimes_K \widehat{K}$ for $L = K[T]/(Q) \in \mathbf{F\acute{E}t}(K)$. Since $\mathbf{F\acute{E}t}(A) \to \mathbf{F\acute{E}t}(K)$ is essentially surjective by Lemma 2.2.3 plus Theorem 1.2.8, this proves the claim. □

Lemma 2.2.5. — *Let K be an analytic field with norm α, and let L be a finite extension of K. Then the unique multiplicative extension of α to L (Definition 2.2.1) is also the unique power-multiplicative extension of α to L.*

Proof. — Let β be the multiplicative extension of α to L, and let γ be a power-multiplicative extension of α to L. Note that for $x \in K^\times, y \in L$, we have

$$\gamma(xy) \leqslant \gamma(x)\gamma(y) = \gamma(x^{-1})^{-1}\gamma(y) \leqslant \gamma(xy),$$

so $\gamma(xy) = \gamma(x)\gamma(y)$.

Given $x \in L^\times$, let $P \in K[T]$ be the minimal polynomial of x over K; since K is complete, the Newton polygon of P consists of a single segment. In other words, if we write $P(T) = \sum_{i=0}^n P_i T^i$ with $P_n = 1$, then $|P_{n-i}|^{1/i} \leqslant |P_0|^{1/n} = \beta(x)$ for $i = 1, \dots, n$. (See [**85**, §2.1] for more discussion of Newton polygons.)

If $\gamma(x) > |P_0|^{1/n}$, then under γ the sum $0 = \sum_{i=0}^n P_i x^i$ would be dominated by the term $P_n x^n$, a contradiction. Hence $\gamma(x) \leqslant |P_0|^{1/n} = \beta(x)$ and similarly $\gamma(x^{-1}) \leqslant \beta(x^{-1})$; by writing

$$1 = \gamma(x \cdot x^{-1}) \leqslant \gamma(x)\gamma(x^{-1}) \leqslant \beta(x)\beta(x^{-1}) = 1,$$

we see that $\gamma(x) = \beta(x)$ as desired. □

Before moving on to Banach modules, we make one observation about modules over a Banach ring.

Lemma 2.2.6. — *Let R be a Banach ring.*

(a) *For any finite R-module M, the quotient seminorm defined by a surjection $\pi : R^n \to M$ of R-modules does not depend, up to equivalence, on the choice of the surjection.*

(b) *Let $R \to S$ be a bounded homomorphism of Banach rings. Let M be a finite R-module, let N be a finite S-module, and let $M \to N$ be an additive R-linear map. Then this map becomes bounded if we equip M and N with seminorms as described in (a).*

Proof. — To prove (a), let $\pi' : R^m \to M$ be a second surjection, and combine π and π' to obtain a third surjection $\pi'' : R^{n+m} \to M$. It is enough to check that the quotient seminorms $|\cdot|, |\cdot|''$ induced by π, π'' are equivalent, as then the same argument will apply with π and π' interchanged.

Let $\mathbf{e}_1, \dots, \mathbf{e}_{n+m}$ be the standard basis of R^{n+m}. On one hand, we clearly have $|\cdot|'' \leqslant |\cdot|$ because lifting an element of M to R^n also gives a lift to R^{n+m}. On the other hand, for $j = n+1, \dots, n+m$, we can write $\pi(\mathbf{e}_j) = \sum_{i=1}^n A_{ij}\pi(\mathbf{e}_i)$ for some $A_{ij} \in R$. If an element of M lifts to $\sum_{i=1}^{n+m} c_i \mathbf{e}_i \in R^{n+m}$, it also lifts to

$$\sum_{i=1}^{n} \left(c_i + \sum_{j=n+1}^{n+m} A_{ij} c_j \right) \mathbf{e}_i \in R^n.$$

Consequently, we have $|\cdot| \leqslant \max\{1, |A|\} |\cdot|''$. This yields (a).

To prove (b), choose surjections $R^m \to M$, $S^n \to N$ of R-modules. We may then lift the composition $R^m \to M \to N$ to a homomorphism $R^m \to S^n$ which is evidently bounded. This proves the claim. □

Definition 2.2.7. — Let R be a Banach ring. A *Banach module* over R is an R-module M whose additive group is complete for a norm $|\cdot|_M$ for which for some $c > 0$, we have $|r\mathbf{v}|_M \leqslant c|r||\mathbf{v}|_M$ for all $r \in R, \mathbf{v} \in M$. In particular, any Banach algebra over R is a Banach module over R.

One has an analogue of the Banach-Schauder open mapping theorem in the nonarchimedean setting. (Note that this result fails completely without a restriction on the base ring.)

Theorem 2.2.8. — *Let R be a Banach ring containing a topologically nilpotent unit. Let $\varphi : V \to W$ be a bounded surjective homomorphism of Banach modules over R. Then φ is open and strict.*

Proof. — For R an analytic field, see [**21**, §I.3.3, Théorème 1] or [**110**, Proposition 8.6]. For the more general case, see [**71**]. □

Lemma 2.2.9. — *Let V, W, X be Banach modules over an analytic field K.*

(a) *The map $V \otimes_K W \to V \widehat{\otimes}_K W$ is injective.*

(b) *Let $f : V \to W$ be a bounded homomorphism and let $f_X : V \widehat{\otimes}_K X \to W \widehat{\otimes}_K X$ be the induced map. Then the natural map* $\ker(f)\widehat{\otimes}_K X \to \ker(f_X)$ *is a bijection.*

(c) *In (b), if f is strict, then so is f_X.*

Proof. — All three parts reduce immediately to the case where V, W, X contain dense K-vector subspaces of at most countable dimension (*i.e.*, they are *separable* Banach modules). In this setting, (a) follows from the existence of Schauder bases for V and W; see for instance [**85**, Lemma 1.3.11]. Similarly, (b) and (c) follow from the existence of a Schauder basis for X. □

Definition 2.2.10. — Let R be a Banach ring. A *finite Banach module/algebra* over R is a Banach module/algebra M over R admitting a strict surjection $R^n \to M$ of Banach modules over R for some nonnegative integer n (for the supremum norm on

R^n defined by the canonical basis). By Lemma 2.2.6, the equivalence class of the norm on M is determined by the underlying R-module.

Remark 2.2.11. — Let R be a Banach ring and let M be a finite R-module. Lemma 2.2.6 equips M with a distinguished equivalence class of seminorms, but M need not be separated or complete under such a seminorm. In fact, by Theorem 2.2.8, M is separate and complete if and only if it is a finite Banach module over R. Moreover, in the case when R contains a topologically nilpotent unit, the following conditions on R are equivalent (see [**18**, Propositions 3.7.3/2, 3.7.3/3] for the case where R is a Banach algebra over an analytic field, then modify the arguments using Theorem 2.2.8).

(a) The ring R is noetherian. (Note that it does not suffice to exhibit a dense noetherian subring of R; see [**23**, Proposition 12].)

(b) Every ideal of R is closed.

(c) The forgetful functor from finite Banach R-modules to finite R-modules is an equivalence of categories.

For Banach rings which are not noetherian, as noted in Remark 2.2.11, we cannot equip arbitrary finite modules over R with natural Banach module structures. However, we can do so for finite projective R-modules.

Lemma 2.2.12. — *Let R be a Banach ring. Let P be a finite projective R-module. Choose a finite projective R-module Q and an isomorphism $P\oplus Q \cong R^n$ of R-modules, for n a suitable nonnegative integer. Equip R^n with the supremum norm defined by the canonical basis.*

(a) *The subspace norm on P for the inclusion into R^n is equivalent to the quotient norm for the projection from R^n, and gives P the structure of a finite Banach module over R.*

(b) *The equivalence class of the norms described in (a) is independent of the choice of Q and of the presentation $P \oplus Q \cong R^n$.*

(c) *The above construction defines a fully faithful functor from finite projective R-modules to finite Banach modules over R whose underlying R-modules are projective, which is a section of the forgetful functor.*

Proof. — Let P', Q' be copies of P, Q, respectively. Note that the supremum norms $|\cdot|_1, |\cdot|_2$ on $P \oplus P' \oplus Q \oplus Q'$ defined by the presentations

$$(P \oplus Q) \oplus (P' \oplus Q') \cong R^n \oplus R^n, \qquad (P \oplus Q') \oplus (P' \oplus Q) \cong R^n \oplus R^n$$

are equivalent by Lemma 2.2.6.

It is clear that the subspace and quotient norms on $P \oplus Q$ induced by $|\cdot|_1$ are identical, and that $P \oplus Q$ is complete under these norms. Consequently, the subspace and quotient norms on $P \oplus Q$ induced by $|\cdot|_2$ are equivalent, and $P \oplus Q$ is complete under these norms. Restricting to P yields the subspace and quotient norms induced by the original presentation, so these two are equivalent. Moreover, P is the intersection of the closed subspaces $P \oplus Q$ and $P \oplus Q'$ of $P \oplus P' \oplus Q \oplus Q'$. This proves (a). Parts (b) and (c) follow from Lemma 2.2.6. □

Lemma 2.2.13. — *Let P be a finite projective module over a Banach ring R, and choose a norm on P as in Lemma 2.2.12. Let $\mathbf{e}_1, \dots, \mathbf{e}_n$ be a finite set of generators of P as an R-module. Then there exists $c > 0$ such that any $\mathbf{e}'_1, \dots, \mathbf{e}'_n \in P$ with $|\mathbf{e}'_i - \mathbf{e}_i| < c$ for $i = 1, \dots, n$ also form a set of generators of P as an R-module.*

Proof. — The conclusion does not depend on the choice of the norm (only the constant c does), so we may use the restriction of the supremum norm on R^n along the homomorphism $R^n \to P$ defined by $\mathbf{e}_1, \dots, \mathbf{e}_n$. In this case, the claim is evident with $c = 1$, as then the matrix A defined by $\mathbf{e}'_j = \sum_i A_{ij}\mathbf{e}_i$ satisfies $|A - 1| < 1$ and hence is invertible. □

Definition 2.2.14. — For A a Banach ring and $B \in \mathbf{F\acute{E}t}(A)$, view B as a finite Banach module over A *via* Lemma 2.2.12. The multiplication map $\mu : B \otimes_A B \to B$ is then bounded by Lemma 2.2.12 again; consequently, we can find an equivalent norm on B which is submultiplicative, and thus view B as a finite Banach algebra over A. We will frequently do so without further comment.

The analogue of a polynomial extension for Banach rings is the following construction.

Definition 2.2.15. — For $r_1, \dots, r_n > 0$, define the *Tate algebra* over the Banach ring A with radii $r_1, \dots, r_n$ to be the ring

$$A\{T_1/r_1, \dots, T_n/r_n\} = \left\{ f = \sum_I a_I T^I : a_I \in A, \lim_{I \to \infty} |a_I| r^I = 0 \right\},$$

where $I = (i_1, \dots, i_n)$ runs over n-tuples of nonnegative integers, $T^I = T_1^{i_1} \cdots T_n^{i_n}$, and $r^I = r_1^{i_1} \cdots r_n^{i_n}$. (That is, the series in question converge on the closed polydisc defined by the conditions $|T_i| \leqslant r_i$ for $i = 1, \dots, n$.) The set $A\{T_1/r_1, \dots, T_n/r_n\}$ is a subring of $A[\![T_1, \dots, T_n]\!]$ complete for the *Gauss norm*

$$\left| \sum_I a_I T^I \right|_r = \sup_I \{ |a_I| r^I \},$$

which is easily seen to be submultiplicative (resp. power-multiplicative, multiplicative) if the seminorm on A is; see [**87**, Lemma 1.7]. In case $r_1 = \cdots = r_n = 1$, we contract the notation to $A\{T_1, \dots, T_n\}$.

A bounded homomorphism $A \to B$ of Banach rings is *affinoid* if it factors as $A \to A\{T_1, \dots, T_n\} \to B$ for some positive integer n and some strict surjection $A\{T_1, \dots, T_n\} \to B$. We also say that B is an *affinoid algebra* over A.

We say that A is *strongly noetherian* if every affinoid algebra over A is noetherian, or equivalently the rings $A\{T_1, \dots, T_n\}$ are noetherian for all $n \geqslant 0$. It appears to be unknown whether every noetherian Banach algebra is strongly noetherian; that is, there is no known analogue of the Hilbert basis theorem for Banach algebras.

Remark 2.2.16. — In Berkovich's theory, what we call an *affinoid homomorphism* is more commonly called a *strictly affinoid homomorphism*; by contrast, an *affinoid homomorphism* would be allowed to have the form $A \to A\{T_1/r_1, \dots, T_n/r_n\} \to B$ for some positive integer n, some $r_1, \dots, r_n > 0$, and some strict surjection $A\{T_1/r_1, \dots, T_n/r_n\} \to B$. This extra generality is important in Berkovich's theory especially when A carries the trivial norm, but is incompatible with Huber's adic constructions.

2.3. The Gel'fand spectrum of a Banach ring

We now introduce one type of topological space corresponding to a Banach ring, as considered by Berkovich [**14, 15**].

Hypothesis 2.3.1. — Throughout §2.3, let A be a Banach ring with norm denoted by $|\cdot|$. Note that we do not yet impose any extra conditions on A, but see Remark 2.3.9.

Definition 2.3.2. — The *Gel'fand spectrum* $\mathcal{M}(A)$ of A is the set of multiplicative seminorms α on A bounded above by $|\cdot|$ (or equivalently, dominated by $|\cdot|$). We topologize $\mathcal{M}(A)$ as a closed subspace of the product $\prod_{a \in A}[0, |a|]$; hence $\mathcal{M}(A)$ is compact by Tikhonov's theorem [**19**, §1.9.5, Théorème 3] (see also [**14**, Theorem 1.2.1]). A subbasis of the topology on $\mathcal{M}(A)$ is given by the sets $\{\alpha \in \mathcal{M}(A) : \alpha(f) \in I\}$ for each $f \in A$ and each open interval $I \subseteq \mathbb{R}$. Any bounded homomorphism $\varphi : A \to B$ between Banach rings induces a continuous map $\varphi^* : \mathcal{M}(B) \to \mathcal{M}(A)$ by restriction.

Remark 2.3.3. — One can use Definition 2.3.2 to define the spectrum $\mathcal{M}(A)$ more generally for any ring A equipped with a submultiplicative seminorm. However, this will provide no useful additional generality, because the map $A \to \widehat{A}$ always induces a homeomorphism $\mathcal{M}(\widehat{A}) \to \mathcal{M}(A)$.

Berkovich's first main theorem about the spectrum is the following.

***Theorem 2.3.4* (Berkovich).** — *For A nonzero, $\mathcal{M}(A) \neq \varnothing$.*

Proof. — See [**14**, Theorem 1.2.1]. □

Corollary 2.3.5. — *For any nontrivial ideal I of A, there exists $\alpha \in \mathcal{M}(A)$ such that $\alpha(f) = 0$ for all $f \in I$.*

Proof. — Let J be the closure of I. By Lemma 2.2.2, A/J is nonzero, so $\mathcal{M}(A/J) \neq \varnothing$ by Theorem 2.3.4. Any element of $\mathcal{M}(A/J)$ restricts to an element $\alpha \in \mathcal{M}(A)$ of the desired form. (Compare [**14**, Corollary 1.2.4].) □

Corollary 2.3.6. — *A finite set $f_1, \dots, f_n$ of elements of A generates the unit ideal if and only if for each $\alpha \in \mathcal{M}(A)$, there exists an index $i \in \{1, \dots, n\}$ for which $\alpha(f_i) > 0$.*

Proof. — If there exist $u_1, \dots, u_n \in A$ for which $u_1 f_1 + \cdots + u_n f_n = 1$, then for each $\alpha \in \mathcal{M}(A)$, we have $\max_i\{\alpha(u_i)\alpha(f_i)\} \geqslant 1$ and so $\alpha(f_i) > 0$ for some index i. Conversely, suppose that $f_1, \dots, f_n$ generate a nontrivial ideal I; then by Corollary 2.3.5, we can choose $\alpha \in \mathcal{M}(A)$ such that $\alpha(f) = 0$ for all $f \in I$. □

Corollary 2.3.7. — *An element $f \in A$ is a unit if and only if $\alpha(f) > 0$ for all $\alpha \in \mathcal{M}(A)$.*

Definition 2.3.8. — For $\alpha \in \mathcal{M}(A)$, define the prime ideal $\mathfrak{p}_\alpha = \alpha^{-1}(0)$; then $\alpha \in \mathcal{M}(A)$ induces a multiplicative norm on $A/\mathfrak{p}_\alpha$. The completion of $\operatorname{Frac}(A/\mathfrak{p}_\alpha)$ for the unique multiplicative extension of this norm is called the *residue field* of α, and denoted $\mathcal{H}(\alpha)$. The image of the map $\mathcal{M}(A) \to \operatorname{Spec}(A)$ taking α to $\mathfrak{p}_\alpha$ contains all maximal ideals, by Corollary 2.3.5; see Lemma 2.3.12 for some consequences of this observation.

The *Gel'fand transform* of A is the map $A \to \prod_{\alpha \in \mathcal{M}(A)} \mathcal{H}(\alpha)$; it is bounded for the supremum norm on the product (or more precisely, on the subring of the product on which the supremum is finite).

Remark 2.3.9. — Starting in §2.4, we will require A to contain a topologically nilpotent unit z. To put this in context, consider the following conditions on A.

(a) We may view A as a Banach algebra over some analytic field.

(b) There exists a topologically nilpotent unit $z \in A$ such that $|z|_{\mathrm{sp}} \left|z^{-1}\right|_{\mathrm{sp}} = 1$. We will refer to any such z as a *uniform unit* in A. Note that for any $\alpha \in \mathcal{M}(A)$,
$$1 = \alpha(z)\alpha(z^{-1}) \leqslant |z|_{\mathrm{sp}} \left|z^{-1}\right|_{\mathrm{sp}} = 1$$
and so $\alpha(z) = |z|_{\mathrm{sp}}$.

(c) There exists a topologically nilpotent unit $z \in A$.

(d) The ring A is *free of trivial spectrum*: there exists no $\alpha \in \mathcal{M}(A)$ such that the norm on $\mathcal{H}(\alpha)$ is trivial, or equivalently (thanks to Corollary 2.3.6) that the ideal generated by $\mathfrak{m}_A$ is trivial.

We record the following observations.

- These conditions occur in increasingly weaker order: (a) implies (b) implies (c) implies (d). To see that (a) implies (b), note that if A is a Banach algebra over an analytic field K, then any $z \in K$ with $|z| \in (0, 1)$ has the desired form.
- If A is of characteristic p, then conditions (a) and (b) are equivalent: the existence of a uniform unit z forces A to be a Banach algebra over the analytic field $\mathbb{F}_p((z))$. This will be important in the study of perfectoid algebras, and is ultimately the reason why we can avoid allowing analytic fields to carry the trivial norm.
- If A is not of characteristic p, then (b) does not imply (a). See Remark 2.5.23 (in the case $rs = 1$) and Remark 5.1.8 for examples.
- Conditions (b) and (c) are not equivalent. For example, let k be any field, choose $c_1, c_2 \in (0, +\infty)$ which are linearly independent over $\mathbb{Q}$, normalize the z_i-adic

norm on $k((z_i))$ by putting $|z_i| = e^{-c_i}$, and put $A = k((z_1)) \oplus k((z_2))$ with the supremum norm (which is power-multiplicative). Then (z_1, z_2) is a topologically nilpotent unit, but there is no element $z \in A$ satisfying $|z|_{\mathrm{sp}}|z^{-1}|_{\mathrm{sp}} = 1$ and $|z|_{\mathrm{sp}} \neq 1$. However, when A is uniform (as in this example) this problem can be remedied by changing the norm without changing the norm topology; see Remark 2.8.18.
- We do not know whether (c) and (d) are equivalent.

Berkovich's second main theorem about the spectrum is the following result.

***Theorem 2.3.10* (Berkovich).** — *The restriction of the supremum norm on* $\prod_{\alpha \in \mathcal{M}(A)} \mathcal{H}(\alpha)$ *along the Gel'fand transform is the spectral seminorm on A.*

Proof. — See [**14**, Corollary 1.3.2]. □

Remark 2.3.11. — We collect several remarks about Theorem 2.3.10.

(a) Theorem 2.3.10 implies Theorem 2.3.4: if A is nonzero, the spectral seminorm of $1 \in A$ equals 1.

(b) The supremum norm in Theorem 2.3.10 is always achieved if A is nonzero: for each $f \in A$, the map $f \mapsto \alpha(f)$ is continuous on the compact space $\mathcal{M}(A)$, and so achieves its maximum. Consequently, Theorem 2.3.10 may be viewed as a form of the *maximum modulus principle* in nonarchimedean analytic geometry. For an analogous result in rigid analytic geometry, see [**18**, Proposition 6.2.1/4].

(c) It is not generally true that A is complete under its spectral seminorm even when the latter is a norm; this observation is related to the definition of *uniformization* (see Definition 2.8.13). One exception is for affinoid algebras over an analytic field; see Corollary 2.5.6.

(d) For any function $g : \mathcal{M}(A) \to \mathbb{R}^+$ whose image is bounded away from 0 and ∞, the norm $\sup\{\alpha^{g(\alpha)} : \alpha \in \mathcal{M}(A)\}$ defines the same topology on A as the spectral seminorm.

Lemma 2.3.12. — *A homomorphism $M \to N$ of A-modules, with N a finite A-module, is surjective if and only if $M \otimes_A \mathcal{H}(\alpha) \to N \otimes_A \mathcal{H}(\alpha)$ is surjective for all $\alpha \in \mathcal{M}(A)$.*

Proof. — Suppose that $M \otimes_A \mathcal{H}(\alpha) \to N \otimes_A \mathcal{H}(\alpha)$ is surjective for all $\alpha \in \mathcal{M}(A)$. For each maximal ideal $\mathfrak{p}$ of A, choose $\alpha \in \mathcal{M}(A)$ with $\mathfrak{p}_\alpha = \mathfrak{p}$. Then $A/\mathfrak{p} \to \mathcal{H}(\alpha)$ is an extension of fields, so surjectivity of $M \otimes_A \mathcal{H}(\alpha) \to N \otimes_A \mathcal{H}(\alpha)$ implies surjectivity of $M \otimes_A A/\mathfrak{p} \to N \otimes_A A/\mathfrak{p}$. This in turn implies surjectivity of $M \otimes_A A_\mathfrak{p} \to N \otimes_A A_\mathfrak{p}$ by Nakayama's lemma, and hence surjectivity of $M \to N$. □

Lemma 2.3.13. — *For $A \to B$, $A \to C$ homomorphisms of Banach rings, the map $\mathcal{M}(B \widehat{\otimes}_A C) \to \mathcal{M}(B) \times_{\mathcal{M}(A)} \mathcal{M}(C)$ is surjective.*

Proof. — This reduces to the case where A, B, C are all analytic fields, for which we may apply Lemma 2.2.9 (a) and Theorem 2.3.4. See also [**87**, Lemma 1.20]. □

Lemma 2.3.14. — *For A a Banach ring and B a faithfully finite étale A-algebra viewed as a Banach algebra over A as per Definition 2.2.14, the map $\mathcal{M}(B) \to \mathcal{M}(A)$ is surjective. (It is also open; see Lemma 2.4.17 (c).)*

Proof. — The hypothesis on B ensures that for each $\alpha \in \mathcal{M}(A)$, $B \otimes_A \mathcal{H}(\alpha) = B\widehat{\otimes}_A\mathcal{H}(\alpha)$ is a nonzero direct sum of finite extensions of $\mathcal{H}(\alpha)$, and so $\mathcal{M}(B \otimes_A \mathcal{H}(\alpha))$ is nonempty. This proves the claim. □

Remark 2.3.15. — When studying spectra, it is helpful to use general facts about compact topological spaces. Here are a few that we will need.

(a) The image of a quasicompact topological space under a continuous map is quasicompact [**19**, §I.9.4, Théorème 2]. Consequently, any continuous map $f : Y \to X$ from a quasicompact topological space to a Hausdorff topological space is closed [**19**, §I.9.4, Corollaire 2].

(b) With notation as in (a), if V is open in Y, then $W = X \setminus f(Y \setminus V)$ is open. One consequence is that if Z is closed in X and V is an open neighborhood of $f^{-1}(Z)$, then W is an open neighborhood of Z and $f^{-1}(W) \subseteq V$. Another consequence is that the quotient and subspace topologies on image(f) coincide: if $U \subseteq$ image(f) and $V = f^{-1}(U)$ is open in Y, then $U =$ image$(f) \cap W$ is open in image(f). That is, any continuous surjection (resp. bijection) from a quasicompact space to a Hausdorff space is a quotient map (resp. a homeomorphism).

(c) If X is the inverse limit of an inverse system $\{X_i\}_{i \in I}$ of nonempty compact spaces, then X is nonempty and compact. This follows from Tikhonov's theorem, or see [**19**, §1.9.6, Proposition 8]. As a corollary, for $i \in I$ and Z a closed subset of X_i, Z has empty inverse image in X if and only if there exists an index $j \geqslant i$ for which Z has empty inverse image in X_j.

(d) With notation as in (c), for any $i \in I$ and any open subsets $V_{1,i}, \dots, V_{n,i}$ of X_i whose inverse images in X form a covering, there exists an index $j \geqslant i$ for which the inverse images $V_{1,j}, \dots, V_{n,j}$ of $V_{1,i}, \dots, V_{n,i}$ in X_j form a covering of X_j itself: apply (c) to the closed set $X_i \setminus (V_{1,i} \cup \dots \cup V_{n,i})$. As a corollary, any finite open covering of X is refined by the pullback of a finite open covering of some X_i.

(e) With notation as in (c), any disconnection of X (*i.e.*, any partition of X into two disjoint closed-open subsets U_1, U_2) is the inverse image of a disconnection of some X_j, by the following argument. Choose any $i \in I$. By (a), the images $V_{1,i}, V_{2,i}$ of U_1, U_2 in X_i are closed and disjoint; they may thus be covered by disjoint open neighborhoods $W_{1,i}, W_{2,i}$. By (d), we can find an index $j \geqslant i$ such that X_j is covered by the inverse images $W_{1,j}, W_{2,j}$ of $W_{1,i}, W_{2,i}$ in X_j. Since $W_{1,j}, W_{2,j}$ are open and disjoint, they form a disconnection of X_j which pulls back to the given disconnection of X.

2.4. The adic spectrum of an adic Banach ring

We next introduce a second type of topological space corresponding to a Banach ring, as considered by Huber [**77, 78, 79**]. The natural levels of generality of the Berkovich and Huber constructions are incompatible; we work at reduced levels of generality where the two constructions can be compared. We begin with the base algebraic objects of Huber's construction.

Before proceeding, we recall that from now on, we only consider Banach rings containing a topologically nilpotent unit (see Remark 2.3.9).

Definition 2.4.1. — For A a ring equipped with a submultiplicative norm, let A° denote the subring of power-bounded elements of A. Note that $A^\circ \neq \mathfrak{o}_A$ in general unless the norm on A is power-multiplicative.

An *adic Banach ring* is a pair (A, A^+) in which A is a Banach ring (which from now on must be a Banach algebra containing a topologically nilpotent unit) and A^+ is a subring of A° which is open and integrally closed in A. These conditions ensure that every topologically nilpotent element of A must belong to A^+.

A *morphism* of adic Banach rings $(A, A^+) \to (B, B^+)$ is a bounded homomorphism $\varphi : A \to B$ of Banach rings such that $\varphi(A^+) \subseteq B^+$. With this definition, the correspondence $A \mapsto (A, A^\circ)$ defines a functor from the category of Banach rings to the category of adic Banach rings.

For $(A, A^+) \to (B, B^+), (A, A^+) \to (C, C^+)$ two morphisms of adic Banach rings, their coproduct in the category of adic Banach rings will be denoted by $(B, B^+)\widehat{\otimes}_{(A,A^+)}(C, C^+)$. It consists of (D, D^+) where $D = B\widehat{\otimes}_A C$ and D^+ is the completion of the integral closure of $B^+ \otimes_{A^+} C^+$ in D.

Remark 2.4.2. — For (A, A^+) an adic Banach ring and $B \in \mathbf{F\acute{E}t}(A)$, view B as a finite Banach A-algebra as in Definition 2.2.14. Then let B^+ be the integral closure of A^+ in B; in this way, we obtain a morphism $(A, A^+) \to (B, B^+)$. This construction will be used in the definition of étale morphisms on adic spaces in §8.

We now associate topological spaces to adic Banach rings. We will discuss the special topological properties of these spaces in more detail in §8.

Definition 2.4.3. — Let Γ be a totally ordered abelian group, and let Γ_0 be the pointed monoid $\Gamma \cup \{0\}$ with $0 \cdot \Gamma_0 = 0$ ordered so that $0 < \gamma$ for all $\gamma \in \Gamma$. A *semivaluation* on a ring A with values in Γ is a function $v : A \to \Gamma_0$ satisfying the following conditions.

(a) For all $a, b \in A$, we have $v(a - b) \leqslant \max\{v(a), v(b)\}$.

(b) For all $a, b \in A$, we have $v(ab) = v(a)v(b)$.

(c) We have $v(0) = 0$ and $v(1) = 1$. If moreover $v^{-1}(0) = \{0\}$, we say that v is a *valuation*.

For example, if $\Gamma = \mathbb{R}^+$, then a (semi)valuation is the same as a multiplicative (semi)norm.

For A a Banach ring, we declare two semivaluations on A (possibly valued in different ordered groups) to be *equivalent* if they define the same order relation on A. It is clear that this defines an equivalence relation and that the equivalence classes form a set (rather than a larger class). Denote the latter set by $\mathrm{Spv}(A)$. For linguistic convenience, we identify each equivalence class in $\mathrm{Spv}(A)$ with a particular representative in an arbitrary but fixed manner.

A semivaluation v on A is *continuous* if for every nonzero γ in the value group of v (*i.e.*, the subgroup of Γ generated by the nonzero images of v) there is a neighborhood U of 0 in A such that $v(u) < \gamma$ for all $u \in U$.

The *adic spectrum* of (A, A^+) is the subset $\mathrm{Spa}(A, A^+)$ of $\mathrm{Spv}(A)$ consisting of the equivalence classes of continuous semivaluations on A bounded by 1 on A^+. Since A^+ is integrally closed, we have the following equality analogous to Theorem 2.3.10:

$$A^+ = \{x \in A : v(x) \leqslant 1 \quad (v \in \mathrm{Spa}(A, A^+))\}. \tag{2.4.3.1}$$

(See [**78**, Proposition 1.6] for more details.) We equip $\mathrm{Spa}(A, A^+)$ with the topology generated by sets of the form

$$\{v \in \mathrm{Spv}(A, A^+) : v(a) \leqslant v(b) \neq 0\} \qquad (a, b \in A).$$

A *rational subspace* of $\mathrm{Spa}(A, A^+)$ is one of the form

$$\{v \in \mathrm{Spa}(A, A^+) : v(f_i) \leqslant v(g) \neq 0 \quad (i = 1, \dots, n)\} \tag{2.4.3.2}$$

for some $f_1, \dots, f_n, g \in A$ generating the unit ideal. One gets the same definition if one only requires that $f_1, \dots, f_n$ generate the unit ideal, since it is harmless to append g as an extra generator. One may also drop the condition $v(g) \neq 0$; see Remark 2.4.7.

Note that any morphism $\psi : (A, A^+) \to (B, B^+)$ induces a continuous map $\psi^* : \mathrm{Spa}(B, B^+) \to \mathrm{Spa}(A, A^+)$. Under this map, the inverse image of any rational subspace is again a rational subspace.

Remark 2.4.4. — Huber's definition of $\mathrm{Spa}(A, A^+)$ applies to more general topological rings than Banach rings. Namely, Huber defines an *f-adic ring* to be a topological ring A containing an open subring A_0 which is adic with a finitely generated ideal of definition (called a *ring of definition* of A). He then says that A is *Tate* if it contains a topologically nilpotent unit z. In this case, for any $c \in (0, 1)$, the norm

$$\alpha(x) = \inf\{c^n : n \in \mathbb{Z}, z^{-n}x \in A_0\}$$

gives A the structure of a Banach ring. One can show in addition (see [**77**, §1]) that the category of Banach rings containing topologically nilpotent units is equivalent to the category of Tate f-adic rings, except that one must allow morphisms of Banach rings which are continuous but not necessarily bounded (*e.g.*, see Remark 2.8.18).

For z a topologically nilpotent unit in A, a semivaluation v on A bounded by 1 on A^+ is continuous if and only if for every $x \in A$ with $v(x) \neq 0$, there exists $n \in \mathbb{Z}$ with $v(z^n) < v(x)$. In case A is a Banach ring over some analytic field K, one may say additionally that a semivaluation on A bounded by 1 on A^+ is continuous if and only if its restriction to K is equivalent to the norm on K. This implies that any continuous morphism of Banach rings over K is bounded; it also arises in the comparison with Gel'fand spectra in Definition 2.4.6.

For the remainder of §2.4, let (A, A^+) be any adic Banach ring. The analogue of the compactness of the Gel'fand spectrum is the following result. A more precise statement is that adic spectra are *spectral spaces*; see §8.1.

***Theorem 2.4.5* (Huber).** — *The space* $\mathrm{Spa}(A, A^+)$ *is quasicompact and the rational subspaces form a topological basis consisting of quasicompact open subsets.*

Proof. — See [**77**, Theorem 3.5 (i)-(ii)]. □

We now relate this construction back to the Gel'fand spectrum.

Definition 2.4.6. — There is a natural map $\mathcal{M}(A) \to \mathrm{Spa}(A, A^+)$ taking each $\alpha \in \mathcal{M}(A)$ to the equivalence class of α as a semivaluation. Beware that this map is not continuous.

Now suppose that A contains a uniform unit z; then there is a map $\mathrm{Spa}(A, A^+) \to \mathcal{M}(A)$ defined as follows. Given a semivaluation $v \in \mathrm{Spa}(A, A^+)$, define the multiplicative seminorm $\alpha = \alpha(v) \in \mathcal{M}(A)$ by the formula

$$\alpha(x) = \inf\{|z|_{\mathrm{sp}}^{r/s} : r \in \mathbb{Z},\ s \in \mathbb{Z}_{>0},\ v(z^r) > v(x^s)\}.$$

The composition $\mathcal{M}(A) \to \mathrm{Spa}(A, A^+) \to \mathcal{M}(A)$ is the identity. In particular, the map $\mathcal{M}(A) \to \mathrm{Spa}(A, A^+)$ is injective, and by Theorem 2.3.4, $\mathrm{Spa}(A, A^+) \neq \varnothing$ whenever $A \neq 0$.

Remark 2.4.7. — Given a rational subspace U of $\mathrm{Spa}(A, A^+)$ as in (2.4.3.2), let $\overline{U}$ be the image of U under the projection $\mathrm{Spa}(A, A^+) \to \mathcal{M}(A)$; since U is quasicompact, $\overline{U}$ is compact by Remark 2.3.15 (a). One has $\alpha(g) > 0$ for all $\alpha \in \overline{U}$, so by compactness $c = \inf\{\alpha(g) : \alpha \in \overline{U}\}$ is positive. For $0 < \epsilon < c$, any $f'_1, \dots, f'_n, g' \in A$ satisfying $|f'_i - f_i| < \epsilon, |g' - g| < \epsilon$ generate the unit ideal and satisfy

$$U = \{v \in \mathrm{Spa}(A, A^+) : v(f'_i) \leqslant v(g') \quad (i = 1, \dots, n)\}.$$

Compare [**87**, Remark 1.15], [**18**, Proposition 7.2.4/1].

Definition 2.4.8. — We define a *rational subspace* of $\mathcal{M}(A)$ as the intersection of $\mathcal{M}(A)$ with a rational subspace of $\mathrm{Spa}(A, A^+)$. For the rational subspace U of $\mathrm{Spa}(A, A^+)$ defined in (2.4.3.2), the corresponding rational subspace of $\mathcal{M}(A)$ is

$$\{\alpha \in \mathcal{M}(A) : \alpha(f_i) \leqslant \alpha(g) \quad (i = 1, \dots, n)\} \tag{2.4.8.1}$$

and the image of U in $\mathcal{M}(A)$ is equal to the intersection $U \cap \mathcal{M}(A)$. As a corollary, we see that every nonempty rational subspace of $\mathrm{Spa}(A, A^+)$ meets $\mathcal{M}(A)$, so $\mathcal{M}(A)$ is dense in $\mathrm{Spa}(A, A^+)$. (See however Remark 2.4.9 below.)

Rational subspaces of $\mathcal{M}(A)$ are closed, not open; as a result, not every rational subspace containing some $\alpha \in \mathcal{M}(A)$ is a neighborhood of α. However, those which are neighborhoods form a neighborhood basis of α in $\mathcal{M}(A)$; we say that such rational subspaces *encircle* α.

By the previous paragraph, $\mathrm{Spa}(A, A^+) \to \mathcal{M}(A)$ is continuous and hence a quotient map by Remark 2.3.15 (b). In fact, $\mathcal{M}(A)$ is the maximal Hausdorff quotient of $\mathrm{Spa}(A, A^+)$: for any continuous map $\mathrm{Spa}(A, A^+) \to U$ with U Hausdorff, any $v \in \mathrm{Spa}(A, A^+)$ projecting to $\alpha \in \mathcal{M}(A)$ is a specialization of α, so v and α must have the same image in U. An immediate consequence is that disconnections of $\mathrm{Spa}(A, A^+)$ and $\mathcal{M}(A)$ correspond, since they define maps to a two-point space.

Remark 2.4.9. — A rational subspace of $\mathrm{Spa}(A, A^+)$ need not be determined by its intersection with $\mathcal{M}(A)$ except in some restricted circumstances (*e.g.*, see Corollary 2.5.13). A typical example is $(A, A^+) = (K\{T\}, \mathfrak{o}_K + \mathfrak{m}_K \cdot \mathfrak{o}_K\{T\})$ for K an analytic field: for

$$U = \{v \in \mathrm{Spa}(A, A^+) : v(T) \geqslant 1\}$$
$$V = \{v \in \mathrm{Spa}(A, A^+) : v(T) \leqslant 1\},$$

one has $U \cap \mathcal{M}(A) = U \cap V \cap \mathcal{M}(A)$ but $U \not\subseteq V$ in general. See [**114**, Example 2.20] for a pictorial representation of this example.

Remark 2.4.10. — As per Remark 2.2.16, what we are calling a *rational subspace* of $\mathcal{M}(A)$ would be called a *strictly rational subspace* in Berkovich's setup. An arbitrary rational subspace would have the form

$$\{\alpha \in \mathcal{M}(A) : \alpha(f_1) \leqslant p_1\alpha(g), \dots, \alpha(f_n) \leqslant p_n\alpha(g)\}$$

for some $f_1, \dots, f_n, g \in A$ generating the unit ideal and some $p_1, \dots, p_n > 0$; such subspaces are needed to obtain a neighborhood basis when A is not required to contain a topologically nilpotent unit.

The analogue of the residue field $\mathcal{H}(\alpha)$ of a point α in a Gel'fand spectrum is the following construction.

Definition 2.4.11. — An *adic field* is an adic Banach ring (K, K^+) in which K is an analytic field and K^+ is a valuation ring in $\mathfrak{o}_K$ (*i.e.*, a subring containing either x or $1/x$ for each $x \in K^\times$). The space $\mathrm{Spa}(K, K^+)$ is not a point unless $K^+ = \mathfrak{o}_K$; however, the valuation corresponding to K^+ defines the generic point of $\mathrm{Spa}(K, K^+)$.

Given $v \in \mathrm{Spa}(A, A^+)$, let $(\mathcal{H}(v), \mathcal{H}(v)^+)$ be the adic field with $\mathcal{H}(v) = \mathcal{H}(\alpha(v))$ and $\mathcal{H}(v)^+$ equal to the valuation ring of the continuous multiplicative extension of v to $\mathcal{H}(\alpha(v))$. By construction, there is a canonical morphism $(A, A^+) \to (\mathcal{H}(v), \mathcal{H}(v)^+)$ under which the generic point of $\mathrm{Spa}(\mathcal{H}(v), \mathcal{H}(v)^+)$ maps to v.

Definition 2.4.12. — Let U be a quasicompact open subset of $\mathrm{Spa}(A, A^+)$. We say that U is an *affinoid subdomain* of $\mathrm{Spa}(A, A^+)$ if there exists an affinoid homomorphism $\varphi : (A, A^+) \to (B, B^+)$ which is initial among morphisms $\psi : (A, A^+) \to (C, C^+)$ of adic Banach rings for which $\psi^*(\mathrm{Spa}(C, C^+)) \subseteq U$. We refer to the representing morphism $(A, A^+) \to (B, B^+)$ as an *affinoid localization*.

In general, the structure of affinoid subdomains is quite mysterious (see Theorem 2.5.11 for an exception). However, every rational subspace U is an affinoid subdomain and the map $\mathrm{Spa}(B, B^+) \cong U$ is a homeomorphism (see Lemma 2.4.13 below). We thus refer to U also as a *rational subdomain* and to the corresponding affinoid localization also as a *rational localization.*

Note that the completed tensor product of two affinoid (resp. rational) localizations is again such a localization, corresponding to the intersection of affinoid (resp. rational) subdomains.

Lemma 2.4.13. — *Let U be a rational subspace of* $\mathrm{Spa}(A, A^+)$ *defined as in* (2.4.3.2).

(a) *The subspace U is an affinoid subdomain represented by $\varphi : (A, A^+) \to (B, B^+)$, where B is the quotient of $A\{T_1, \dots, T_n\}$ for the closure of the ideal $(gT_1 - f_1, \dots, gT_n - f_n)$, equipped with the quotient norm, and B^+ is the completion of the integral closure of the image of $A^+[T_1, \dots, T_n]$ in B.*

(b) *The map $\varphi^* : \mathrm{Spa}(B, B^+) \to \mathrm{Spa}(A, A^+)$ induces a homeomorphism $\mathrm{Spa}(B, B^+) \cong U$. More precisely, the rational subspaces of* $\mathrm{Spa}(B, B^+)$ *correspond to the rational subspaces of* $\mathrm{Spa}(A, A^+)$ *contained in U.*

Proof. — For (a), see [**78**, Proposition 1.3]. To check (b), note that φ^* by definition gives a continuous map from $\mathrm{Spa}(B, B^+)$ to U. To see that the map is bijective, choose any $v \in U$. The map $(A, A^+) \to (\mathcal{H}(v), \mathcal{H}(v)^+)$ factors uniquely through a bounded homomorphism $(B, B^+) \to (\mathcal{H}(v), \mathcal{H}(v)^+)$; the generic point of $(\mathcal{H}(v), \mathcal{H}(v)^+)$ maps to the unique point of $\mathrm{Spa}(B, B^+)$ in the preimage of v.

To see that the induced morphism $\mathrm{Spa}(B, B^+) \to U$ is a homeomorphism, it suffices to check the final assertion, *i.e.*, that any rational subspace of $\mathrm{Spa}(B, B^+)$ is also a rational subspace of $\mathrm{Spa}(A, A^+)$. This follows from Remark 2.4.7: any rational subspace of $\mathrm{Spa}(B, B^+)$ can be described using generators in $A[f_1/g, \dots, f_n/g]$, and such a description can be translated into a description using generators in A. (See also [**18**, Theorem 7.2.4/2] and [**77**, Lemma 1.5].) □

To obtain building blocks for the theory of adic spaces, we must define structure sheaves on adic Banach rings. We postpone the globalization step until §8.

Definition 2.4.14. — By a *rational covering* (resp. *affinoid covering*) of $\mathrm{Spa}(A, A^+)$, we will mean either a finite collection $\{U_i\}_i$ of rational (resp. affinoid) subdomains of $\mathrm{Spa}(A, A^+)$ forming a set-theoretic covering, or the corresponding collection $\{\mathrm{Spa}(A, A^+) \to \mathrm{Spa}(B_i, B_i^+)\}_i$ of rational (resp. affinoid) localizations, depending on context.

Note that a rational covering of $\mathrm{Spa}(A, A^+)$ induces a set-theoretic covering of $\mathcal{M}(A)$ by rational subspaces, but not conversely in general. However, a finite collection of rational subspaces whose relative interiors cover $\mathcal{M}(A)$ does induce a rational covering of $\mathrm{Spa}(A, A^+)$; we call such a covering a *strong rational covering* of $\mathrm{Spa}(A, A^+)$ (or of $\mathcal{M}(A)$).

Definition 2.4.15. — Define the *structure presheaf* $\mathcal{O}$ on $\mathrm{Spa}(A, A^+)$ as the functor taking each open subset U to the inverse limit of B over all rational localizations $(A, A^+) \to (B, B^+)$ for which $\mathrm{Spa}(B, B^+) \subseteq U$. In particular, for any rational localization $(A, A^+) \to (B, B^+)$, we have $\Gamma(\mathrm{Spa}(B, B^+), \mathcal{O}) = B$. The stalks of $\mathcal{O}$ are henselian local rings (see Lemma 2.4.17 below).

We say that (A, A^+) is *sheafy* if the structure presheaf is a sheaf; an equivalent condition (*e.g.*, see Proposition 2.4.21) is that for any rational localization $(A, A^+) \to (B, B^+)$, the map $B \to H^0(\mathrm{Spa}(B, B^+), \mathcal{O})$ is an isomorphism. In this case, $(\mathrm{Spa}(A, A^+), \mathcal{O})$ is a locally ringed space. This is not true in general; see [**77**, §1] or [**23**, §4.1] for failures of injectivity, and Example 2.8.7 for a failure of surjectivity.

***Proposition 2.4.16* (Huber).** — *Let (A, A^+) be an adic Banach ring such that A is strongly noetherian (see Definition 2.2.15). Then (A, A^+) is sheafy.*

Proof. — See [**78**, Theorem 2.2]. □

Lemma 2.4.17. — *The following statements hold.*

(a) *For $v \in \mathrm{Spa}(A, A^+)$, the stalk $\mathcal{O}_v$ is a henselian local ring whose residue field is dense in $\mathcal{H}(v)$.*

(b) *For $\alpha \in \mathcal{M}(A)$, let A_α be the direct limit of B over all rational localizations $(A, A^+) \to (B, B^+)$ encircling α. (We call this ring the* Hausdorff localization *at α to distinguish it from the stalk $\mathcal{O}_\alpha$.) Then A_α is a henselian local ring whose residue field is dense in $\mathcal{H}(\alpha)$.*

(c) *With notation as in Lemma 2.3.14, the map $\mathcal{M}(B) \to \mathcal{M}(A)$ is open.*

Proof. — In (a) and (b), the local property follows from Corollary 2.3.7 and the henselian property follows from Lemma 2.2.3 (a).

To check (c), we may work locally around $\alpha \in \mathcal{M}(A)$; by (b) and Theorem 1.2.8, we reduce to the case where $\mathcal{M}(B)$ contains a unique point β lifting α, $\mathcal{H}(\beta)$ is a Galois extension of $\mathcal{H}(\alpha)$ with group G, and G acts on B. In this case, for any open subset V of $\mathcal{M}(B)$ with image U in $\mathcal{M}(A)$, the inverse image of U in $\mathcal{M}(B)$ is the open set $\cup_{g \in G} g(V)$; since $\mathcal{M}(B) \to \mathcal{M}(A)$ is a quotient map by Remark 2.3.15 (b), U is open. □

We conclude this section by introducing the key formal arguments in the proofs of the theorems of Tate and Kiehl (Theorem 2.5.20), which allow us to reduce certain questions about coverings (namely sheaf, acyclicity, and glueing properties) to coverings of a very simple form.

Definition 2.4.18. — For $f_1, \dots, f_n \in A$ generating the unit ideal, the *standard rational covering* of $\mathrm{Spa}(A, A^+)$ generated by $f_1, \dots, f_n$ is the covering by the rational subspaces

$$U_i = \{v \in \mathrm{Spa}(A, A^+) : v(f_j) \leqslant v(f_i) \quad (j = 1, \dots, n)\} \qquad (i = 1, \dots, n).$$

For $f_1, \dots, f_n \in A$ arbitrary, the *standard Laurent covering* generated by $f_1, \dots, f_n$ is the covering by the rational subspaces

$$S_e = \bigcap_{i=1}^{n} S_{i,e_i} \qquad (e = (e_1, \dots, e_n) \in \{-, +\}^n),$$

where

$$S_{i,-} = \{v \in \mathrm{Spa}(A, A^+) : v(f_i) \leqslant 1\}, \qquad S_{i,+} = \{v \in \mathrm{Spa}(A, A^+) : v(f_i) \geqslant 1\}.$$

A standard Laurent covering with $n = 1$ is also called a *simple Laurent covering.*

We will use the following observations.

Lemma 2.4.19. — *The following statements hold.*

(a) *Any rational covering can be refined by a standard rational covering.*

(b) *For any standard rational covering $\mathfrak{U}$ of X, there exists a standard Laurent covering $\mathfrak{V}$ of X such that for each $V = \mathrm{Spa}(B, B^+) \in \mathfrak{V}$, the restriction of $\mathfrak{U}$ to V (omitting empty intersections) is a standard rational covering generated by units in B.*

(c) *Any standard rational covering generated by units can be refined by a standard Laurent covering generated by units.*

Proof. — To prove (a), we follow [**18**, Lemma 8.2.2/2]. Given a rational covering $U_1, \dots, U_n$ where U_i is generated by the parameter set $S_i = \{f_{i1}, \dots, f_{in_i}, g_i\}$, let S be the set of products of the form $s_1 \cdots s_n$ where $s_i \in S_i$ for all i. Let S' be the subset of S consisting of products $s_1 \cdots s_n$ for which $s_i = g_i$ for at least one i. Note that S' generates the unit ideal: for any $v \in \mathrm{Spa}(A, A^+)$, for each i we can find $s_i \in S_i$ not vanishing at v, taking $s_i = g_i$ for any i for which $v \in U_i$. Thus the parameter set S' defines a standard rational covering. To see that this refines the original covering, note that the rational subspace with final parameter $s_1 \cdots s_n$ does not change if we add $S \setminus S'$ to the set of parameters (again because the U_i form a covering), which makes it clear that this subspace is contained in U_i for any index i for which $s_i = g_i$ (because we now have parameters obtained from $s_1, \dots, s_n$ by replacing s_i with each of the other elements of S_i).

To prove (b), we follow [**18**, Lemma 8.2.2/3]. Let $\mathfrak{U}$ be the standard rational covering defined by the parameters $f_1, \dots, f_n$. We argue as in Remark 2.4.7: since $f_1, \dots, f_n$ generate the unit ideal, by Corollary 2.3.6 the quantity

$$c = \inf\{\max_i\{\alpha(f_i)\} : \alpha \in \mathcal{M}(A)\}$$

is positive. Since A contains a topologically nilpotent unit, we may rescale $f_1, \dots, f_n$ to reduce to the case $c > 1$. In this case, the standard Laurent covering $\mathfrak{V}$ defined by $f_1, \dots, f_n$ has the desired property: on the subspace where $|f_1|, \dots, |f_s| \leqslant 1$ and $|f_{s+1}|, \dots, |f_n| \geqslant 1$, the restriction of $\mathfrak{U}$ is the standard rational covering generated by $f_{s+1}, \dots, f_n$ plus some empty intersections.

To prove (c), we follow [**18**, Lemma 8.2.2/4]. Consider the standard rational covering generated by the units $f_1, \dots, f_n$. This cover is refined by the standard Laurent covering generated by $f_i f_j^{-1}$ for $1 \leqslant i < j \leqslant n$, by an elementary combinatorics argument (any total ordering on a finite set has a maximal element). □

Using these observations, we obtain the following criterion.

Proposition 2.4.20. — *Let $\mathcal{P}$ be a property of rational coverings of rational subdomains of* $\mathrm{Spa}(A, A^+)$ *satisfying the following conditions.*

(a) *The property $\mathcal{P}$ is local: if it holds for a refinement of a given covering, it also holds for the original covering.*

(b) *The property $\mathcal{P}$ is transitive: if it holds for a covering $\{(B, B^+) \to (C_i, C_i^+)\}_i$ and for some coverings $\{(C_i, C_i^+) \to (D_{ij}, D_{ij}^+)\}_j$ for each i, then it holds for the composite covering $\{(B, B^+) \to (D_{ij}, D_{ij}^+)\}_{i,j}$.*

(c) *The property $\mathcal{P}$ holds for any simple Laurent covering.*

Then the property $\mathcal{P}$ holds for any rational covering of any rational subdomain of $\mathrm{Spa}(A, A^+)$.

Proof. — We make the following observations.

(i) We may deduce $\mathcal{P}$ for any standard Laurent covering generated by units by writing it as a composition of simple Laurent coverings generated by units, then invoking (b) and (c).

(ii) We may deduce $\mathcal{P}$ for any standard rational covering generated by units by applying Lemma 2.4.19 (c) to refine the covering by a standard Laurent covering generated by units, then invoking (a) and (i).

(iii) Given a standard rational covering $\{(B, B^+) \to (C_i, C_i^+)\}_i$, we obtain, using Lemma 2.4.19 (b), a standard Laurent covering $\{(B, B^+) \to (D_j, D_j^+)\}_j$ such that for each j, the covering $\{(D_j, D_j^+) \to (D_j, D_j^+)\widehat{\otimes}_{(B,B^+)}(C_i, C_i^+)\}_i$ is a standard rational covering generated by units in D_j. We may thus deduce $\mathcal{P}$ for the covering $\{(B, B^+) \to (D_j, D_j^+)\widehat{\otimes}_{(B,B^+)}(C_i, C_i^+)\}_{i,j}$ by invoking (ii) and (b), and then deduce $\mathcal{P}$ for the original covering by invoking (a).

(iv) We may deduce $\mathcal{P}$ for any covering by applying Lemma 2.4.19 (b) to refine the covering by a standard rational covering, then invoking (i).

These observations prove the claim. □

We will apply Proposition 2.4.20 to two general purposes: construction of acyclic sheaves of rings, and comparison of certain modules over such sheaves with their global sections. In the latter case, the conditions of Proposition 2.4.20 will be easy to verify directly. In the former case, one must be slightly more careful; we package an extra argument into the following proposition modeled on [**18**, Proposition 8.2.2/5].

Proposition 2.4.21. — *Let $\mathcal{F}$ be a presheaf of abelian groups on* $\mathrm{Spa}(A, A^+)$. *Suppose that for every rational subdomain* $U = \mathrm{Spa}(B, B^+)$ *of* $\mathrm{Spa}(A, A^+)$ *and every simple Laurent covering* V_1, V_2 *of* U, *we have*

$$\check{H}^0(U, \mathcal{F}; \mathfrak{V}) = \mathcal{F}(U), \quad \textit{resp.} \quad \check{H}^i(U, \mathcal{F}; \{V_1, V_2\}) = \begin{cases} \mathcal{F}(U) & i = 0 \\ 0 & i = 1. \end{cases} \tag{2.4.21.1}$$

Then for every rational subdomain U *of* $\mathrm{Spa}(A, A^+)$ *and every rational covering* $\mathfrak{V}$ *of* U,

$$H^0(U, \mathcal{F}) = \check{H}^0(U, \mathcal{F}; \mathfrak{V}) = \mathcal{F}(U), \quad \textit{resp.} \quad H^i(U, \mathcal{F}) = \check{H}^i(U, \mathcal{F}; \mathfrak{V}) = \begin{cases} \mathcal{F}(U) & i = 0 \\ 0 & i > 0. \end{cases}$$

Proof. — Throughout this argument, let U be an arbitrary rational subdomain of $\mathrm{Spa}(A, A^+)$ and let $\mathfrak{V}$ be a rational covering of U. We identify a series of properties of $\mathfrak{V}$ which satisfy the criteria of Proposition 2.4.20, and hence hold for all U and $\mathfrak{V}$.

First, the property that $\mathcal{F}(U) \to \check{H}^0(U, \mathcal{F}; \mathfrak{V})$ is injective satisfies (a) and (b) formally and (c) by (2.4.21.1).

Next, the property that $\mathcal{F}(U) \to \check{H}^0(U, \mathcal{F}; \mathfrak{V})$ is bijective satisfies (b) formally and (c) by (2.4.21.1). To check (a), let $\mathfrak{V}'$ be a refinement of $\mathfrak{V}$ such that $\mathcal{F}(U) \to \check{H}^0(U, \mathcal{F}; \mathfrak{V}')$ is bijective. The map $\check{H}^0(U, \mathcal{F}; \mathfrak{V}) \to \check{H}^0(U, \mathcal{F}; \mathfrak{V}')$ is then surjective, but it is also injective by the previous paragraph (applied to each element of $\mathfrak{V}$).

From now on, assume we are in the second situation. Next, we say that $\mathfrak{V}$ is *Čech-acyclic* if $\check{H}^i(U, \mathcal{F}; \mathfrak{V}) = 0$ for all $i > 0$. This property satisfies criteria (b) formally and (c) by (2.4.21.1), but not (a).

Instead, we say that $\mathfrak{V}$ is *universally Čech-acyclic* if its pullback to any rational subdomain of U is Čech-acyclic. This property formally also satisfies criteria (b) and (c) of Proposition 2.4.20. However, it also satisfies (a) by a spectral sequence argument; see [**18**, Corollary 8.1.4/3].

We thus deduce that every rational covering of every rational subdomain is Čech-acyclic. Acyclicity for sheaf cohomology then follows by a standard homological algebra argument (see [**122**, Tag 01EW]). □

As a first application of this argument, we have the following result which asserts that for an arbitrary adic Banach ring, the only obstruction to the analogue of Tate's acyclicity theorem is the failure of the structure presheaf to be a sheaf.

Lemma 2.4.22. — *Let* S_-, S_+ *be the simple Laurent covering of* $\mathrm{Spa}(A, A^+)$ *defined by some* $f \in A$. *Let* $(A, A^+) \to (B_1, B_1^+), (A, A^+) \to (B_2, B_2^+), (A, A^+) \to (B_{12}, B_{12}^+)$ *be the rational localizations corresponding to* $S_-, S_+, S_- \cap S_+$. *Then the map* $B_1 \oplus B_2 \to B_{12}$ *taking* (b_1, b_2) *to* $b_1 - b_2$ *is surjective.*

Proof. — By Lemma 2.4.13, we obtain strict surjections

$$A\{T\} \longrightarrow B_1, \quad A\{U\} \longrightarrow B_2, \quad A\{T, U\} \longrightarrow B_{12}$$

taking T to f and U to f^{-1}. In particular, any $b \in B_{12}$ can be lifted to some $\sum_{i,j=0}^{\infty} a_{ij} T^i U^j \in A\{T, U\}$. Let a'_n be the sum of a_{ij} over all $i, j \geqslant 0$ with $i - j = n$; note that this sum converges in A. Let b_1 be the image of $\sum_{n=0}^{\infty} a'_n T^n$ in B_1. Let b_2 be the image of $-\sum_{n=1}^{\infty} a'_{-n} U^n$ in B_2. Then $(b_1, b_2) \in B_1 \oplus B_2$ maps to $b \in B_{12}$, proving the desired exactness. □

Theorem 2.4.23. — *Suppose that (A, A^+) is sheafy. Then for every rational covering $\mathfrak{U}$ of* $\mathrm{Spa}(A, A^+)$,

$$H^i(\mathrm{Spa}(A, A^+), \mathcal{O}) = \check{H}^i(\mathrm{Spa}(A, A^+), \mathcal{O}; \mathfrak{U}) = \begin{cases} A & i = 0 \\ 0 & i > 0. \end{cases}$$

Proof. — By Proposition 2.4.21, it suffices to check Čech-acyclicity for simple Laurent coverings. Since the sheafy condition propagates to rational subspaces, we may as well consider only simple Laurent coverings of $\mathrm{Spa}(A, A^+)$ itself. In the notation of Lemma 2.4.22, the sequence

$$0 \longrightarrow A \longrightarrow B_1 \oplus B_2 \longrightarrow B_{12} \longrightarrow 0$$

is exact at B_{12}; by the sheafy hypothesis, it is also exact at A and $B_1 \oplus B_2$. Thus Proposition 2.4.21 yields the claim. □

In some cases, one can apply Proposition 2.4.20 to prove properties of individual inclusions of rational subdomains, by taking $\mathcal{P}$ to be the condition that every subdomain in a covering has the desired property. However, in some cases it is not straightforward to verify locality, in which case the following alternate reduction process may be preferable.

Proposition 2.4.24. — *Let $\mathcal{P}$ be a property of inclusions $V \subseteq U$ of rational subdomains of* $\mathrm{Spa}(A, A^+)$ *satisfying the following conditions.*

(a) *The property $\mathcal{P}$ is transitive: if it holds for $V \subseteq U$ and $W \subseteq V$, then it holds for $W \subseteq U$.*

(b) *The property $\mathcal{P}$ holds for any inclusion $V \subseteq U$ which is part of a simple Laurent covering of U.*

Then the property $\mathcal{P}$ holds for any inclusion of rational subdomains of $\mathrm{Spa}(A, A^+)$.

Proof. — To check that $\mathcal{P}$ holds for $V \subseteq U$, write $U = \mathrm{Spa}(B, B^+)$ and suppose that V is defined by the parameters $f_1, \dots, f_n, g \in B$. Let $z \in B$ be a topologically nilpotent unit. By Remark 2.4.7, for any sufficiently large m the set

$$V_0 = \{v \in U : v(gz^{-m}) \geqslant 1\}$$

is contained in V. For $i = 1, \dots, n$ in turn, define

$$V_i = \{v \in V_{i-1} : v(f_i g^{-1}) \leqslant 1\}.$$

Since each of the inclusions $V = V_n \subseteq \cdots \subseteq V_0 \subseteq U$ is part of a simple Laurent covering, we may deduce the claim from (a) and (b). □

2.5. Coherent sheaves on affinoid spaces

We now restrict to the setting of affinoid spaces over an analytic field, where a good theory of coherent sheaves is available thanks to the work of Tate and Kiehl. However, since we are working in the framework of adic spectra, we must be a bit careful to ensure that our statements do indeed follow from the classical ones. (If one returns to the classical results, these can mostly be extended to the most general setting of Berkovich; see Remark 2.5.22.)

Hypothesis 2.5.1. — Throughout §2.5, let (A, A^+) be an adic Banach ring in which A is an affinoid algebra over an analytic field K. We refer to any such object as an *adic affinoid algebra* over K. Unless specified we do not assume $A^+ = A^\circ$; when this does occur, we get an affinoid K-algebra of tft (topologically finite type) in the terminology of [**114**].

Lemma 2.5.2. — *The ring A is noetherian, so (by Remark 2.2.11) any ideal of A is closed. Moreover, any finite A-module may be viewed as a finite Banach A-module in a canonical way, under which any A-linear homomorphism of finite A-modules is continuous and strict.*

Proof. — For the first assertion, see [**18**, Proposition 6.1.1/3]. For the other assertions (which apply to any noetherian Banach ring), see [**18**, §3.7.3]. □

Remark 2.5.3. — A refinement of Lemma 2.5.2 is that A is *excellent* in the sense of Grothendieck [**38**, Théorème 2.6], and hence catenary.

***Lemma 2.5.4* (Noether normalization).** — *For $A \neq 0$, there exists a finite strict monomorphism $K\{T_1, \dots, T_n\} \to A$ for some $n \geqslant 0$.*

Proof. — See [**18**, Corollary 6.1.2/2]. □

Corollary 2.5.5. — *The following statement are true.*

(a) *Every maximal ideal of A has residue field finite over K.*

(b) *The formula $\alpha \mapsto \mathfrak{p}_\alpha$ defines a bijection from* $\mathrm{Maxspec}(A)$ *to the points of $\mathcal{M}(A)$ with residue field finite over K. (We will hereafter identify* $\mathrm{Maxspec}(A)$ *with a subspace of $\mathcal{M}(A)$ and of* $\mathrm{Spa}(A, A^+)$.*)*

(c) *If A is nonzero, then the sets in (b) are nonempty.*

Proof. — Assertion (a) follows from Lemma 2.5.4. For (b), note that on one hand, if $\mathcal{H}(\alpha)$ is finite over K, then $A \to \mathcal{H}(\alpha)$ is surjective because its image generates a dense subfield containing K. Consequently, $\mathfrak{p}_\alpha$ is a maximal ideal of A. Conversely, if $\mathfrak{m}$ is a maximal ideal of A, then $A/\mathfrak{m}$ is a finite extension of K, and so is complete for the unique multiplicative extension of the norm on K. It thus may be identified with $\mathcal{H}(\alpha)$ for some $\alpha \in \mathcal{M}(A)$. This proves (b); since any nonzero ring has a maximal ideal, (b) implies (c). □

Corollary 2.5.6. — *For A reduced, the spectral seminorm on A is a norm equivalent to the given norm.*

Proof. — See [**18**, Theorem 6.2.4/1]. □

Corollary 2.5.7. — *Let $A \to B$ be a finite morphism. Then for the spectral seminorms on A and B, the induced map $\kappa_A \to \kappa_B$ is finite.*

Proof. — See [**18**, Theorem 6.3.4/2]. □

We next relate affinoid subdomains of adic spectra with the corresponding notion in rigid analytic geometry.

Definition 2.5.8. — An *affinoid subdomain* of $\mathrm{Maxspec}(A)$ is a subset U of $\mathrm{Maxspec}(A)$ for which there exists a morphism $\varphi : A \to B$ of affinoid algebras over K which is initial for the property that $\varphi^* : \mathrm{Maxspec}(B) \to \mathrm{Maxspec}(A)$ factors through U. For example, the intersection of $\mathrm{Maxspec}(A)$ with a rational subspace of $\mathrm{Spa}(A, A^+)$ is an affinoid subdomain [**18**, Proposition 7.2.3/4]; any such subspace is called a *rational subdomain* of $\mathrm{Maxspec}(A)$.

Lemma 2.5.9. — *Let U be a rational subspace of* $\mathrm{Spa}(A, A^+)$ *defined as in* (2.4.3.2). *Let $(A, A^+) \to (B, B^+)$ be the representing morphism.*

(a) *We have $B \cong A\{T_1, \dots, T_n\}/(gT_1 - f_1, \dots, gT_n - f_n)$.*

(b) *The space $U \cap \mathrm{Maxspec}(A)$ is a rational subdomain of* $\mathrm{Maxspec}(A)$ *represented by the morphism $A \to B$.*

(c) *If A is reduced, then so is B.*

(d) *If $A^+ = A^\circ$, then $B^+ = B^\circ$.*

Proof. — By Lemma 2.5.2 the ideal $(gT_1 - f_1, \dots, gT_n - f_n)$ in $A\{T_1, \dots, T_n\}$ is closed, so (a) follows from Lemma 2.4.13. We deduce (b) from (a) plus [**18**, Proposition 7.2.3/4]. We deduce (c) from (b) plus [**18**, Corollary 7.3.2/10]. To deduce (d), apply Corollary 2.5.7 to the finite morphism $A\{T_1, \dots, T_n\} \to B$ to deduce that κ_B is integral over $\kappa_A[T_1, \dots, T_n]$ *via* the map taking T_i to f_i/g. □

Lemma 2.5.10. — *Any finite covering of* $\mathrm{Maxspec}(A)$ *by rational subdomains is refined by a covering induced by a rational covering of* $\mathrm{Spa}(A, A^+)$.

Proof. — By [**18**, Lemma 8.2.2/2], any finite covering of $\mathrm{Maxspec}(A)$ by rational subdomains is refined by standard rational covering in the sense of Definition 2.4.18. □

The Gerritzen-Grauert theorem in rigid analytic geometry can be interpreted as follows. (See also Temkin's proof in the context of Berkovich's theory [**123**, Theorem 3.1].)

Theorem 2.5.11. — *Let U be an affinoid subdomain of* $\mathrm{Maxspec}(A)$. *Then there exists a rational covering $V_1, \dots, V_n$ of* $\mathrm{Spa}(A, A^+)$ *such that $U \cap V_i$ is a rational subdomain in $V_i \cap \mathrm{Maxspec}(A)$ for $i = 1, \dots, n$. In particular, U can be written as a finite union of rational subdomains of* $\mathrm{Maxspec}(A)$ *(but not conversely: not every finite union of rational subdomains is an affinoid subdomain).*

Proof. — By [**18**, Theorem 7.3.5/1], there exists a finite collection of rational subdomains $W_1, \dots, W_m$ of $\mathrm{Maxspec}(A)$ with the property that $U \cap W_i$ is a rational subdomain in W_i for $i = 1, \dots, m$. By Lemma 2.5.10, this covering can be refined to a covering induced by a rational covering of $\mathrm{Spa}(A, A^+)$. □

Lemma 2.5.12. — *Suppose that $A^+ = A^\circ$ (and hence A is reduced; see Definition 2.8.1). For any rational subdomain* $\mathrm{Spa}(B, B^+)$ *of* $\mathrm{Spa}(A, A^+)$, *a finite collection of rational subdomains $\{U_i\}_i$ of* $\mathrm{Spa}(B, B^+)$ *is a rational covering if and only if $\{U_i \cap \mathrm{Maxspec}(B)\}_i$ is a covering of* $\mathrm{Maxspec}(B)$.

Proof. — We first check the special case of a one-element covering. Suppose that $V \subseteq U$ are rational subspaces of $\mathrm{Spa}(A, A^\circ)$ such that U is rational and $V \cap \mathrm{Maxspec}(A) = U \cap \mathrm{Maxspec}(A)$. Let $(A, A^\circ) \to (B, B^+) \to (C, C^+)$ be the corresponding rational localizations. By Lemma 2.5.9, $B^+ = B^\circ$. Since $B \to C$ is a rational localization, it induces isomorphisms at completed (algebraic) local rings, and hence is an open immersion of rigid analytic spaces [**18**, Proposition 7.3.3/5]. But by assumption $\mathrm{Maxspec}(B) = \mathrm{Maxspec}(C)$, so we obtain an isomorphism of rigid analytic spaces (see the discussion after [**18**, Corollary 7.3.3/6]) and $B \to C$ is itself an isomorphism of rings. Since $B^\circ = C^\circ \subseteq C^+ \subseteq C^\circ$, we have $C^+ = C^\circ$ and hence $V = U$.

To prove the statement of the lemma, note that $B^+ = B^\circ$ by Lemma 2.5.9; we may thus assume $(B, B^+) = (A, A^\circ)$. Let $\{U_i\}_i$ be a finite collection of rational subdomains of $\mathrm{Spa}(A, A^\circ)$ such that $\{U_i \cap \mathrm{Maxspec}(A)\}_i$ is a covering of $\mathrm{Maxspec}(A)$. By Lemma 2.5.10, there exists a rational covering $\{V_j\}_j$ of $\mathrm{Spa}(A, A^\circ)$ such that the covering $\{V_j \cap \mathrm{Maxspec}(A)\}_{j=1}^n$ of $\mathrm{Maxspec}(A)$ refines the covering $\{U_i \cap \mathrm{Maxspec}(A)\}_i$. That is, for each $j = 1, \dots, n$, there exists some i such that $V_j \cap \mathrm{Maxspec}(A) \subseteq U_i \cap \mathrm{Maxspec}(A)$, or in other words $(V_j \cap U_i) \cap \mathrm{Maxspec}(A) = V_j \cap \mathrm{Maxspec}(A)$. By the previous paragraph, this implies that $V_j \cap U_i = V_j$, or in other words $V_j \subseteq U_i$. Hence $\{U_i\}_i$ is a rational covering as claimed. □

Corollary 2.5.13. — *Suppose that A is reduced. Then no two distinct rational subspaces of* $\mathrm{Spa}(A, A^+)$ *have the same intersection with* $\mathrm{Maxspec}(A)$.

Proof. — By Corollary 2.5.6, we may reduce to the case $A^+ = A^\circ$. For $U = \mathrm{Spa}(B, B^+)$ a rational subspace of $\mathrm{Spa}(A, A^\circ)$, $U = \varnothing$ iff $B = 0$ (Theorem 2.3.4) iff $\mathrm{Maxspec}(B) = 0$. We may thus deduce the claim from Lemma 2.5.12. □

Proposition 2.5.14. — *Let U be an affinoid subdomain of* $\mathrm{Spa}(A, A^+)$ *represented by* $(A, A^+) \to (B, B^+)$.

(a) *The space* $U \cap \mathrm{Maxspec}(A)$ *is an affinoid subdomain of* $\mathrm{Maxspec}(A)$ *represented by the morphism* $A \to B$.

(b) *If* A *is reduced, then so is* B.

(c) *If* $A^+ = A^\circ$, *then* $B^+ = B^\circ$.

Proof. — To check (a), let $A \to C$ be a morphism of affinoid algebras over K such that $\mathrm{Maxspec}(C)$ maps into $U \cap \mathrm{Maxspec}(A)$; we must show that this morphism factors through B. Form a morphism $(A, A^+) \to (C, C^+)$ of adic Banach rings by taking $C^+ = C^\circ$. To factor $A \to C$ through B, it suffices to check that $\mathrm{Spa}(C, C^+)$ maps into U; for this purpose, we may assume that A is reduced, then pass from (A, A^+) to (A, A°) by base extension (using Corollary 2.5.6) and thus assume that $A^+ = A^\circ$. Apply Theorem 2.5.11 to obtain a rational covering $\{V_i\}_i$ of $\mathrm{Spa}(A, A^\circ)$ such that for each i, $U \cap V_i \cap \mathrm{Maxspec}(A)$ is a rational subspace of $V_i \cap \mathrm{Maxspec}(A)$. Let $(A, A^+) \to (D_i, D_i^+)$ be the rational localization representing V_i; by Lemma 2.5.9, $D_i^+ = D_i^\circ$. Put $(E_i, E_i^+) = (B, B^+)\widehat{\otimes}_{(A,A^+)}(D_i, D_i^+)$ and $(F_i, F_i^+) = (C, C^+)\widehat{\otimes}_{(A,A^+)}(D_i, D_i^+)$; it now suffices to check that the image of $\mathrm{Spa}(F_i, F_i^+) \to \mathrm{Spa}(D_i, D_i^+)$ is contained in $U \cap V_i$. Since $(D_i, D_i^+) \to (E_i, E_i^+)$ and $(C, C^+) \to (F_i, F_i^+)$ are rational localizations, Lemma 2.5.9 implies that $E_i^+ = E_i^\circ$, $F_i^+ = F_i^\circ$, and $U \cap V_i \cap \mathrm{Maxspec}(A)$ is a rational subdomain of $\mathrm{Maxspec}(D_i) = V_i \cap \mathrm{Maxspec}(A)$. Since $\mathrm{Maxspec}(F_i)$ maps into $U \cap V_i \cap \mathrm{Maxspec}(A)$, it follows that $D_i \to F_i$ factors through E_i and so $(D_i, D_i^\circ) \to (F_i, F_i^\circ)$ factors through (E_i, E_i°). Hence the image of $\mathrm{Spa}(F_i, F_i^+) \to \mathrm{Spa}(D_i, D_i^+)$ is contained in $U \cap V_i$, yielding (a).

Given (a), we may deduce (b) and (c) by the same arguments as in the proofs of parts (c) and (d) of Lemma 2.5.9. □

Proposition 2.5.15. — *Suppose that* $A^+ = A^\circ$ *(and hence* A *is reduced). For any affinoid subdomain* $\mathrm{Spa}(B, B^+)$ *of* $\mathrm{Spa}(A, A^\circ)$, *a finite collection of affinoid subdomains* $\{U_i\}_i$ *of* $\mathrm{Spa}(B, B^+)$ *is an affinoid covering if and only if* $\{U_i \cap \mathrm{Maxspec}(B)\}_i$ *is a covering of* $\mathrm{Maxspec}(B)$.

Proof. — This follows by the same proof as Lemma 2.5.12, but using Proposition 2.5.14 in lieu of Lemma 2.5.9. □

Remark 2.5.16. — Proposition 2.5.14 and Proposition 2.5.15 allow us to assert statements about affinoid subdomains of $\mathrm{Spa}(A, A^+)$ by invoking the corresponding statements about affinoid subdomains of $\mathrm{Maxspec}(A)$ from rigid analytic geometry. We will do this without further comment in what follows.

Lemma 2.5.17. — *For any affinoid localization* $(A, A^+) \to (B, B^+)$, *the morphism* $A \to B$ *is flat.*

Proof. — See [**18**, Corollary 7.3.2/6]. □

Proposition 2.5.18. — *Let* $\{(A, A^+) \to (B_i, B_i^+)\}_{i=1}^n$ *be an affinoid covering. Then the ring homomorphism* $A \to B_1 \oplus \cdots \oplus B_n$ *is faithfully flat.*

Proof. — The homomorphism is flat by Lemma 2.5.17. It is faithful by Lemma 1.1.4 and the fact that every maximal ideal of A is closed (by Corollary 2.3.5). □

Corollary 2.5.19. — *Let* $\{(A, A^+) \to (B_i, B_i^+)\}_{i=1}^n$ *be an affinoid covering.*

(a) *A finite A-module M is locally free if and only if $M \otimes_A B_i$ is a locally free B_i module for* $i = 1, \dots, n$.

(b) *A finite A-algebra R is étale if and only if $R \otimes_A B_i$ is an étale B_i-algebra for* $i = 1, \dots, n$.

Proof. — This follows from Proposition 2.5.18 plus Theorem 1.3.5. □

Theorem 2.5.20. — *Let $\mathfrak{U}$ be an affinoid covering of* $\operatorname{Spa}(A, A^+)$.

(a) *For any finite A-module M, let $\tilde{M}$ be the sheaf of $\mathcal{O}$-modules on* $\operatorname{Spa}(A, A^+)$ *induced by M. Then $H^i(\operatorname{Spa}(A, A^+), \tilde{M}; \mathfrak{U}) = M$ for $i = 0$ and 0 for $i > 0$. In particular, (A, A^+) is sheafy and $H^i(\operatorname{Spa}(A, A^+), \tilde{M}) = M$ for $i = 0$ and 0 for* $i > 0$.

(b) *The functor $M \mapsto \tilde{M}$ defines a tensor equivalence between finite A-modules and coherent sheaves of $\mathcal{O}$-modules on* $\operatorname{Spa}(A, A^+)$. *In particular, for $\{(A, A^+) \to (B_i, B_i^+)\}_{i=1}^n$ the morphisms representing $\mathfrak{U}$, the homomorphism $A \to B_1 \oplus \cdots \oplus B_n$ is an effective descent morphism for finite Banach modules over Banach rings.*

Proof. — Part (a) is due to Tate; see [**18**, Corollary 8.2.1/5]. Part (b) is due to Kiehl; see [**18**, Theorem 9.4.3/3]. □

Corollary 2.5.21. — *Let U be a closed-open subset of* $\operatorname{Spa}(A, A^+)$. *Then there exists a unique idempotent element $e \in A$ whose image in $\mathcal{H}(v)$ is 1 if $v \in U$ and 0 if $v \notin U$. In particular, the projection $A \to eA$ taking $x \in A$ to ex induces a homeomorphism* $\operatorname{Spa}(eA, eA^+) \cong U$.

Note that the analogous result for $\mathcal{M}(A)$ also holds because the closed-open subsets of $\mathcal{M}(A)$ and $\operatorname{Spa}(A, A^+)$ correspond; see Definition 2.4.8. Note also that this result generalizes to arbitrary adic Banach rings; see Proposition 2.6.4.

Proof. — Cover U with finitely many rational subdomains $U_1, \dots, U_m$ and the complement of U with finitely many rational subdomains $V_1, \dots, V_n$. Let $(A, A^+) \to (B_i, B_i^+)$ and $(A, A^+) \to (C_j, C_j^+)$ be the morphisms representing U_i and V_j, respectively. By Theorem 2.5.20 (a) applied with $M = A$, the element $((1, \dots, 1), (0, \dots, 0))$ of $(B_1 \oplus \cdots \oplus B_m) \oplus (C_1 \oplus \cdots \oplus C_n)$ determines an idempotent element $e \in A$ with the desired property.

To verify uniqueness, let $e' \in A$ be another idempotent of the desired form. Then $1 - e - e'$ maps to 1 or -1 in $\mathcal{H}(v)$ for each $v \in \operatorname{Spa}(A, A^+)$, and so is a unit in A by Corollary 2.3.7. Now $(e - e')(1 - e - e') = 0$, so $e - e' = 0$ as desired. □

Remark 2.5.22. — As described in [**14**, §2], Berkovich extends the preceding results (excluding Lemma 2.5.4 and Corollary 2.5.5) in three ways: the base field K is permitted to carry the trivial norm; affinoid algebras are defined as in Remark 2.2.16; and rational subspaces are defined as in Remark 2.4.10. The basic idea is that for any affinoid algebra A over K in the sense of Berkovich, one can construct a nontrivially normed analytic field L containing K such that $A_L = A\widehat{\otimes}_K L$ is an affinoid algebra over L in the classical sense (by adjoining some transcendentals with prescribed norms). The most nontrivial points are that the map $A \to A_L$ is strict (Lemma 2.2.9) and faithfully flat [**15**, Lemma 2.1.2] and the corresponding restriction map $\mathcal{M}(A_L) \to \mathcal{M}(A)$ is surjective (Lemma 2.3.13).

Remark 2.5.23. — In a different direction, note that for any $r \in (0,1), s \in (1,+\infty)$ with $rs \geqslant 1$, the ring $\mathbb{Z}((z))$ equipped with the norm

$$\left|\sum_{i\in\mathbb{Z}} c_i z^i\right| = \max\left\{\max\{r^i : i \geqslant 0,\, c_i \neq 0\}, \max\{s^i : i > 0,\, c_{-i} \neq 0\}\right\}$$

is strongly noetherian: we may construct $\mathbb{Z}((z))\{T_1,\dots,T_n\}$ by taking the z-adic completion of $\mathbb{Z}[z,T_1,\dots,T_n]$ and then inverting z. This means that many of the preceding results apply also to affinoid algebras over $\mathbb{Z}((z))$, including Lemma 2.5.17 (see [**79**, Lemma 1.7.6]), Proposition 2.5.18 (as a corollary of Lemma 2.5.17), Corollary 2.5.19 (as a corollary of Proposition 2.5.18), Theorem 2.5.20 (a) (see [**78**, Theorem 2.5]), and Corollary 2.5.21 (as a corollary of Theorem 2.5.20 (a)). One may also extend Theorem 2.5.20 (b): we are unaware of a precise reference, but given the previous results one may directly emulate the proof of [**18**, Theorem 9.4.3/3]. In this paper we will only apply Theorem 2.5.20 (b) in the case of finite projective modules, in which case one may instead appeal to Theorem 2.7.7; note that the proof of that theorem does not depend on any results from §2.6, so there is no vicious circle.

2.6. Affinoid systems

To get some handle on Banach algebras which are not affinoid algebras over a field, we use an analogue of the observation that every ring is a direct limit of noetherian subrings (namely its finitely generated $\mathbb{Z}$-subalgebras). As usual, we restrict to classical affinoid algebras and note in passing that similar arguments can be derived in Berkovich's framework.

Definition 2.6.1. — By an *affinoid system,* we will mean a directed system $\{((A_i, A_i^+), \alpha_i)\}_{i\in I}$ in the category of adic affinoid algebras over $\mathbb{Z}((z))$ and submetric (not just bounded) morphisms. Note that each ring A_i is strongly noetherian (Remark 2.5.23).

Given an affinoid system, equip the direct limit A of the A_i in the category of rings with the submultiplicative seminorm α given by taking the infimum of the quotient norms induced by the α_i, and let A^+ be the direct limit of the A_i^+. We will refer to the completion of (A, A^+) with respect to α as the *completed direct limit* of the system.

Lemma 2.6.2. — *Let (R, R^+) be an adic Banach algebra. Then there exists an affinoid system with completed direct limit (R, R^+).*

Proof. — By choosing a topologically nilpotent unit $z \in R$, we may view R as a Banach algebra over $\mathbb{Z}((z))$ for a suitable norm as in Remark 2.5.23. Let I be the set of finite subsets of R^+. For each $S \in I$, let A_i be the quotient of $\mathbb{Z}((z))\{S\}$ by the kernel of the map to R taking s (as a generator of the ring) to s (as an element of R); this is an affinoid algebra over $\mathbb{Z}((z))$. Equip A_i with the supremum of the quotient norm and the subspace norm; since this is again a norm under which A_i is a Banach algebra over $\mathbb{Z}((z))$, it is equivalent to the quotient norm by the open mapping theorem (Theorem 2.2.8). Let A_i^+ be the image of $\mathbb{Z}[\![z]\!]\{S\}$ in A_i. This gives the desired affinoid system. □

The previous observation has some strong consequences for Banach algebras.

Remark 2.6.3. — Let $\{((A_i, A_i^+), \alpha_i)\}_{i \in I}$ be an affinoid system. For $((A, A^+), \alpha)$ the direct limit and (R, R^+) the completion, the restriction map $\mathrm{Spa}(R, R^+) \to \varprojlim_i \mathrm{Spa}(A_i, A_i^+)$ is continuous, and also bijective because specifying a compatible system of semivaluations on each A_i bounded by α_i is equivalent to specifying a semivaluation on A bounded by α. Moreover, every rational subspace of $\mathrm{Spa}(R, R^+)$ arises from some $\mathrm{Spa}(A_i, A_i^+)$ by Remark 2.4.7. We thus obtain a homeomorphism $\mathrm{Spa}(R, R^+) \cong \varprojlim_i \mathrm{Spa}(A_i, A_i^+)$.

We obtain the following extension of Corollary 2.5.21. For an alternate approach that also includes the case of a Banach algebra over a trivially normed field (without a topologically nilpotent unit), see [**14**, Theorem 7.4.1].

Proposition 2.6.4. — *Let (R, R^+) be an adic Banach algebra, and let U be a closed-open subset of* $\mathrm{Spa}(R, R^+)$. *Then there exists a unique idempotent element $e \in R$ whose image in $\mathcal{H}(v)$ is 1 if $v \in U$ and 0 if $v \notin U$. In particular, the projection $R \to eR$ taking $x \in R$ to ex induces a homeomorphism* $\mathrm{Spa}(eR, eR^+) \cong U$.

Proof. — By Lemma 2.6.2, we can ensure that there exists an affinoid system $\{((A_i, A_i^+), \alpha_i)\}_{i \in I}$ with completed direct limit $\mathrm{Spa}(R, R^+)$. By Remark 2.3.15 (e) and Remark 2.6.3, U is the inverse image of a closed-open subset of some $\mathrm{Spa}(A_i, A_i^+)$, and hence is induced by some idempotent element of some A_i by Corollary 2.5.21 (as extended by Remark 2.5.23). □

We next relate rational localizations of the completed direct limit of an affinoid system with the corresponding objects defined on individual terms of the system.

Lemma 2.6.5. — *Let $\{((A_i, A_i^+), \alpha_i)\}_{i\in I}$ be an affinoid system with direct limit $((A, A^+), \alpha)$. Let (R, R^+) be the completion of (A, A^+).*

(a) *For any rational localization $(R, R^+) \to (S, S^+)$, there exist an index $i \in I$ and a rational localization $(A_i, A_i^+) \to (B_i, B_i^+)$ such that $(S, S^+) \cong (B_i, B_i^+)\widehat{\otimes}_{(A_i,A_i^+)}(R, R^+)$. The same is then true for each $j \geqslant i$ for $(B_j, B_j^+) = (B_i, B_i^+)\widehat{\otimes}_{(A_i,A_i^+)}(A_j, A_j^+)$; in fact, the (B_j, B_j^+) form another affinoid system with completed direct limit (S, S^+).*

(b) *With notation as in (a), for any $v \in \mathrm{Spa}(R, R^+)$ restricting to $v_i \in \mathrm{Spa}(A_i, A_i^+)$, v belongs to $\mathrm{Spa}(S, S^+)$ if and only if v_i belongs to $\mathrm{Spa}(B_i, B_i^+)$.*

(c) *With notation as in (a), for any $\beta \in \mathcal{M}(R)$ restricting to $\beta_i \in \mathcal{M}(A_i)$, $(R, R^+) \to (S, S^+)$ encircles β if and only if there exists an index $j \geqslant i$ for which $(A_j, A_j^+) \to (B_j, B_j^+)$ encircles β_j.*

Proof. — Part (a) is immediate from Remark 2.4.7. Part (b) is immediate from Remark 2.6.3. Part (c) follows by taking maximal Hausdorff quotients in Remark 2.6.3 to view $\mathcal{M}(R)$ as the inverse limit of the $\mathcal{M}(A_i)$. □

Corollary 2.6.6. — *Let $\{((A_i, A_i^+), \alpha_i)\}_{i\in I}$ be an affinoid system with direct limit $((A, A^+), \alpha)$. Let (R, R^+) be the completion of (A, A^+).*

(a) *For $v \in \mathrm{Spa}(A, A_+)$ and $i \in I$, let $\mathcal{O}_{i,v}$ be the stalk of the structure sheaf of $\mathrm{Spa}(A_i, A_i^+)$ at v. Then $\varinjlim_{i\in I} \mathcal{O}_{i,v}$ is a local ring whose residue field is dense in $\mathcal{H}(v)$.*

(b) *For $\beta \in \mathcal{M}(R)$ and $i \in I$, let $A_{i,\beta}$ denote the Hausdorff localization of A_i at the restriction of β. Then $\varinjlim_{i\in I} A_{i,\beta}$ is a local ring whose residue field is dense in $\mathcal{H}(\beta)$.*

Remark 2.6.7. — With notation as in Lemma 2.6.5, note that by Remark 2.3.15 (d), any rational covering of $\mathrm{Spa}(A, A^+)$ is defined over some (A_i, A_i^+).

We have the following extension of Lemma 2.2.4.

Proposition 2.6.8. — *Let $\{((A_i, A_i^+), \alpha_i)\}_{i\in I}$ be an affinoid system with direct limit $((A, A^+), \alpha)$. Put $I = \ker(\alpha)$, $\overline{A} = A/I$, and $R = \widehat{A}$. Then the base change functors $\mathbf{F\acute{E}t}(A) \to \mathbf{F\acute{E}t}(\overline{A}) \to \mathbf{F\acute{E}t}(R)$ are tensor equivalences.*

Proof. — The base change functor $\mathbf{F\acute{E}t}(A) \to \mathbf{F\acute{E}t}(\overline{A})$ is a tensor equivalence by Lemma 2.2.3 (a) and Theorem 1.2.8. The functor $\mathbf{F\acute{E}t}(\overline{A}) \to \mathbf{F\acute{E}t}(R)$ is rank-preserving and fully faithful by Lemma 2.2.4 (a). It is thus enough to check that $\mathbf{F\acute{E}t}(A) \to \mathbf{F\acute{E}t}(R)$ is essentially surjective.

Choose any $V \in \mathbf{F\acute{E}t}(R)$. For each $\beta \in \mathcal{M}(R)$, for $A_{i,\beta}$ the Hausdorff localization of A_i at the restriction of β, the functor $\mathbf{F\acute{E}t}(\varinjlim_{i\in I} A_{i,\beta}) \to \mathbf{F\acute{E}t}(R_\beta)$ is an equivalence by Corollary 2.6.6 (to see that both $\varinjlim_{i\in I} A_{i,\beta}$ and R_β have dense images in $\mathcal{H}(\beta)$) and Lemma 2.2.4 (b). We can thus choose an index $i \in I$ and a rational localization $(A_i, A_i^+) \to (B_i, B_i^+)$ encircling β such that for $S = R\widehat{\otimes}_{A_i} B_i$, the object $V \otimes_R S$ in $\mathbf{F\acute{E}t}(S)$ descends to an object in $\mathbf{F\acute{E}t}(B_i)$. By the compactness of $\mathcal{M}(R)$, we can find an index $i \in I$ and a strong rational covering $\{(A_i, A_i^+) \to (B_{i,j}, B_{i,j}^+)\}_{j=1}^n$ such that for $S_j = R\widehat{\otimes}_{A_i} B_{i,j}$, the object $V \otimes_R S_j$ in $\mathbf{F\acute{E}t}(S_j)$ descends to an object $U_{i,j}$ in $\mathbf{F\acute{E}t}(B_{i,j})$. If write $B_{i,jl}$ and S_{jl} for $B_{i,j}\widehat{\otimes}_{A_i} B_{i,l}$ and $S_j\widehat{\otimes}_R S_l$, the functor $\mathbf{F\acute{E}t}(\varinjlim_{i\in I} B_{i,jl}) \to \mathbf{F\acute{E}t}(S_{jl})$ is fully faithful; we thus obtain (after suitably increasing i) isomorphisms among the $U_{i,j}$ on overlaps satisfying the cocycle condition. By Theorem 2.5.20 (b) and Corollary 2.5.19 (as extended by Remark 2.5.23), the $U_{i,j}$ glue to an object in $\mathbf{F\acute{E}t}(A_i)$, and hence in $\mathbf{F\acute{E}t}(A)$. This proves the claim. □

Using Lemma 2.6.5, we obtain a weak extension of Theorem 2.5.20 to arbitrary Banach algebras. A better result would be to glue finite projective modules, but this is more difficult; see §2.7.

Theorem 2.6.9. — *Let (R, R^+) be an adic Banach algebra. Let $\{(R, R^+) \to (R_i, R_i^+)\}_{i=1}^n$ be a rational covering. Then the homomorphism $R \to R_1 \oplus \cdots \oplus R_n$ is an effective descent morphism for finite étale algebras over Banach rings.*

Proof. — By Lemma 2.6.2, we can construct an affinoid system $\{((A_i, A_i^+), \alpha_i)\}_{i\in I}$ with completed direct limit (R, R^+). By Remark 2.6.7, for each sufficiently large j, the given covering family is induced by a covering family $\{(A_j, A_j^+) \to (B_{j,i}, B_{j,i}^+)\}_{i=1}^n$. By Proposition 2.6.8, any descent datum for finite étale algebras over Banach rings with respect to $R \to S_1 \oplus \cdots \oplus S_n$ arises from a descent datum with respect to $A_j \to B_{j,1} \oplus \cdots \oplus B_{j,n}$ for some j. This descent datum is effective by Theorem 2.5.20 (b) (as extended by Remark 2.5.23, to uniquely glue the underlying finite flat algebras) and Corollary 2.5.19 (to show that the resulting algebra is finite étale). □

Corollary 2.6.10. — *With notation as in Theorem 2.6.9, the morphism $R \to R_1 \oplus \cdots \oplus R_n$ is an effective descent morphism for étale $\mathbb{Z}_p$-local systems over Banach rings.*

Proof. — This follows from Theorem 2.6.9 and Remark 1.4.3. □

2.7. Glueing of finite projective modules

We now turn to the problem of glueing finite modules over Banach rings, using the formalism of §1.3 as a starting point. We begin with a cautionary note.

Remark 2.7.1. — For $(R, R^+) \to (S, S^+)$ a rational localization of adic Banach rings, the map $R \to S$ is flat when R is an affinoid algebra over an analytic field by Lemma 2.5.17, but need not be flat in general. For instance, flatness almost always fails for perfectoid algebras. Guided by this observation and by the analogy with the Beauville-Laszlo theorem (Proposition 1.3.6), we limit our glueing ambitions to cases where the modules being glued are themselves flat.

Taking Remark 2.7.1 into account, we will be interested in the categories of sheaves of locally free modules of finite rank over various sheaves of rings on adic spectra; in particular, we will want to know when these categories are equivalent to the categories of finite projective modules over the ring of global sections. The guiding principle at work is that the only obstructions to obtaining such results (analogous to Kiehl's theorem) are failures of acyclicity of the base rings (analogous to Tate's theorem).

Lemma 2.7.2. — *Let $R_1 \to S$, $R_2 \to S$ be bounded homomorphisms of Banach rings (not necessarily containing topologically nilpotent units) such that the sum homomorphism $\psi : R_1 \oplus R_2 \to S$ of groups is strict surjective. Then there exists a constant $c > 0$ such that for every positive integer n, every matrix $U \in \mathrm{GL}_n(S)$ with $|U - 1| < c$ can be written in the form $\psi(U_1)\psi(U_2)$ with $U_i \in \mathrm{GL}_n(R_i)$. Moreover, if ψ is almost optimal, this holds with $c = 1$.*

Proof. — By hypothesis, there exists a constant $d \geqslant 1$ such that every $x \in S$ lifts to some pair $(y_1, y_2) \in R_1 \oplus R_2$ with $|y_1|, |y_2| \leqslant d|x|$. It will suffice to prove the claim for $c = d^{-2}$.

Given $U \in \mathrm{GL}_n(S)$ with $|U - 1| < c$, put $V = U - 1$, and lift each entry V_{ij} to a pair $(X_{ij}, Y_{ij}) \in R_1 \oplus R_2$ with $|X_{ij}|, |Y_{ij}| \leqslant d|V_{ij}|$. Then the matrix $U' = \psi(1 - X)U\psi(1 - Y)$ satisfies $|U' - 1| \leqslant d|U - 1|^2$. If $|U - 1| \leqslant d^{-l}$ for some integer $l \geqslant 2$, then $|U' - 1| \leqslant d^{-l-1}$, so we may construct the desired matrices by iterating the construction. (See [**51**, Lemma 4.5.3] for a similar argument or [**85**, Theorem 2.2.2] for a more general result.) □

Definition 2.7.3. — Let

$$\begin{array}{ccc} R & \longrightarrow & R_1 \\ \downarrow & & \downarrow \\ R_2 & \longrightarrow & R_{12} \end{array}$$

be a commutative diagram of Banach rings. (For the purposes of this definition, it is not necessary to assume the presence of topologically nilpotent units.) We call this diagram a *glueing square* if the following conditions hold.

(a) The sequence

$$0 \longrightarrow R \longrightarrow R_1 \oplus R_2 \longrightarrow R_{12} \longrightarrow 0$$

of R-modules, in which the last nontrivial arrow takes (s_1, s_2) to $s_1 - s_2$, is strict exact.

(b) The map $R_2 \to R_{12}$ has dense image.

(c) The map $\mathcal{M}(R_1 \oplus R_2) \to \mathcal{M}(R)$ is surjective.

We define *glueing data* on a glueing square as in Definition 1.3.7.

The following argument is a variant of Lemma 2.2.13.

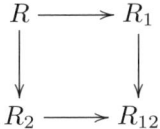

Lemma 2.7.4. — *Consider a glueing square as in Definition 2.7.3, and let M_1, M_2, M_{12} be a finite glueing datum. Let M be the kernel of the map $M_1 \oplus M_2 \to M_{12}$ taking (m_1, m_2) to $\psi_1(m_1) - \psi_2(m_2)$.*

(a) *For $i = 1, 2$, the natural map $M \otimes_R R_i \to M_i$ is surjective.*

(b) *The map $M_1 \oplus M_2 \to M_{12}$ is surjective.*

Proof. — We follow [**51**, Lemmas 4.5.4 and 4.5.5]. Choose generating sets $\mathbf{v}_1, \dots, \mathbf{v}_n$ and $\mathbf{w}_1, \dots, \mathbf{w}_n$ of M_1 and M_2, respectively, of the same cardinality. We may then choose $n \times n$ matrices A, B over R_{12} such that $\psi_2(\mathbf{w}_j) = \sum_i A_{ij}\psi_1(\mathbf{v}_i)$ and $\psi_1(\mathbf{v}_j) = \sum_i B_{ij}\psi_2(\mathbf{w}_i)$.

By hypothesis, the map $R_2 \to R_{12}$ has dense image. We may thus choose an $n \times n$ matrix B' over R_2 so that $A(B' - B)$ has norm less than the constant c of Lemma 2.7.2. We may then write $1 + A(B' - B) = C_1 C_2^{-1}$ with $C_i \in \mathrm{GL}_n(R_i)$.

We now may define elements $\mathbf{x}_j \in M_1 \oplus M_2$ by the formula

$$\mathbf{x}_j = (\mathbf{x}_{j,1}, \mathbf{x}_{j,2}) = \left(\sum_i (C_1)_{ij}\mathbf{v}_i, \sum_i (B'C_2)_{ij}\mathbf{w}_i\right) \qquad (j = 1, \dots, n).$$

Then

$$\psi_1(\mathbf{x}_{j,1}) - \psi_2(\mathbf{x}_{j,2}) = \sum_i (C_1 - AB'C_2)_{ij}\psi_1(\mathbf{v}_i) = \sum_i ((1 - AB)C_2)_{ij}\psi_1(\mathbf{v}_i) = 0,$$

so $\mathbf{x}_j \in M$. Since $C_1 \in \mathrm{GL}_n(R_1)$, the $\mathbf{x}_{i,1}$ generate M_1 over R_1, so the map $M \otimes_R R_1 \to M_1$ is surjective. We may now apply Lemma 1.3.8 to deduce (a) and (b). □

Proposition 2.7.5. — *Consider a glueing square as in Definition 2.7.3, and let M_1, M_2, M_{12} be a finite projective glueing datum. Let M be the kernel of the map $M_1 \oplus M_2 \to M_{12}$ taking (m_1, m_2) to $\psi_1(m_1) - \psi_2(m_2)$. Then M is a finite projective R-module and the natural maps $M \otimes_R R_i \to M_i$ are isomorphisms.*

Proof. — By Lemma 2.7.4, the hypotheses of Lemma 1.3.8 are satisfied. It thus suffices to check that the additional hypothesis of Lemma 1.3.9 (b) is satisfied, *i.e.*, that the image of $\mathrm{Spec}(R_1 \oplus R_2) \to \mathrm{Spec}(R)$ contains $\mathrm{Maxspec}(R)$. Given $\mathfrak{p} \in \mathrm{Maxspec}(R)$, choose $\alpha \in \mathcal{M}(R)$ with $\mathfrak{p}_\alpha = \mathfrak{p}$ (see Definition 2.3.8). By assumption, α lifts to some $\beta \in \mathcal{M}(R_1 \oplus R_2)$; then $\mathfrak{p}_\beta$ is a prime ideal of $\mathrm{Spec}(R_1 \oplus R_2)$ lifting $\mathfrak{p}$. □

Definition 2.7.6. — Let (A, A^+) be an adic Banach ring. Let $\mathcal{R}$ be a presheaf of topological rings on $\mathrm{Spa}(A, A^+)$. We say that $\mathcal{R}$ satisfies the *Tate sheaf property* if for every rational localization $(A, A^+) \to (B, B^+)$ and every rational covering $\mathfrak{V}$ of $U = \mathrm{Spa}(B, B^+)$,

$$H^i(U, \mathcal{R}) = \check{H}^i(U, \mathcal{R}; \mathfrak{V}) = \begin{cases} \mathcal{R}(U) & i = 0 \\ 0 & i > 0. \end{cases} \tag{2.7.6.1}$$

In particular, this implies that $\mathcal{R}$ is a sheaf. By Proposition 2.4.21, it suffices to check (2.7.6.1) for simple Laurent coverings.

We say that $\mathcal{R}$ satisfies the *Kiehl glueing property* if for every rational subdomain U of $\mathrm{Spa}(A, A^+)$, the functor from the category of finite projective $\mathcal{R}(U)$-modules to the category of sheaves of $\mathcal{R}$-modules over U which are locally free of finite rank is an equivalence of categories.

Theorem 2.7.7. — *Let (A, A^+) be a sheafy adic Banach ring. Then the structure sheaf on* $\mathrm{Spa}(A, A^+)$ *satisfies the Tate sheaf property and the Kiehl glueing property.*

Proof. — The Tate sheaf property is a consequence of Theorem 2.4.23. The Kiehl glueing property follows from acyclicity plus Proposition 2.4.20 and Proposition 2.7.5. □

Remark 2.7.8. — By Theorem 2.5.20, any adic affinoid algebra over an analytic field is sheafy. Some additional cases in which we will establish the sheafy property, and hence the Tate and Kiehl properties, will be perfect uniform Banach $\mathbb{F}_p$-algebras (Theorem 3.1.13), perfectoid algebras (Theorem 3.6.15), and preperfectoid algebras (Theorem 3.7.4). For some additional examples of presheaves of rings satisfying the Tate and Kiehl properties, see §5.3.

Remark 2.7.9. — As an aside, we use the glueing formalism to produce a new proof of the Beauville-Laszlo theorem as formulated in Proposition 1.3.6. (Note that [**8**] also includes a somewhat stronger result which we do not treat here.)

Set

$$R_1 = \widehat{R}, \qquad R_2 = R[t^{-1}], \qquad R_{12} = \widehat{R}[t^{-1}].$$

Since t is not a zero divisor in R, the maps $R \to R_2, R_1 \to R_{12}$ are both injective. It is clear that $R_1 \oplus R_2 \to R_{12}$ is surjective, so we obtain a glueing square.

Given a finite glueing datum, set notation as in the first paragraph of the proof of Lemma 2.7.4. Since $R_2 \to R_{12}$ has dense image for the t-adic topology, we may choose a matrix B' over R_2 so that $A(B' - B)$ has entries in tR_1. Put $C_1 = 1 + A(B' - B) \in \mathrm{GL}_n(R_1)$ and $C_2 = 1 \in \mathrm{GL}_n(R_2)$; we may then continue as in Lemma 2.7.4 to conclude that $M \otimes_R R_1 \to M_1$ is surjective.

Since $\mathrm{Spec}(R_1 \oplus R_2) \to \mathrm{Spec}(R)$ is surjective, the hypotheses of Lemma 1.3.9 are satisfied. Proposition 1.3.6 follows at once.

2.8. Uniform Banach rings

In the classical theory of Banach algebras over $\mathbb{R}$ or $\mathbb{C}$, an important role is played by the class of *uniform function algebras* (*i.e.*, algebras of continuous functions on compact spaces topologized using the supremum norm). We now introduce the analogous objects in the nonarchimedean setting.

Definition 2.8.1. — The following conditions on a Banach ring A are equivalent.

(a) The norm on A is equivalent to some power-multiplicative norm.

(b) The norm on A is equivalent to its spectral seminorm (which we therefore also call the *spectral norm*).

(c) There exists $c > 0$ such that $|x^2| \geqslant c|x|^2$ for all $x \in A$. (One gets another equivalent condition by replacing 2 with any larger integer.)

(d) The subring A° of A is bounded.

If these conditions hold, we say A is *uniform*. Any uniform Banach ring is reduced; the converse is false in general, but any reduced affinoid algebra over an analytic field is uniform by Corollary 2.5.6. We say that an adic Banach ring (A, A^+) is *uniform* if A is a uniform Banach ring.

Example 2.8.2. — For A a uniform Banach ring and $r_1, \dots, r_n > 0$, the Tate algebra

$$A\{T_1/r_1, \dots, T_n/r_n\}$$

is again uniform; see Definition 2.2.15.

Remark 2.8.3. — For A a uniform Banach ring, by Theorem 2.3.10 the spectral norm on A is equal to the restriction of the supremum norm on $\prod_{\alpha \in \mathcal{M}(A)} \mathcal{H}(\alpha)$ along the Gel'fand transform. Here are some notable consequences.

(a) Any bounded homomorphism $A \to B$ of uniform Banach rings equipped with their spectral norms is submetric in general, and isometric if and only if $\mathcal{M}(B) \to \mathcal{M}(A)$ is surjective.

(b) For (A, A^+) a uniform adic Banach ring, the map $A \to H^0(\mathrm{Spa}(A, A^+), \mathcal{O})$ is injective. That is, (A, A^+) can only fail to be sheafy if local sections fail to glue.

For uniform Banach rings, we have the following criterion for projectivity of finitely generated modules, analogous to criterion (d) in Definition 1.1.2 for reduced rings.

Proposition 2.8.4. — *Let A be a uniform Banach ring and let M be a finitely generated A-module. Then the following conditions are equivalent.*

(a) *The module M is projective.*

(b) *The rank function $\beta \mapsto \dim_{\mathcal{H}(\beta)}(M \otimes_A \mathcal{H}(\beta))$ on $\mathcal{M}(A)$ is continuous.*

(c) *There exists a bounded homomorphism $A \to B$ of uniform Banach rings such that $\mathcal{M}(B) \to \mathcal{M}(A)$ is surjective and $M \otimes_A B$ is a projective B-module.*

Proof. — If (a) holds, then the function $\mathfrak{p} \to \dim_{A/\mathfrak{p}}(M \otimes_A (A/\mathfrak{p}))$ on $\mathrm{Spec}(A)$ is continuous. By restricting along the map $\mathcal{M}(A) \to \mathrm{Spec}(A)$, we obtain (b).

If (b) holds, then the function $\beta \mapsto \dim_{\mathcal{H}(\beta)}(M \otimes_A \mathcal{H}(\beta))$ is constant on each set in some finite disconnection of $\mathcal{M}(A)$. By Proposition 2.6.4 and the relationship between disconnections of $\mathcal{M}(A)$ and $\mathrm{Spa}(A, A^\circ)$ (Definition 2.4.8), this disconnection descends to $\mathrm{Spec}(A)$, so we may reduce to the case where $\dim_{\mathcal{H}(\beta)}(M \otimes_A \mathcal{H}(\beta))$ is equal to a constant value n. For each maximal ideal $\mathfrak{p}$ of A, we may choose $\alpha \in \mathcal{M}(A)$ with $\mathfrak{p}_\alpha = \mathfrak{p}$ (see Definition 2.3.8). Choose elements $\mathbf{v}_1, \dots, \mathbf{v}_n$ of M whose images in $M \otimes_A \mathcal{H}(\alpha)$ are linearly independent. Then $\mathbf{v}_1, \dots, \mathbf{v}_n$ form a basis of $M \otimes_A A/\mathfrak{p}$, so they also generate $M \otimes_A A_\mathfrak{p}$ by Nakayama's lemma. We may then choose $f \in A \setminus \mathfrak{p}$ so that $\mathbf{v}_1, \dots, \mathbf{v}_n$ generate $M \otimes_A A[f^{-1}]$. Suppose that $\mathbf{v}_1, \dots, \mathbf{v}_n$ fail to form a basis of $M \otimes_A A[f^{-1}]$; then there must exist $a_1, \dots, a_n \in A$ not all mapping to zero in $A[f^{-1}]$ and a nonnegative integer m such that $f^m a_1 \mathbf{v}_1 + \cdots + f^m a_n \mathbf{v}_n = 0$. For each $\beta \in \mathcal{M}(A)$, if $\beta(f) = 0$, then obviously $\beta(f^m a_i) = 0$ for $i = 1, \dots, n$. Otherwise, $\mathfrak{p}_\beta \in \operatorname{Spec}(A[f^{-1}])$ and so $\mathbf{v}_1, \dots, \mathbf{v}_n$ generate $M \otimes_A A/\mathfrak{p}_\beta$, again without relations because $\dim_{\mathcal{H}(\beta)}(M \otimes_A \mathcal{H}(\beta)) = n$. Hence $\beta(f^m a_i) = 0$ for $i = 1, \dots, n$ again. By Theorem 2.3.10, we deduce that $f^m a_i = 0$ for $i = 1, \dots, n$, a contradiction. We conclude that $M \otimes_A A[f^{-1}]$ is a free $A[f^{-1}]$-module; in other words, M is free over a distinguished open subset of $\operatorname{Spec}(A)$ containing the original maximal ideal $\mathfrak{p}$ as well as all other prime ideals contained in $\mathfrak{p}$. We may thus cover $\operatorname{Spec}(A)$ by such open subsets, so (a) holds.

If (a) holds, then (c) is evident. Conversely, if (c) holds, then the function $\gamma \mapsto \dim_{\mathcal{H}(\gamma)}(M \otimes_A \mathcal{H}(\gamma))$ on $\mathcal{M}(B)$ is continuous by the previous paragraph. This function factors through the function $\beta \mapsto \dim_{\mathcal{H}(\beta)}(M \otimes_A \mathcal{H}(\beta))$ on $\mathcal{M}(A)$; the latter is forced to be continuous because $\mathcal{M}(B) \to \mathcal{M}(A)$ is a surjective continuous map of compact spaces and hence a quotient map (Remark 2.3.15 (b)). We thus deduce (b). Hence all three conditions are equivalent. □

Remark 2.8.5. — Unfortunately, the class of uniform Banach rings is not stable under some key operations.

- For $A \to B, A \to C$ morphisms of uniform Banach rings, the completed tensor product $B \widehat{\otimes}_A C$ need not be uniform. A simple example is $A = \mathbb{Q}_p$ and $B = C = \mathbb{C}_p$ (the completion of an algebraic closure of $\mathbb{Q}_p$). For a special case where uniformity is preserved, see Lemma 2.8.6.
- For (A, A^+) a uniform adic Banach ring, a rational localization of $\operatorname{Spa}(A, A^+)$ need not be uniform; see Example 2.8.7. When this is always true, we say that (A, A^+) is *stably uniform*; see Theorem 2.8.10 for an example of this condition.

One operation under which the class of uniform Banach rings does turn out to be stable is the formation of finite étale extensions; see Proposition 2.8.16 and Remark 2.8.19.

Lemma 2.8.6. — *Let $k \subseteq \ell$ be an extension of perfect fields of characteristic p. Put $K = W(k)[p^{-1}]$ and $L = W(\ell)[p^{-1}]$. Let A be a uniform Banach algebra over K equipped with the spectral norm. Then the tensor product seminorm on $A \otimes_K L$ is power-multiplicative; in particular, $A \widehat{\otimes}_K L$ is again uniform.*

Proof. — Let S be a basis of ℓ over k; we can then write each element $b \in A \otimes_K L$ uniquely in the form $\sum_{s \in S} a_s \otimes [s]$ for some $a_s \in A$, all but finitely of which are zero. In terms of such a representation, we have

$$|b| = \max\{|a_s| : s \in S\}$$

and hence

$$\left| b^p - \sum_{s \in S} a_s^p \otimes [s^p] \right| \leqslant p^{-1} |b|^p .$$

Since k and ℓ are perfect, $\{s^p : s \in S\}$ is also a basis of ℓ over k, so

$$\left|\sum_{s\in S} a_s^p \otimes [s^p]\right| = \max\{|a_s^p| : s \in S\}.$$

It follows that $|b^p| = c^p$, proving the claim. □

The following example is due to Mihara [**99**].

Example 2.8.7. — Let K be an analytic field and pick any $r > 0$. Let A be the closure of the K-subalgebra of $K\{X/r, U\}$ generated by $U^n X^{\lceil \log_2 n \rceil}$ for $n = 1, 2, \ldots$. Then A is uniform because it is a subring of the Tate algebra $K\{X/r, U\}$ (see Example 2.8.2). However, for $(A, A^\circ) \to (B, B^+)$ the rational localization corresponding to the set $\{v \in \operatorname{Spa}(A, A^\circ) : v(X) \leqslant 1\}$, the Banach ring B is not uniform [**99**, Theorem 3.11].

It is not known whether (A, A°) is sheafy. However, for B the subset of the infinite product $A \times A \times \cdots$ on which the supremum norm is bounded, Mihara shows that $\operatorname{Spa}(B, B^\circ)$ is not sheafy [**99**, Theorem 3.15]. For another example, see [**23**, §4.6].

The following lemma is [**99**, Proposition 2.3].

Lemma 2.8.8. — *For any uniform Banach ring A and any $f \in A$, the ideals $(T - f)$ and $(1 - fT)$ in $A\{T\}$ and $A\{T, T^{-1}\}$ are closed.*

Proof. — Equip A with the spectral norm. For each $\alpha \in \mathcal{M}(A)$, let $\tilde{\alpha} \in \mathcal{M}(A\{T\})$ be the Gauss norm relative to α. For any $g \in A\{T\}$, we then have $\tilde{\alpha}(T - f) = \max\{1, \alpha(f)\} \geqslant 1$, so

$$\tilde{\alpha}(g) \leqslant \tilde{\alpha}(T - f)\tilde{\alpha}(g) = \tilde{\alpha}((T - f)g).$$

Since $\tilde{\alpha}$ equals the spectral norm on $\mathcal{H}(\alpha)\{T\}$, taking the supremum of the $\tilde{\alpha}$ computes the spectral norm on $A\{T\}$ by Theorem 2.3.10. We thus deduce that

$$|g| \leqslant |(T - f)g|,$$

so multiplication by $T - f$ defines a strict endomorphism of $A\{T\}$. This proves the claim for the ideal $(T - f)$ in $A\{T\}$; the other cases are similar. □

The following consequence is a special case of [**23**, Corollary 4], but our proof is slightly different.

Corollary 2.8.9. — *Let (A, A^+) be a uniform adic Banach ring. Let $\{(A, A^+) \to (B_i, B_i^+)\}_{i=1}^2$ be the standard Laurent covering defined by some $f \in A$, and put $B_{12} = B_1 \widehat{\otimes}_A B_2$. Then the sequence*

$$0 \longrightarrow A \longrightarrow B_1 \oplus B_2 \longrightarrow B_{12} \longrightarrow 0$$

is exact.

Proof. — In the diagram

(2.8.9.1)

$$\begin{array}{ccccccccc}
 & & & & 0 & & 0 & & \\
 & & & & \downarrow & & \downarrow & & \\
 & & & & (T-f)A\{T\} \oplus (1-fU)A\{U\} & \longrightarrow & (T-f)A\{T,T^{-1}\} & \longrightarrow & 0 \\
 & & & & \downarrow & & \downarrow & & \\
0 & \longrightarrow & A & \longrightarrow & A\{T\} \oplus A\{U\} & \longrightarrow & A\{T,T^{-1}\} & \longrightarrow & 0 \\
 & & \downarrow & & \downarrow & & \downarrow & & \\
0 & \longrightarrow & A & \longrightarrow & B_1 \oplus B_2 & \longrightarrow & B_{12} & \longrightarrow & 0 \\
 & & \downarrow & & \downarrow & & \downarrow & & \\
 & & 0 & & 0 & & 0 & &
\end{array}$$

the first and second rows and the first column are evidently exact, while the second and third columns are exact thanks to Lemma 2.4.13 and Corollary 2.8.9. By Theorem 2.3.10, the third row is exact at A; by diagram chasing, it is also exact at the other positions. □

The following theorem is due to Buzzard and Verberkmoes [**23**, Theorem 7].

***Theorem 2.8.10* (Buzzard-Verberkmoes).** — *A stably uniform adic Banach ring is sheafy.*

Proof. — Let (A, A^+) be a uniform adic Banach ring. If (A, A^+) is stably uniform, then by Corollary 2.8.9, the structure presheaf satisfies the criterion of Proposition 2.4.21. Hence (A, A^+) is sheafy. □

Remark 2.8.11. — We do not know whether conversely to Theorem 2.8.10, any sheafy uniform Banach ring is necessarily stably uniform. To check this, it would suffice by Proposition 2.4.24 to check that for any uniform sheafy adic Banach ring (A, A^+) and any $f \in A$, the quotient $A\{T\}/(T-f)$ is again uniform.

One easy but important special case is that if (A, A^+) is sheafy and admits a rational covering by stably uniform adic Banach rings, then A is stably uniform. This is of particular interest when the covering spaces are perfect (Proposition 3.1.16) or perfectoid (Remark 3.6.27).

Remark 2.8.12. — Let A be a Banach algebra over an analytic field K, let $K \to L$ be a morphism of analytic fields, and put $A_L = A\widehat{\otimes}_K L$. Since $K \to L$ is an isometric inclusion, it is strict; by Lemma 2.2.9 (c), the map $A \to A_L$ is also strict. Using criterion (c) of Definition 2.8.1, we see that if A_L is uniform, then so is A (but not conversely; see Remark 2.8.5). Moreover, if A_L is stably uniform, then any rational localization

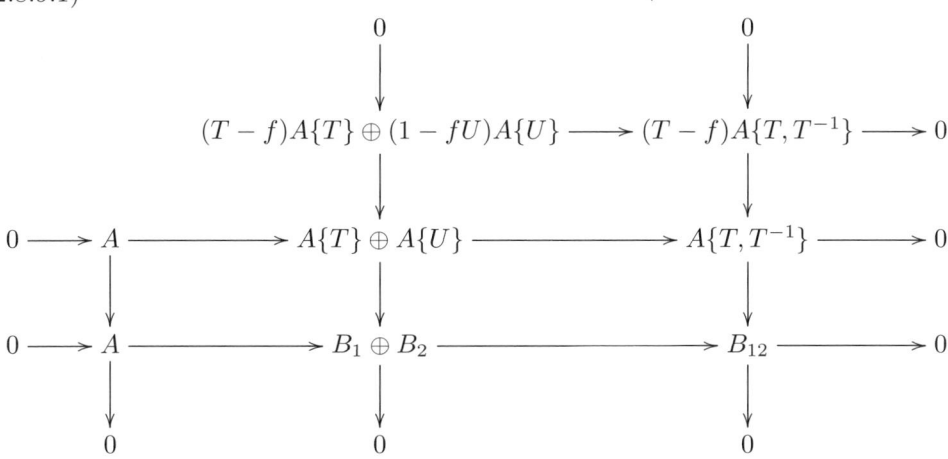

$(A, A^+) \to (B, B^+)$ gives rise to a rational localization $(A_L, A_L^+) \to (B_L, B_L^+)$, and so A is also stably uniform.

A closely related observation is that if $(A, A^+) \to (B, B^+)$ is a bounded morphism of adic Banach rings such that B is stably uniform and $A \to B$ splits in the category of Banach modules over A, then A is stably uniform. Namely, the splitting persists under rational localizations, so we need only check that A is uniform; this again follows from criterion (c) of Definition 2.8.1.

In light of Remark 2.8.5, we are compelled to introduce the following construction.

Definition 2.8.13. — For any Banach ring A, the separated completion of A for the spectral seminorm is a uniform Banach ring, called the *uniformization* of A and denoted A^u. Note that the natural map $A \to A^u$ induces a continuous bijection $\mathcal{M}(A^u) \to \mathcal{M}(A)$, which is thus a homeomorphism by Remark 2.3.15 (b). Using this observation plus Theorem 2.3.10 (as in Remark 2.8.3), we see that uniformization defines a left adjoint to the forgetful functor from uniform Banach rings to arbitrary Banach rings.

For (A, A^+) an adic Banach ring, the *uniformization* of (A, A^+) is defined as (A^u, A^{u+}), where A^{u+} is the completion of the image of A^+ in A^u. Again, this construction defines a left adjoint to the forgetful functor from uniform adic Banach rings to arbitrary adic Banach rings, and the map $\mathrm{Spa}(A^u, A^{u+}) \to \mathrm{Spa}(A, A^+)$ is a homeomorphism which identifies rational subspaces on both sides.

Lemma 2.8.14. — *Let A be a uniform Banach ring. Then for $B \in \mathbf{F\acute{E}t}(A)$, there is a maximal power-multiplicative seminorm on B for which the homomorphism $A \to B$ is bounded.*

Proof. — Let α be the spectral norm on A. When A and B are analytic fields, the claim is clear: since the homomorphism $A \to B$ is nonzero, it is bounded if and only if it is isometric, and the only power-multiplicative extension of α to S is the multiplicative extension (Lemma 2.2.5).

When A is an analytic field and B is arbitrary, we may split B as a direct sum $B_1 \oplus \cdots \oplus B_n$ of finite separable field extensions of A. The maximal power-multiplicative seminorm in this case is the supremum of the maximal seminorms on the B_i.

For general A, we may write $\alpha = \sup\{\gamma : \gamma \in \mathcal{M}(A)\}$ by Theorem 2.3.10. We then take the supremum of the restrictions to B of the maximal power-multiplicative seminorms on the rings $B \otimes_R \mathcal{H}(\alpha)$; this is maximal by Theorem 2.3.10 again. □

Lemma 2.8.15. — *For any Banach ring A and any $B \in \mathbf{F\acute{E}t}(A)$, there exists a strong rational covering $((A, A^\circ) \to (A_i, A_i^+))_i$ such that for each i, $B \otimes_A A_i$ splits as a direct sum of monogenic A-algebras (*i.e.*, algebras of the form $A[T]/(P)$ for some monic polynomial P).*

Proof. — By compactness, it suffices to check that for each $\alpha \in \mathcal{M}(A)$, there exists a rational localization encircling α with the desired property. Using Lemma 2.4.17 and compactness, we may reduce to the case where $B \otimes_A \mathcal{H}(\alpha)$ is connected. By the primitive element theorem and Lemma 2.2.13, the primitive elements of $B \otimes_A \mathcal{H}(\alpha)$ form an open subset; we may thus choose a primitive element f belonging to $\operatorname{Frac}(B)$. By Lemma 2.4.17 again, we may reduce to the case where the minimal polynomial of f over $\operatorname{Frac}(A)$ has coefficients in A. Let $P \in A[T]$ be this polynomial; its resultant is nonzero at α, so we may reduce to the case where the resultant has nonzero norm on all of $\mathcal{M}(A)$, and hence is a unit by Corollary 2.3.7. Hence $A[T]/(P)$ is a finite étale A-algebra which becomes isomorphic to B upon base extension from A to $\mathcal{H}(\alpha)$; by Lemma 2.4.17 again, after replacing A by some localization we obtain an isomorphism $B \cong A[T]/(P)$ as desired. □

Proposition 2.8.16. — *Let A be a Banach ring.*

(a) *The functor $\mathbf{F\acute{E}t}(A) \to \mathbf{F\acute{E}t}(A^u)$ is a tensor equivalence.*

(b) *For $B \in \mathbf{F\acute{E}t}(A^u)$, the power-multiplicative seminorm given by Lemma 2.8.14 is equivalent to any norm on B provided by Definition 2.2.14, and thus provides B with the structure of a uniform Banach algebra over A^u.*

Proof. — To prove (a), apply Lemma 2.6.2 to construct an affinoid system $\{((A_i, A_i^+), \alpha_i)\}_{i \in I}$ with completed direct limit A. Define another affinoid system $\{((B_i, B_i^+), \beta_i)\}_{i \in I}$ in which B_i is the reduced quotient of A_i and β_i is the spectral norm, under which B_i is complete by Corollary 2.5.6. We have $\mathbf{F\acute{E}t}(A_i) \cong \mathbf{F\acute{E}t}(B_i)$ by Theorem 1.2.8. The completed direct limit of the new affinoid system is A^u, so by Proposition 2.6.8 we have tensor equivalences

$$\mathbf{F\acute{E}t}(A) \cong \mathbf{F\acute{E}t}(\varinjlim_i A_i) \cong \mathbf{F\acute{E}t}(\varinjlim_i B_i) \cong \mathbf{F\acute{E}t}(A^u).$$

To prove (b), we may assume A is uniform with norm α. By Theorem 2.3.10, we have $\alpha = \sup\{\gamma : \gamma \in \mathcal{M}(A)\}$ and $\beta \in \sup\{\gamma : \gamma \in \mathcal{M}(B)\}$. Suppose first that A is an analytic field; then B is a direct sum of analytic fields containing A (see Definition 2.2.1), so in particular B is complete under β. The equivalence of norms follows from the open mapping theorem (Theorem 2.2.8).

Suppose next that B is a direct sum of monogenic extensions; we immediately reduce to the case where $B = A[T]/(P)$ is monogenic of degree $d > 0$. In this case, we compare β to the supremum norm β' defined by the basis $1, T, \dots, T^{d-1}$. Since β is equal to the spectral seminorm of β', we need only check that $\beta' \leqslant c\beta$ for some $c > 0$. For $\gamma \in \mathcal{M}(A)$, choose an algebraic closure L of $\mathcal{H}(\gamma)$ and let $z_1, \dots, z_d$ be the roots of P in L. For $a_0, \dots, a_{d-1} \in A$, the supremum of $Q = \sum_{i=0}^{d-1} a_i T^i \in A[T]$ over those $\delta \in \mathcal{M}(B)$ lying over γ may be computed as the supremum of $|Q(z_j)|$ for $j = 1, \dots, d$. Let V be the Vandermonde matrix in $z_1, \dots, z_d$; then

$$\max\{\gamma(a_i) : i = 0, \dots, d-1\} \leqslant \left|V^{-1}\right| \max\{|Q(z_j)| : j = 1, \dots, d\}.$$

For $P = T^d + \sum_{i=0}^{d-1} P_i T^i$, we have

$$|z_i| \leqslant c_1, \qquad c_1 = \max\{\alpha(P_{d-i})^{1/i} : i = 1, \dots, d\}.$$

Let f be the resultant of P; since $B \in \mathbf{F\acute{E}t}(A)$, f is a unit in A. By writing V^{-1} as $\det(V)^{-1}$ times the cofactor matrix of V, we compute that

$$\beta' \leqslant c\beta, \qquad c = c_1^{d-1}\alpha(f^{-1}),$$

proving the claim.

For general A and B, apply Lemma 2.8.15 to construct a strong covering family $\{(A, A^\circ) \to (A_i, A_i^+)\}_{i=1}^n$ such that for $i = 1, \ldots, n$, $B \otimes_A A_i$ splits a direct sum of monogenic extensions of A_i. Let β' be any norm derived as in Lemma 2.2.12. To show that $\beta' \leqslant c\beta$ for some $c > 0$, we reduce to the previous paragraph: by Theorem 2.3.10, both β and β' may be computed by taking suprema over the corresponding norms on $B \otimes_A A_i^u$ for $i = 1, \ldots, n$. □

Remark 2.8.17. — The conclusions of Proposition 2.8.16 continue to hold without assuming that A contains a topologically nilpotent unit. The primary modification needed is to redefine the Hausdorff localization A_α in terms of Berkovich rational subspaces (Remark 2.4.10).

Remark 2.8.18. — For A a uniform Banach ring, recall (Remark 2.3.11) that for any function $g : \mathcal{M}(A) \to \mathbb{R}^+$ whose image is bounded away from 0 and ∞, the function $|\cdot|_g = \sup\{\alpha^{g(\alpha)} : \alpha \in \mathcal{M}(A)\}$ is a norm defining the topology on A. In particular, for any topologically nilpotent z in A, we can choose g so that $|z|_g \left|z^{-1}\right|_g = 1$. That is, for uniform Banach rings, conditions (b) and (c) of Remark 2.3.9 become equivalent if we only keep track of the norm topology, rather than the equivalence class of the norm.

Remark 2.8.19. — With notation as in Remark 2.4.2, if $A^+ = A^\circ$, then $B^+ = B^\circ$: for $b \in B^\circ$, the characteristic polynomial of multiplication by b as an A-linear endomorphism of B is monic, has b as a root (by Cayley-Hamilton), and has coefficients in A° (by Theorem 2.3.10 to reduce to the case of an analytic field, plus usual properties of Newton polygons as in [**85**, Chapter 2]).

CHAPTER 3

PERFECT RINGS AND STRICT p-RINGS

Recall that there is a natural way to lift perfect rings of characteristic p to *strict p-rings* of characteristic 0 (see Definition 3.2.1), and that these can be used to describe étale local systems on perfect rings of characteristic p using a nonabelian generalization of Artin-Schreier-Witt theory (see Proposition 3.2.7). We take a first step towards exploiting this description in p-adic Hodge theory by setting up a correspondence between certain highly ramified analytic fields of mixed characteristics and perfect analytic fields of characteristic p. This correspondence, which has also been described recently by Scholze [**114**], generalizes the *field of norms* construction of Fontaine-Wintenberger [**49**] but with a rather different proof. We then extend the correspondence to Banach algebras, in the direction of generalizing Faltings's *almost purity theorem*; however, this will not be completed until we introduce extended Robba rings (see §5.5).

Convention 3.0.1. — We will refer frequently to *(adic) Banach $\mathbb{F}_p$-algebras* even though $\mathbb{F}_p$ cannot be viewed as an analytic field. What we will mean are (adic) Banach rings whose underlying rings are of characteristic p.

3.1. Perfect $\mathbb{F}_p$-algebras

We begin with some observations about perfect rings of characteristic p, which we may more briefly characterize as *perfect $\mathbb{F}_p$-algebras*.

Definition 3.1.1. — For R an $\mathbb{F}_p$-algebra, let $\overline{\varphi} : R \to R$ denote the p-th power map, *i.e.*, the *Frobenius endomorphism*. We say R is *perfect* if $\overline{\varphi}$ is a bijection; this forces R to be reduced. Note that any localization of a perfect $\mathbb{F}_p$-algebra is also perfect.

By a *perfect uniform Banach $\mathbb{F}_p$-algebra* (resp. a *perfect uniform adic Banach $\mathbb{F}_p$-algebra*), we will mean a uniform Banach algebra R (resp. a uniform adic Banach algebra (R, R^+)) over $\mathbb{F}_p$ such that R is a perfect ring. In the adic case, the fact that R^+ is integrally closed forces it to also be perfect. We will see shortly that any perfect uniform adic Banach $\mathbb{F}_p$-algebra is stably uniform (Proposition 3.1.7).

We will sometimes have need to pass from an $\mathbb{F}_p$-algebra to an associated perfect $\mathbb{F}_p$-algebra.

Definition 3.1.2. — Let R be an $\mathbb{F}_p$-algebra. The *perfect closure* of R is the limit R^{perf} of the direct system

$$R \xrightarrow{\overline{\varphi}} R \xrightarrow{\overline{\varphi}} \cdots,$$

viewed as an R-algebra *via* the map to the first factor. (The map $R \to R^{\mathrm{perf}}$ induces an injection of the reduced quotient of R into R^{perf}.) Any power-multiplicative seminorm on R extends uniquely to a power-multiplicative seminorm on R^{perf}; in particular, given a power-multiplicative norm on R, we may extend it to R^{perf} and then obtain homeomorphisms $\mathcal{M}(R^{\mathrm{perf}}) \to \mathcal{M}(R)$ and $\mathrm{Spa}(R^{\mathrm{perf}}, R^{+,\mathrm{perf}}) \to \mathrm{Spa}(R, R^+)$. We will also call R^{perf} the *direct perfection* of R, to distinguish it from the *inverse perfection* in which one takes the arrows in the opposite direction; we will have more use for the latter construction in §3.4.

Lemma 3.1.3. — *Let R be a perfect $\mathbb{F}_p$-algebra. If $e \in R$ satisfies $e^p = e$, then*

$$e_i = \prod_{j \in \mathbb{F}_p \setminus \{i\}} \frac{e-j}{i-j} \qquad (i \in \mathbb{F}_p)$$

is an idempotent in R and $\sum_{i \in \mathbb{F}_p} i e_i = e$.

Proof. — Note that $e_i(e-i)$ is divisible by $e^p - e$ and thus equals 0. Hence

$$e_i^2 = e_i \prod_{j \in \mathbb{F}_p \setminus \{i\}} \frac{e-j}{i-j} = e_i \prod_{j \in \mathbb{F}_p \setminus \{i\}} \frac{i-j}{i-j} = e_i.$$

The identity $\sum_{i \in \mathbb{F}_p} i e_i = e$ arises from Lagrange interpolation of the polynomial $T \in \mathbb{F}_p[T]$ at the points of $\mathbb{F}_p$. □

Corollary 3.1.4. — *The ring $R^{\overline{\varphi}}$ is the $\mathbb{F}_p$-algebra generated by the idempotents of R. In particular, this ring equals $\mathbb{F}_p$ if and only if R is connected.*

Lemma 3.1.5. — *For R a perfect $\mathbb{F}_p$-algebra, any $S \in \mathbf{F\acute{E}t}(R)$ is also perfect.*

Proof. — Since S is étale over the reduced ring R, it is also reduced [**60**, Proposition 17.5.7]; hence $\overline{\varphi} : S \to S$ is injective. Since $\Omega_{S/R} = 0$ by [**60**, Proposition 17.2.1], S is generated over R by S^p by [**59**, Proposition 0.21.1.7]. Combining this with the surjectivity of $\overline{\varphi} : R \to R$ yields surjectivity of $\overline{\varphi} : S \to S$. Hence S is perfect, as desired. □

The study of perfect uniform Banach $\mathbb{F}_p$-algebras is greatly simplified by the following observations. For some related results in characteristic 0, see §3.6.

Remark 3.1.6. — Let R, S, T be perfect uniform Banach $\mathbb{F}_p$-algebras.

(a) Any strict homomorphism $f : R \to S$ is almost optimal (but not necessarily optimal).

(b) If $f_1, f_2 : R \to S$ are homomorphisms such that $f_1 - f_2$ is strict, then $f_1 - f_2$ is almost optimal.

(c) For any bounded homomorphisms $T \to R$, $T \to S$, the completed tensor product $R\widehat{\otimes}_T S$ is again a perfect uniform Banach $\mathbb{F}_p$-algebra. (By contrast, the completed tensor product of uniform Banach rings is not guaranteed to be uniform in general.)

(d) If I is a closed and perfect ideal of R, then R/I is also a perfect and uniform $\mathbb{F}_p$-algebra.

The proofs of (a)–(c) are all similar, so we only describe (a) in detail. Choose $c > 0$ such that every $x \in \text{image}(f)$ lifts to some $y \in R$ with $|y| \leqslant c|x|$. For any positive integer n, x^{p^n} then lifts to some $y_n \in R$ with $|y_n| \leqslant c|x^{p^n}|$; we may then lift x to $y_n^{p^{-n}}$ and note that

$$|y_n^{p^{-n}}| = |y_n|^{p^{-n}} \leqslant c^{p^{-n}} |x^{p^n}|^{p^{-n}} = c^{p^{-n}} |x|.$$

Since $c^{p^{-n}}$ can be made arbitrarily close to 1, this proves the claim.

To prove (d), note that the perfectness is obvious; it remains to show that the spectral seminorm on R/I is a norm. Suppose the spectral seminorm for some $\bar{y} \in R/I$ is 0. That is, for any $\epsilon > 0$, there exist some $n \in \mathbb{N}$ and $y_n \in R$ lifting $\bar{y}^{p^n}$ such that $|y_n|^{p^{-n}} = |y_n^{p^{-n}}| \leqslant \epsilon$. Since I is perfect, $y_n^{p^{-n}}$ lifts $\bar{y}$. Hence the quotient norm for $\bar{y}$ is less than ϵ, yielding that $\bar{y} = 0$.

Proposition 3.1.7. — *Let (R, R^+) be a perfect uniform adic Banach $\mathbb{F}_p$-algebra. Then any rational localization of (R, R^+) is again perfect uniform; in particular, (R, R^+) is stably uniform.*

Proof. — Suppose $(R, R^+) \to (S, S^+)$ corresponds to a rational subdomain U as in (2.4.3.2). By Lemma 2.4.13, we may view S as the quotient of $R\{T_1, \dots, T_n\}$ by the closure of the ideal $(gT_1 - f_1, \dots, gT_n - f_n)$. Equip $S^{1/p}$ with the norm given by $|x|_{S^{1/p}} = |x^p|_S^{1/p}$. By applying $\overline{\varphi}^{-1}$, raising norms to the p-th power, and using that R is perfect uniform (so its norm remains unchanged), we obtain another rational localization $(R, R^+) \to (S^{1/p}, (S^+)^{1/p})$ representing U. By Lemma 2.4.13 again, we may view $S^{1/p}$ as the quotient of $R\{T_1^{1/p}, \dots, T_n^{1/p}\}$ by the closure of the ideal $(g^{1/p}T_1^{1/p} - f_1^{1/p}, \dots, g^{1/p}T_n^{1/p} - f_n^{1/p})$. The inclusion $R\{T_1, \dots, T_n\} \to R\{T_1^{1/p}, \dots, T_n^{1/p}\}$ then induces a morphism $(S, S^+) \to (S^{1/p}, (S^+)^{1/p})$ of adic Banach algebras over (R, R^+), which must be an isomorphism by the universal property of rational localizations. It follows that S is perfect and uniform. □

Remark 3.1.8. — Retain notation as in Proposition 3.1.7. Let R' be the completed perfect closure of $R\{T_1, \dots, T_n\}$. Let I' be the closure of the ideal of R' generated by $(gT_i - f_i)^{p^{-h}}$ for $i = 1, \dots, n$ and $h = 0, 1, \dots$. Put $S' = R'/I'$; by Remark 3.1.6 (d), S' is perfect uniform. By the universal property of rational localizations, we obtain a morphism $S \to S'$ inducing a bijection $\mathcal{M}(S') \cong \mathcal{M}(S)$. Since S is uniform, the map $S \to S'$ is injective by Theorem 2.3.10.

Choose $h_1, \dots, h_n, k \in R$ such that $h_1 f_1 + \cdots + h_n f_n + kg = 1$, then note that any element

$$y = \sum_{i_1,\dots,i_n} y_{i_1,\dots,i_n} T_1^{i_1} \cdots T_n^{i_n} \in R\{T_1,\dots,T_n\}^{\mathrm{perf}}$$

represents the same element of the quotient as

$$z = (k + h_1 T_1 + \cdots + h_n T_n)^n \times$$
$$\times \sum_{i_1,\dots,i_n} y_{i_1,\dots,i_n} f_1^{i_1 - \lfloor i_1 \rfloor} \cdots f_n^{i_n - \lfloor i_n \rfloor} g^{n-(i_1 - \lfloor i_1 \rfloor + \cdots + i_n - \lfloor i_n \rfloor)} T_1^{\lfloor i_1 \rfloor} \cdots T_n^{\lfloor i_n \rfloor},$$

which satisfies

$$|z| \leqslant c\,|y|, \qquad c = \max\{|h_1|, \dots, |h_n|, |k|\}^n\, |f_1| \cdots |f_n|\, |g|^n.$$

Thus the map $S \to S'$ is also surjective, hence an isomorphism.

An additional consequence of this calculation is that while the map $R\{T_1,\dots,T_n\} \to S$ is not almost optimal, for any $c > 1$ we can arrange for the quotient norm to be at most c times the spectral norm on S, by running the construction with $f_1, \dots, f_n, g$ replaced by suitable p-th power roots.

For perfect uniform Banach $\mathbb{F}_p$-algebras, we have the following refinement of Proposition 2.8.16.

Lemma 3.1.9. — *Let R be a perfect uniform Banach $\mathbb{F}_p$-algebra with norm α. Let S be a finite perfect R-algebra admitting the structure of a finite Banach module over R for some norm β. (Such β exists when S is projective as an R-module by Lemma 2.2.12, and hence when $S \in$ **FÉt**(R).) Then S is a perfect uniform Banach algebra.*

Proof. — Equip $S \otimes_R S$ with the product seminorm induced by β. By Lemma 2.2.6, the multiplication map $\mu : S \otimes_R S \to S$ is bounded. Consequently, there exists $c > 0$ such that

(3.1.9.1) $$\beta(xy) \leqslant c\beta(x)\beta(y) \qquad (x, y \in S).$$

Rewrite (3.1.9.1) as $c\beta(xy) \leqslant (c\beta(x))(c\beta(y))$, then apply Fekete's lemma to deduce that the limit $\gamma(x) = \lim_{n\to\infty}(c\beta(x^n))^{1/n}$ exists. From (3.1.9.1) again, we see that γ is a power-multiplicative seminorm on S and that $\gamma(x) \leqslant c\beta(x)$.

Let R' be a copy of R equipped with the norm α^p; the homomorphism $\overline{\varphi}^{-1} : R \to R'$ is isometric because R is uniform. Let S' be a copy of S equipped with the norm β^p; then S' is a finite Banach module over R' and the map $\overline{\varphi}^{-1} : S \to S'$ is semilinear with respect to $\overline{\varphi}^{-1} : R \to R'$. By Lemma 2.2.6 again, $\overline{\varphi}^{-1} : S \to S'$ is bounded; that is, there exists $d > 0$ such that for all $x \in S$, $\beta(x^{1/p})^p \leqslant d\beta(x)$. Equivalently, for all $x \in S$, $\beta(x^p) \geqslant d^{-1}\beta(x)^p$. By induction on the positive integer n, we have $c\beta(x^{p^n}) \geqslant cd^{-1-p-\cdots-p^{n-1}}\beta(x)^{p^n}$; by taking p^n-th roots and then taking the limit as $n \to \infty$, we deduce that $\gamma(x) \geqslant d^{-1/(p-1)}\beta(x)$.

From the preceding paragraphs, γ is equivalent to β; it is thus a norm on S under which S is a perfect uniform Banach algebra. $\square$

Lemma 3.1.10. — *Let R be a perfect uniform Banach $\mathbb{F}_p$-algebra, and let γ be an isometric automorphism of R extending to an automorphism of $S \in \mathbf{F\acute{E}t}(R)$. Then γ is also isometric on S for the norm provided by Lemma 3.1.9 (or equivalently Proposition 2.8.16).*

Proof. — Suppose first that $R = L$ is an analytic field. Given $y \in S$, let $P = \sum_i P_i T^i \in L[T]$ be the minimal polynomial of y. As in the proof of Lemma 2.2.5, we have $|y| = |P_0|^{1/d}$ for $d = \deg(P)$ and $|\gamma(y)| = |\gamma(P_0)|^{1/d} = |y|$ as desired.

We reduce the general case to the case of an analytic field using Theorem 2.3.10. More precisely, for each $\beta \in \mathcal{M}(R)$, we may use γ to identify $\mathcal{H}(\beta)$ with $\mathcal{H}(\gamma^*(\beta))$, then use the extended action of γ to define an automorphism of $S\widehat{\otimes}_R\mathcal{H}(\beta)$. Since this automorphism is isometric by the previous paragraph, we may apply Theorem 2.3.10 to deduce that the action of γ on S is isometric. □

Remark 3.1.11. — Suppose that R is a perfect uniform $\mathbb{F}_p$-algebra over a perfect analytic field L. For $S \in \mathbf{F\acute{E}t}(R)$, $\mathfrak{o}_S$ is perfect because S is, so $\Omega_{\mathfrak{o}_S/\mathfrak{o}_R} = 0$. This does not imply that $\mathfrak{o}_S$ is finite étale over $\mathfrak{o}_R$, because $\mathfrak{o}_S$ need not be a finite $\mathfrak{o}_R$-module. However, we can say that the quotient of $\mathfrak{o}_S$ by the sum of its finitely generated projective $\mathfrak{o}_R$-submodules is killed by $\mathfrak{m}_L$: the quotient by a single submodule is killed by a nonzero element $\overline{z} \in \mathfrak{m}_L$, then use perfectness to replace $\overline{z}$ by $\overline{z}^{p^{-n}}$ for any nonnegative integer n. A related statement in the language of almost ring theory is that $\mathfrak{o}_S$ is *almost finite étale* over $\mathfrak{o}_R$; see Theorem 5.5.9 for a similar statement and derivation.

As a consequence of these observations, we obtain analogues of the Tate-Kiehl theorems for perfect uniform Banach algebras. For extensions of these results, see §5.3; for an analogue for perfectoid algebras, see Theorem 3.6.15.

Lemma 3.1.12. — *Let (R, R^+) be a perfect uniform adic Banach $\mathbb{F}_p$-algebra. Choose $f \in R$ and let $\{(R, R^+) \to (R_i, R_i^+)\}_{i=1,2,12}$ represent the rational subdomains*

$$\{v \in \mathrm{Spa}(R, R^+) : v(f) \leqslant 1\}, \qquad \{v \in \mathrm{Spa}(R, R^+) : v(f) \geqslant 1\},$$
$$\{v \in \mathrm{Spa}(R, R^+) : v(f) = 1\}$$

of $\mathrm{Spa}(R, R^+)$. *Then the sequence*

$$0 \longrightarrow R \longrightarrow R_1 \oplus R_2 \longrightarrow R_{12} \longrightarrow 0 \tag{3.1.12.1}$$

*is almost optimal exact (*i.e.*, exact with each morphism being almost optimal).*

Proof. — Strict exactness follows from Corollary 2.8.9; almost optimality then follows from Remark 3.1.6. One can also give a more direct proof using affinoid systems and Tate's theorem; we leave this as an exercise. □

Theorem 3.1.13. — *Any perfect uniform adic Banach $\mathbb{F}_p$-algebra (R, R^+) is sheafy, and the structure sheaf on* $\mathrm{Spa}(R, R^+)$ *satisfies the Tate sheaf and Kiehl glueing properties (see Definition 2.7.6).*

Proof. — We may deduce sheafiness either from Theorem 2.8.10 and Proposition 3.1.7 or from Lemma 3.1.12 and Proposition 2.4.21. The properties of the structure sheaf then follow from Theorem 2.7.7. □

Example 3.1.14. — For X an arbitrary (possibly infinite) set and R a ring, let $R[X]$ denote the free commutative R-algebra generated by X, and write $R[X^{p^{-\infty}}]$ for $\cup_{n=1}^{\infty} R[X^{p^{-n}}]$. Then for any $\mathbb{F}_p$-algebra R, we have a natural (in R) identification $R[X]^{\mathrm{perf}} \cong R^{\mathrm{perf}}[X^{p^{-\infty}}]$.

The operation of forming the perfect closure, or the completed perfect closure in case we have a power-multiplicative norm, does not change the étale fundamental group.

Theorem 3.1.15. — *Let R be an $\mathbb{F}_p$-algebra.*

(a) *The base change functor* $\mathbf{F\acute{E}t}(R) \to \mathbf{F\acute{E}t}(R^{\mathrm{perf}})$ *is a tensor equivalence.*

(b) *Suppose that R is a uniform Banach $\mathbb{F}_p$-algebra, and let S be the completion of R^{perf}. Then the base change functor* $\mathbf{F\acute{E}t}(R) \to \mathbf{F\acute{E}t}(S)$ *is a tensor equivalence.*

Proof. — The morphism $\mathrm{Spec}(R^{\mathrm{perf}}) \to \mathrm{Spec}(R)$ is surjective, integral, and radicial, so by [**60**, Corollaire 18.12.11], it is a universal homeomorphism. In particular, it is universally submersive [**62**, Exposè IX, Définition 2.1], so $\mathbf{F\acute{E}t}(R) \to \mathbf{F\acute{E}t}(R^{\mathrm{perf}})$ is fully faithful by [**62**, Exposé IX, Corollaire 3.3]. On the other hand, by Remark 1.2.9, $\mathbf{F\acute{E}t}(R^{\mathrm{perf}})$ is the direct 2-limit of $\mathbf{F\acute{E}t}(T)$ as T runs over all R-subalgebras of R^{perf} of the form $R[x_1^{1/p^m}, \dots, x_n^{1/p^m}]$ for some nonnegative integer m and some $x_1, \dots, x_n \in R$. For each such T, the morphism $\mathrm{Spec}(T) \to \mathrm{Spec}(R)$ is finite, radicial, surjective, and of finite presentation, so $\mathbf{F\acute{E}t}(R) \to \mathbf{F\acute{E}t}(T)$ is essentially surjective by [**62**, Exposé IX, Théorème 4.10]. We deduce that $\mathbf{F\acute{E}t}(R) \to \mathbf{F\acute{E}t}(R^{\mathrm{perf}})$ is also essentially surjective. This proves (a).

To prove (b), note that by Lemma 2.6.2, we can write R as the completion of the direct limit of some affinoid system $\{A_i\}_{i \in I}$. Form a new affinoid system $\{B_i\}_{i \in I}$ by taking $B_i = A_i^u$ (note that by Corollary 2.5.6, B_i is isomorphic to the reduced quotient of A_i). Let B be the direct limit of the second affinoid system; its completion is again R. From the second affinoid system, form a third one by adding $\overline{\varphi}^{-j}(B_i)$ for all nonnegative integers j; the completed directed limit becomes S. Applying $\mathbf{F\acute{E}t}$ to the commutative diagram

$$\begin{array}{ccc} B & \longrightarrow & R \\ \downarrow & & \downarrow \\ B^{\mathrm{perf}} & \longrightarrow & S \end{array}$$

yields tensor equivalences along the horizontal arrows by Proposition 2.6.8 and along the left vertical arrow by (a). Hence the right vertical arrow also becomes a tensor equivalence; this proves (b). □

For attribution of the following result, see Remark 3.1.17.

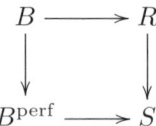

Proposition 3.1.16. — *Let (A, A^+) be a sheafy adic Banach $\mathbb{F}_p$-algebra. Suppose that there exists a rational covering $\{(A, A^+) \to (B_i, B_i^+)\}_i$ such that each B_i is perfect uniform. Then A is perfect.*

Proof. — Treating the desired result as a property of rational coverings, we verify the criteria of Proposition 2.4.20: (a) follows from Proposition 3.1.7, (b) follows because (A, A^+) is stably uniform, and (c) follows from Corollary 2.8.9. Explicitly, if A is uniform and B_1 and B_2 are perfect, then B_{12} is also perfect by Proposition 3.1.7; consequently, any $f \in A$ has a unique p-th root in each of B_1, B_2, B_{12}, so these p-th roots define an element of A *via* the exact sequence in Corollary 2.8.9. □

Remark 3.1.17. — Proposition 3.1.16 is essentially [**23**, Corollary 10], except that we assume that (A, A^+) is sheafy rather than stably uniform. This implies [**23**, Corollary 10] because stably uniform adic Banach rings are sheafy (Theorem 2.8.10), and indeed this is how the proof of [**23**, Corollary 10] proceeds. On the other hand, one may easily deduce Proposition 3.1.16 from [**23**, Corollary 10] because an adic Banach algebra which is both sheafy and locally stably uniform is stably uniform (Remark 2.8.11).

Either of these results may be viewed as a partial resolution of [**114**, Conjecture 2.16]. As originally stated, that conjecture asserts that an adic Banach $\mathbb{F}_p$-algebra (A, A^+) which admits a covering by perfect uniform algebras is itself perfect uniform; however, a counterexample against this conjecture is given in [**23**, Proposition 13]. According to [**23**], Scholze has proposed to amend this conjecture by adding the hypothesis that (A, A^+) be uniform, which avoids the counterexample from [**23**, Proposition 13]. If we further assume that (A, A^+) is stably uniform, we get precisely [**23**, Corollary 10]. The corresponding question for perfectoid algebras is somewhat subtler; see Remark 3.6.27.

3.2. Strict p-rings

Perfect $\mathbb{F}_p$-algebras lift naturally to characteristic zero, as follows. (Our derivations follow [**119**, §5]; see [**80**, §1] for a discussion more explicitly in terms of Witt vectors.)

Definition 3.2.1. — A *strict p-ring* is a p-torsion-free, p-adically complete ring S for which $S/(p)$ is perfect. Given such a ring, for any p-adically complete ring U and any ring homomorphism $\bar{t} : S/(p) \to U/(p)$, $\bar{t}$ lifts uniquely to a multiplicative map $t : S/(p) \to U$; more precisely, for any $\overline{x} \in S/(p)$ and any $y \in U$ lifting $\bar{t}(x^{p^{-n}})$, we have $t(\overline{x}) \equiv y^{p^n} \pmod{p^{n+1}}$. In particular, the projection $S \to S/(p)$ admits a multiplicative section $[\cdot] : S/(p) \to S$, called the *Teichmüller map*; each $x \in S$ can be written uniquely as $\sum_{n=0}^{\infty} p^n [\overline{x}_n]$ with $\overline{x}_n \in S/(p)$.

Lemma 3.2.2. — *Let S be a strict p-ring, let U be a p-adically complete ring, and let $\pi : U \to U/(p)$ be the natural projection. Let $\bar{t} : S/(p) \to U/(p)$ be a ring* *homomorphism, and lift $\overline{t}$ to a multiplicative map $t : S/(p) \to U$ as in Definition 3.2.1. Then the formula*

$$T\left(\sum_{n=0}^{\infty} p^n [\overline{x_n}]\right) = \sum_{n=0}^{\infty} p^n t(\overline{x}_n) \qquad (\overline{x}_0, \overline{x}_1, \dots \in S/(p)) \tag{3.2.2.1}$$

defines a (necessarily unique) homomorphism $T : S \to U$ such that $T \circ [\cdot] = t$.

Proof. — We check by induction that for each positive integer n, T induces an additive map $S/(p^n) \to U/(p^n)$. This holds for $n = 1$ because $\pi \circ t$ is a homomorphism. Suppose the claim holds for some $n \geqslant 1$. For $x = [\overline{x}] + px_1, y = [\overline{y}] + py_1, z = [\overline{z}] + pz_1 \in S$ with $x + y = z$,

$$[\overline{z}] \equiv ([\overline{x}^{p^{-n}}] + [\overline{y}^{p^{-n}}])^{p^n} \pmod{p^{n+1}}$$
$$t(\overline{z}) \equiv (t(\overline{x}^{p^{-n}}) + t(\overline{y}^{p^{-n}}))^{p^n} \pmod{p^{n+1}}$$

as in Definition 3.2.1. In particular,

$$T([\overline{z}]) - T([\overline{x}]) - T([\overline{y}]) \equiv \sum_{i=1}^{p^n-1} \binom{p^n}{i} T([\overline{x}^{ip^{-n}} \overline{y}^{1-ip^{-n}}]) \pmod{p^{n+1}}. \tag{3.2.2.2}$$

On the other hand, since $\frac{1}{p}\binom{p^n}{i} \in \mathbb{Z}$ for $i = 1, \dots, p^n - 1$, we may write

$$z_1 - x_1 - y_1 = \frac{[\overline{x}] + [\overline{y}] - [\overline{z}]}{p} \equiv -\sum_{i=1}^{p^n-1} \frac{1}{p}\binom{p^n}{i} [\overline{x}^{ip^{-n}} \overline{y}^{1-ip^{-n}}] \pmod{p^n},$$

apply T, invoke the induction hypothesis on both sides, and multiply by p to obtain

$$pT(z_1) - pT(x_1) - pT(y_1) \equiv -\sum_{i=1}^{p^n-1} \binom{p^n}{i} T([\overline{x}^{ip^{-n}} \overline{y}^{1-ip^{-n}}]) \pmod{p^{n+1}}. \tag{3.2.2.3}$$

Since $T(x) = T([\overline{x}]) + pT(x_1)$ and so on, we may add (3.2.2.2) and (3.2.2.3) to deduce that $T(z) - T(x) - T(y) \equiv 0 \pmod{p^{n+1}}$, completing the induction. Hence T is additive; multiplicativity of t forces T to also be multiplicative, as desired. □

Remark 3.2.3. — For X an arbitrary set, the p-adic completion S of $\mathbb{Z}[X^{p^{-\infty}}]$ is a strict p-ring with $S/(p) \cong \mathbb{F}_p[X^{p^{-\infty}}]$. If we take $X = \{\overline{x}, \overline{y}\}$, then

$$[\overline{x}] + [\overline{y}] = \sum_{n=0}^{\infty} p^n [P_n(\overline{x}, \overline{y})] \tag{3.2.3.1}$$

for some $P_n(\overline{x}, \overline{y})$ in the ideal $(\overline{x}^{p^{-\infty}}, \overline{y}^{p^{-\infty}}) \subset \mathbb{F}_p[\overline{x}^{p^{-\infty}}, \overline{y}^{p^{-\infty}}]$ and homogeneous of degree 1. For instance, $P_0(\overline{x}, \overline{y}) = \overline{x} + \overline{y}$ and $P_1(\overline{x}, \overline{y}) = -\sum_{i=1}^{p-1} p^{-1}\binom{p}{i} \overline{x}^{i/p} \overline{y}^{1-i/p}$. By Lemma 3.2.2, (3.2.3.1) is also valid for any strict p-ring S and any $\overline{x}, \overline{y} \in S/(p)$. One can similarly derive formulas for arithmetic in a strict p-ring in terms of Teichmüller coordinates; these can also be obtained using Witt vectors (Definition 3.2.5).

Theorem 3.2.4. — *The functor $S \rightsquigarrow S/(p)$ from strict p-rings to perfect $\mathbb{F}_p$-algebras is an equivalence of categories.*

Proof. — Full faithfulness follows from Lemma 3.2.2. To prove essential surjectivity, let R be a perfect $\mathbb{F}_p$-algebra, choose a surjection $\psi : \mathbb{F}_p[X^{p^{-\infty}}] \to R$ for some set X, and put $\overline{I} = \ker(\psi)$. Let S_0 be the p-adic completion of $\mathbb{Z}[X^{p^{-\infty}}]$; this is a strict p-ring with $S_0/(p) \cong \mathbb{F}_p[X^{p^{-\infty}}]$. Put $I = \left\{ \sum_{n=0}^{\infty} p^n[\overline{x}_n] \in S_0 : \overline{x}_0, \overline{x}_1, \dots \in \overline{I} \right\}$; this forms an ideal in S_0 by Remark 3.2.3. Then $S = S_0/I$ is a strict p-ring with $S/(p) \cong R$. □

Definition 3.2.5. — For R a perfect $\mathbb{F}_p$-algebra, let $W(R)$ denote the strict p-ring with $W(R)/(p) \cong R$; this object is unique up to unique isomorphism by Theorem 3.2.4. More concretely, we may identify $W(R)$ with the set of infinite sequences over R so that the sequence $(\overline{x}_0, \overline{x}_1, \dots)$ corresponds to the ring element $\sum_{n=0}^{\infty} p^n[\overline{x}_n]$. This is a special case of the construction of the ring of *p-typical Witt vectors* associated to a ring R, hence the notation. The construction of $W(R)$ is functorial in R, so for instance $\overline{\varphi}$ lifts functorially to an endomorphism φ of $W(R)$. It is common shorthand to write $W_n(R)$ for $W(R)/(p^n)$.

One of the key roles that strict p-rings play in our work is in the classification of local systems over rings of positive characteristic. The central point is a nonabelian version of Artin-Schreier-Witt theory, for which we follow [**81**, Proposition 4.1.1] (see also [**28**, Theorem 2.2]).

Lemma 3.2.6. — *Let R be a perfect $\mathbb{F}_p$-algebra, and let n be a positive integer. Let M be a finite projective $W_n(R)$-module of everywhere positive rank, equipped with a semilinear φ^a-action for some positive integer a. Then there exists a faithfully finite étale R-algebra S such that $M \otimes_{W_n(R)} W_n(S)$ admits a basis fixed by φ^a. More precisely, if $m < n$ is another positive integer and $M \otimes_{W_n(R)} W_m(R)$ admits a φ^a-fixed basis, then S can be chosen so that this basis lifts to a φ^a-fixed basis of $M \otimes_{W_n(R)} W_n(S)$.*

Proof. — We treat the case $n = 1$ first. Suppose first that M is free; choose a basis $\mathbf{e}_1, \dots, \mathbf{e}_m$ of M on which φ^a acts *via* the invertible matrix A over $W_1(R) \cong R$. Let X be the closed subscheme of $\operatorname{Spec}(R[U_{ij} : i, j = 1, \dots, m])$ defined by the matrix equation $\overline{\varphi}^a(U) = A^{-1}U$. The morphism $X \to \operatorname{Spec}(R)$ is finite (evidently) and étale (by the Jacobian criterion), so $X = \operatorname{Spec}(S)$ for some finite étale R-algebra S. The elements $\mathbf{v}_1, \dots, \mathbf{v}_m$ of $M \otimes_R S$ defined by $\mathbf{v}_j = \sum_i U_{ij}\mathbf{e}_i$ form a basis fixed by φ^a. Since the construction is naturally independent of the choice of the original basis, for general M we can glue to obtain a finite étale R-algebra S and a fixed basis of $M \otimes_R S$.

What is left to check is that S has positive rank everywhere as an R-module. This can be checked pointwise on R, and also may be checked after faithfully flat descent, so we may reduce to the case where R is an algebraically closed field. It is enough to check that the map $U \mapsto U^{-1}\overline{\varphi}^a(U)$ on $\mathrm{GL}_m(R)$ is surjective; this observation is due to Lang and is proved as follows (following [**120**, §VI.1, Proposition 4], [**37**, Exposé XXII, Proposition 1.1]). For each $A \in \mathrm{GL}_m(R)$, the map $L_A : U \mapsto U^{-1}A\overline{\varphi}^a(U)$ induces a bijective map from the tangent space at 1, so the image of L_A contains a nonempty

Zariski open subset V_A of $\mathrm{GL}_m(R)$. Since GL_m is a connected group scheme, the open sets V_A and V_1 must intersect in some matrix B, for which

$$B = U_1^{-1}\overline{\varphi}^a(U_1) = U_2^{-1}A\overline{\varphi}^a(U_2)$$

for some $U_1, U_2 \in \mathrm{GL}_m(R)$. We then have $A = U^{-1}\overline{\varphi}^a(U)$ for $U = U_1U_2^{-1}$.

The case $n = 1$ is now complete; we treat the case $n > 1$ by induction on n. We may assume that there exists a basis $\mathbf{e}_1, \dots, \mathbf{e}_m$ of M on which φ^a acts *via* a matrix A congruent to 1 modulo p^{n-1}. We may then take $\mathrm{Spec}(S)$ to be the closed subscheme of $\mathrm{Spec}(R[U_{ij} : i,j = 1,\dots,m])$ defined by the matrix equation $\overline{\varphi}^a(U) - U + p^{1-n}(A-1) = 0$: this subscheme is again finite étale (and hence affine) over R, and the elements $\mathbf{v}_1,\dots,\mathbf{v}_m$ of $M \otimes_{W(R)} W(S)$ defined by $\mathbf{v}_j = \mathbf{e}_j + \sum_i p^{n-1}U_{ij}\mathbf{e}_i$ form a basis fixed by φ^a modulo p^n. □

Proposition 3.2.7. — *For R a perfect $\mathbb{F}_p$-algebra, for each positive integer n, there is a natural (in R and n) tensor equivalence between lisse sheaves of $\mathbb{Z}/p^n\mathbb{Z}$-modules on R and finite projective modules over $W_n(R)$ equipped with semilinear φ-actions.*

One can also weaken the condition on the modules over $W_n(R)$; see Proposition 3.2.13.

Proof. — Let T be a lisse sheaf of $\mathbb{Z}/p^n\mathbb{Z}$-modules on R. Let $\mathrm{Spec}(R_n)$ be the finite étale R-scheme parametrizing trivializations of T. In case T is of constant rank d, R_n carries an action of the group $G = \mathrm{GL}_d(\mathbb{Z}/p^n\mathbb{Z})$, so we may define $M(T) = W_n(R_n)^G$; by faithfully flat descent (Theorems 1.3.4 and 1.3.5), $M(T)$ is projective of constant rank d over $W_n(R)$. The construction extends naturally to general T.

Let M be a finite projective module over $W_n(R)$ equipped with a semilinear φ-action. The assignment

$$S \longmapsto (M \otimes_{W_n(R)} W_n(S))^{\varphi}$$

defines an étale sheaf $T(M)$ on $\mathrm{Spec}(R)$. It is easy to check thanks to Lemma 3.2.6 that the functors $T \rightsquigarrow M(T)$ and $M \rightsquigarrow T(M)$ form an equivalence. □

Remark 3.2.8. — One might like to assert Proposition 3.2.7 with GL_d replaced by other group schemes. The main difficulty is that the analogue of Hilbert's Theorem 90 is not always valid; this is related to the classification of *special groups* by Serre [**118**] and Grothendieck [**55**]. One tractable special case is that of a unipotent group scheme; see Proposition 3.2.9.

Proposition 3.2.9. — *Let d, m, n be integers with $d, m \geqslant 1$ and $n \geqslant 2$ (we may also take $n = 1$ in case $p > 2$). Let $\mathfrak{g}$ be an algebraic Lie subalgebra of the Lie algebra of $d \times d$ matrices over $\mathbb{Q}_p$. Let $\mathfrak{g}_n$ be the intersection of $\mathfrak{g}$ with the Lie algebra of $d \times d$ matrices over $p^n\mathbb{Z}_p$. Let $G_{n,m}$ be the unipotent group scheme defined by the Lie algebra $\mathfrak{g}_n \otimes_{\mathbb{Z}_p} \mathbb{Z}_p/(p^m)$, viewed over $\mathbb{F}_p$ by Greenberg realization (*i.e., *identifying $\mathbb{Z}_p$ with the Witt vectors of $\mathbb{F}_p$). For R a perfect $\mathbb{F}_p$-algebra, define the equivalence* *relation $\sim$ on $G_{n,m}(R)$ by declaring that $g_1 \sim g_2$ if there exists $h \in G_{n,m}(R)$ for which $h^{-1}g_1\varphi(h) = g_2$. Then there is a natural (in G, R, n, m) bijection*

$$G_{n,m}(R)/\sim \longrightarrow H^1_{\text{ét}}(R, G_{n,m}(\mathbb{F}_p)).$$

Proof. — For $g \in G_{n,m}(R)$, as in Lemma 3.2.6, we may construct a faithfully finite étale R-algebra S such that $g = h^{-1}\varphi(h)$ for some $h \in G_{n,m}(S)$. The choice of h then defines an element of $H^1_{\text{ét}}(R, G_{n,m}(\mathbb{F}_p))$ which depends only on g up to equivalence. This gives the claimed map; its injectivity is straightforward. Surjectivity comes down to the fact that $H^1_{\text{ét}}(R, G_{n,m})$ is trivial as a pointed set, which holds because $G_{n,m}$ is unipotent. □

The operation of direct perfection can be extended to certain p-torsion-free rings in order to generate strict p-rings.

Definition 3.2.10. — Let A be a p-torsion-free ring with A/pA reduced, equipped with an endomorphism $\varphi_A : A \to A$ inducing the p^r-power Frobenius map on A/pA for some $r > 0$ and with an identification $(A/pA)^{\text{perf}} \cong R$. Then φ induces a map $s_\varphi : A \to W(R)$ satisfying $\varphi^r \circ s_\varphi = s_\varphi \circ \varphi_A$; this may be seen by using the uniqueness property of $W(R)$ to identify it with the p-adic completion of the limit of the direct system

$$A \xrightarrow{\varphi_A} A \xrightarrow{\varphi_A} \cdots.$$

(For more details, we follow [**80**, (1.3.16)] in suggesting the reference [**95**, VII, §4].) We describe $W(R)$ as the *direct perfection* of A with respect to φ_A.

Example 3.2.11. — Put $A = \mathbb{Z}[T]$, $R = \mathbb{F}_p[\overline{T}]^{\text{perf}}$, and identify $(A/pA)^{\text{perf}}$ with R by mapping the class of T to $\overline{T}$. For the endomorphism $\varphi_A : A \to A$ defined by $\varphi_A(T) = T^p$, the map s_φ takes T to $[\overline{T}]$. However, note for instance that $s_\varphi(T+1) \neq [\overline{T}+1]$.

We next weaken the hypothesis on the modules over $W_n(R)$ in Proposition 3.2.7.

Lemma 3.2.12. — *Let R be a perfect $\mathbb{F}_p$-algebra and let J be a finitely generated ideal of R such that $\overline{\varphi}(J) = J$. Then J is generated by an idempotent element of R; in particular, R/J is a finite projective R-module.*

The hypothesis that J is finitely generated is crucial; otherwise, one could choose any $x \in R$ and take J to be the ideal generated by $\overline{\varphi}^n(x)$ for all $n \in \mathbb{Z}$.

Proof. — Choose generators $x_1, \ldots, x_n$ of J and write $x_i = \sum_j A_{ij}\overline{\varphi}(x_j)$ for some $A_{ij} \in R$. Define the $n \times n$ matrix B over R by setting $B_{ii} = x_i^{p-1}$ and $B_{ij} = 0$ for $i \neq j$; then

$$0 = \sum_i (1 - AB)_{ij} x_j \qquad (j = 1, \ldots, n).$$

For each prime ideal $\mathfrak{p}$ of R, if $x_1, \ldots, x_n$ all map to zero in $\kappa_{\mathfrak{p}}$, then B maps to the zero matrix over $\kappa_{\mathfrak{p}}$ and so $\det(1 - AB)$ maps to 1 in $\kappa_{\mathfrak{p}}$; otherwise, $1 - AB$ maps to a matrix over $\kappa_{\mathfrak{p}}$ with nontrivial kernel and so $\det(1 - AB)$ maps to 0 in $\kappa_{\mathfrak{p}}$. Since R is reduced, this implies that $\det(1 - AB)$ is an idempotent in R.

We may thus reduce to the cases where $\det(1 - AB) \in \{0, 1\}$. If $\det(1 - AB) = 1$, then by the previous paragraph $x_1, \dots, x_n$ map to zero in every $\kappa_{\mathfrak{p}}$ and so $J = 0$. If $\det(1 - AB) = 0$, then by the previous paragraph for every prime ideal $\mathfrak{p}$ of R, at least one of $x_1, \dots, x_n$ has nonzero image in $\kappa_{\mathfrak{p}}$ and so $J = R$. □

Proposition 3.2.13. — *Let R be a perfect $\mathbb{F}_p$-algebra, let n be a positive integer, and let M be a finitely presented $W_n(R)$-module which is flat over $\mathbb{Z}/p^n\mathbb{Z}$ and admits a semilinear φ-action. Then M is a finite projective $W_n(R)$-module.*

Proof. — Since M is flat over $\mathbb{Z}/p^n\mathbb{Z}$, we may reduce to the case $n = 1$. Since M is finitely presented, we may define the Fitting ideals $\mathrm{Fitt}_i(M)$ as in Definition 1.1.2. Since M admits a semilinear $\overline{\varphi}$-action, we have $\overline{\varphi}(\mathrm{Fitt}_i(M)) = \mathrm{Fitt}_i(M)$ for all $i \geqslant 0$. By Lemma 3.2.12, each ideal $\mathrm{Fitt}_i(M)$ is generated by an idempotent element of R; we may thus reduce to the case where for some $n \geqslant 0$ we have $\mathrm{Fitt}_i(M) = 0$ for $i < n$ and $\mathrm{Fitt}_n(M) = R$. In this case, as in Definition 1.1.2, M is a finite projective R-module of constant rank n, as claimed. □

3.3. Norms on strict p-rings

We now take a more metric look at strict p-rings.

Hypothesis 3.3.1. — Throughout §3.3, let R be a perfect $\mathbb{F}_p$-algebra.

We introduce some operations relating the spectra of R and $W(R)$. For variants that do not require the norm on R to be trivial, see Proposition 5.1.2.

Definition 3.3.2. — For α a submultiplicative (resp. power-multiplicative, multiplicative) seminorm on R bounded by the trivial norm,

$$\lambda(\alpha)\left(\sum_{i=0}^{\infty} p^i[\overline{x}_i]\right) = \sup_i\{p^{-i}\alpha(\overline{x}_i)\} \tag{3.3.2.1}$$

is a submultiplicative (resp. power-multiplicative, multiplicative) seminorm on $W(R)$ bounded by the p-adic norm [**87**, Lemma 4.1]. For β a submultiplicative (resp. power-multiplicative, multiplicative) seminorm on $W(R)$ bounded by the p-adic norm,

$$\mu(\beta)(\overline{x}) = \beta([\overline{x}]) \tag{3.3.2.2}$$

is a submultiplicative (resp. power-multiplicative, multiplicative) seminorm on R bounded by the trivial norm [**87**, Lemma 4.2].

Lemma 3.3.3. — *Equip R with the trivial norm and $W(R)$ with the p-adic norm. Then the functions $\lambda : \mathcal{M}(R) \to \mathcal{M}(W(R))$ and $\mu : \mathcal{M}(W(R)) \to \mathcal{M}(R)$ are continuous, and satisfy $(\mu \circ \lambda)(\alpha) = \alpha$ and $(\lambda \circ \mu)(\beta) \geqslant \beta$.*

Proof. — See [**87**, Theorem 4.5]. □

Definition 3.3.4. — Suppose that R is complete with respect to a power-multiplicative norm α bounded above by the trivial norm. An element $z = \sum_{i=0}^{\infty} p^i[\overline{z}_i] \in W(R)$ is *primitive of degree* 1 if

$$\alpha(\overline{x}\overline{z}_0) = p^{-1}\alpha(\overline{x}) \qquad (x \in R) \tag{3.3.4.1}$$

and $\overline{z}_1 \in R^\times$ (or equivalently $z - [\overline{z}_0] \in pW(R)^\times$). This implies that the principal ideal (z) in $W(R)$ is closed (see [**87**, Theorem 5.11]. Note that if (3.3.4.1) holds, then $S = R[\overline{z}_0^{-1}]$ is a Banach algebra over the analytic field $\mathbb{F}_p((\overline{z}_0))$. Conversely, if R is contained in a Banach algebra over $\mathbb{F}_p((\overline{z}_0))$, then (3.3.4.1) holds if and only if $\alpha(\overline{z}_0) = p^{-1}$, as then

$$\alpha(\overline{x}\overline{z}_0) \leqslant \alpha(\overline{x})\alpha(\overline{z}_0) = \alpha(\overline{x})\alpha(\overline{z}_0^{-1})^{-1} \leqslant \alpha(\overline{x}\overline{z}_0).$$

The terminology is modeled on that of [**45**], in which a result similar to our Theorem 3.3.7 can be found; the wording is meant to evoke an analogy with the theory of Weierstrass preparation for analytic power series. Note however that when $R = L$ is an analytic field, our definition is more restrictive than that used in [**45**], in which the condition $\alpha(\overline{z}_0) = p^{-1}$ is relaxed to $\alpha(\overline{z}_0) < 1$.

A key example of the previous definition is the following.

Example 3.3.5. — Suppose R is a uniform Banach ring with spectral norm α. Choose $\overline{\pi} \in R^\times$ with $\alpha(\overline{\pi}) = p^{-p/(p-1)}$ and $\alpha(\overline{\pi}^{-1}) = p^{p/(p-1)}$, and put

$$z = \sum_{i=0}^{p-1} [\overline{\pi} + 1]^{i/p} = \sum_{i=0}^{\infty} p^i [\overline{z}_i].$$

Then $\overline{z}_0 = \overline{\pi}^{(p-1)/p}$, so $\alpha(\overline{z}_0) = p^{-1}$ and $\alpha(\overline{z}_0^{-1}) = p$. We may check that $\overline{z_1} \in \mathfrak{o}_R^\times$ by noting that under the map $W(\mathbb{F}_p[\overline{\pi}]^{\mathrm{perf}}) \to W(\mathbb{F}_p)$ induced by reduction modulo $\overline{\pi}$, the image of $\sum_{i=0}^{p-1}[\overline{\pi} + 1]^{i/p}$ is $\sum_{i=0}^{p-1} 1 = p$. Hence $z \in W(\mathfrak{o}_R)$ is primitive of degree 1.

Lemma 3.3.6. — *Suppose that R is complete with respect to a power-multiplicative norm α and that $z \in W(\mathfrak{o}_R)$ is primitive of degree 1. Then any $x \in W(\mathfrak{o}_R)$ is congruent modulo z to some $y = \sum_{i=0}^{\infty} p^i[\overline{y}_i] \in W(\mathfrak{o}_R)$ with $\alpha(\overline{y}_0) \geqslant \alpha(\overline{y}_i)$ for all $i > 0$.*

Proof. — See [**87**, Lemma 5.5]. □

Theorem 3.3.7. — *Take R, z as in Lemma 3.3.6.*

(a) *For each submultiplicative (resp. power-multiplicative, multiplicative) seminorm γ on $\mathfrak{o}_R$ bounded by the trivial norm, the quotient seminorm $\sigma(\gamma)$ on $W(\mathfrak{o}_R)/(z)$ induced by $\lambda(\gamma)$ is submultiplicative (resp. power-multiplicative, multiplicative) and satisfies $\mu(\sigma(\gamma)) = \gamma$.*

(b) *Equip $W(\mathfrak{o}_R)$ with the power-multiplicative norm $\lambda(\alpha)$. Then the map $\sigma : \mathcal{M}(\mathfrak{o}_R) \to \mathcal{M}(W(\mathfrak{o}_R))$ indicated by (a) is a continuous section of μ, which induces a homeomorphism of $\mathcal{M}(\mathfrak{o}_R)$ with $\mathcal{M}(W(\mathfrak{o}_R)/(z))$. Under this homeomorphism, a subspace of $\mathcal{M}(\mathfrak{o}_R)$ is rational if and only if the corresponding subspace of $\mathcal{M}(W(\mathfrak{o}_R)/(z))$ is rational.*

(c) *The homeomorphism of (b) induces a homeomorphism of $\mathcal{M}(R)$ with $\mathcal{M}(W(\mathfrak{o}_R)[[\overline{z}]^{-1}]/(z))$ under which rational subspaces again correspond.*

For more on the relationship between $\mathcal{M}(R)$ and $\mathcal{M}(W(\mathfrak{o}_R)[[\overline{z}]^{-1}]/(z))$, see §3.6 and §5.4.

Proof. — For (a), see [**87**, Theorem 5.11(a)] (which is itself an easy corollary of Lemma 3.3.6). For (b), see [**87**, Corollary 7.2]. Note that for these results, z need not be primitive of degree 1; it is enough to assume that $\alpha(\overline{z}_0) \leqslant p^{-1}$ and $\overline{z}_1 \in \mathfrak{o}_R^\times$.

By assuming that z is primitive of degree 1, however, we ensure that the quotient norm β on $\mathcal{M}(W(\mathfrak{o}_R)/(z))$ has the property that $\beta(px) = p^{-1}\beta(x)$, so that we may extend β after inverting p. We may then identify

$$\mathcal{M}(R) = \{\gamma \in \mathcal{M}(\mathfrak{o}_R) : \gamma(\overline{z}_0) \geqslant p^{-1}\}$$
$$\mathcal{M}(W(\mathfrak{o}_R)[[\overline{z}]^{-1}]/(z)) = \{\gamma \in \mathcal{M}(W(\mathfrak{o}_R)/(z)) : \gamma([\overline{z}_0]) \geqslant p^{-1}\}.$$

Since these are rational subspaces, we may deduce (c). □

Example 3.3.8. — Let L be a perfect analytic field of characteristic p, let α be the norm on L, and choose $z \in W(\mathfrak{o}_L)$ which is primitive of degree 1. By Theorem 3.3.7 (a), the quotient norm on $W(\mathfrak{o}_L)/(z)$ induced by $\lambda(\alpha)$ is multiplicative. Moreover, by Lemma 3.3.6, every nonzero element of $W(\mathfrak{o}_L)/(z)$ can be lifted to an element of $W(\mathfrak{o}_L)$ which becomes invertible in $W(\mathfrak{o}_L)[[\overline{z}]^{-1}]$. It follows that $W(\mathfrak{o}_L)/(z)$ is the valuation subring of an analytic field $F = W(\mathfrak{o}_L)[[\overline{z}]^{-1}]/(z)$, whose residue field is the same as that of L. (In terms of the rings $\tilde{\mathcal{R}}_L^{\mathrm{int},r}$ to be introduced in Definition 4.2.2 below, we can also realize F as $\tilde{\mathcal{R}}_L^{\mathrm{int},r}/(z)$ for any $r \geqslant 1$. See Lemma 5.5.5.) Two key examples are the following.

- For L the completed perfection of $\mathbb{F}_p((\overline{\pi}))$ and z as in Example 3.3.5, F is the completion of $\mathbb{Q}_p(\mu_{p^\infty})$ for the p-adic norm.
- For L the completed perfection of $\mathbb{F}_p((\overline{\pi}))$ with $\alpha(\overline{\pi}) = p^{-1}$ and $z = [\overline{\pi}] - p$, F is the completion of $\mathbb{Q}_p(p^{p^{-\infty}})$ for the p-adic norm.

Note that

$$\mathfrak{o}_L/(\overline{z}) \cong W(\mathfrak{o}_L)/(p, [\overline{z}]) = W(\mathfrak{o}_L)/(p, z) = \mathfrak{o}_F/(p).$$

This implies that $\overline{\varphi}$ is surjective on $\mathfrak{o}_F/(p)$ and that $\mathfrak{o}_F$ is not discretely valued. These conditions turn out to characterize the fields F which arise in this manner; see Lemma 3.5.2.

The following refinement of [**87**, Lemma 5.16] is useful for some calculations.

Lemma 3.3.9. — *Take R, z as in Lemma 3.3.6. Then for any $\epsilon > 0$ and any nonnegative integer m, every $x \in W(\mathfrak{o}_R)[[\overline{z}]^{-1}]$ is congruent modulo z to some $y = \sum_{n=0}^{\infty} p^n[\overline{y}_n] \in W(\mathfrak{o}_R)[[\overline{z}]^{-1}]$ such that for each $\alpha \in \mathcal{M}(R)$,*

$$\alpha(\overline{y}_1) \leqslant \max\{p^{-p^{-1}-\cdots-p^{-m}}\alpha(\overline{y}_0), \epsilon\} \tag{3.3.9.1}$$

$$\alpha(\overline{y}_n) \leqslant \max\{\alpha(\overline{y}_0), \epsilon\} \quad (n > 1). \tag{3.3.9.2}$$

Proof. — Define the sequence $x = x_0, x_1, \ldots$ as in the proof of [**87**, Lemma 5.5]. That is, let w be the inverse of $p^{-1}(z - [\overline{z}])$ in $W(\mathfrak{o}_R)$, then write $x_i = \sum_{j=0}^{\infty} p^j[\overline{x}_{ij}]$ with $\overline{x}_{ij} \in R$ and put

$$x_{i+1} = x_i - p^{-1}w(x_i - [\overline{x}_{i0}])z = [\overline{x}_{i0}] - p^{-1}w(x_i - [\overline{x}_{i0}])[\overline{z}].$$

The proof of [**87**, Lemma 5.16] shows that there exists i_0 such that for each $\alpha \in \mathcal{M}(R)$,

$$\alpha(\overline{x}_{ij}) \leqslant \max\{\alpha(\overline{x}_{i0}), \epsilon\} \quad (i \geqslant i_0, j > 0).$$

If we take $y = x_{i_0+k}$ for some nonnegative integer k, then (3.3.9.2) is satisfied. If $\alpha(\overline{x}_{i_0 0}) \leqslant \epsilon$, then (3.3.9.1) is also satisfied.

Suppose instead that $\alpha(\overline{x}_{i_0 0}) > \epsilon$; in this case, it will complete the proof to show that (3.3.9.1) is satisfied whenever $k \geqslant m$. It will suffice to check that for each nonnegative integer k,

$$\alpha(\overline{x}_{(i_0+k)1}) \leqslant \max\{p^{-p^{-1}-\cdots-p^{-k}}|\overline{x}_{(i_0+k)0}|, \epsilon\}. \tag{3.3.9.3}$$

We have this for $k = 0$, so we may proceed by induction on k. Given (3.3.9.3) for some k, write

$$x_{i_0+k+1} \equiv [\overline{x}_{(i_0+k)0}] - w[\overline{x}_{(i_0+k)1}\overline{z}] + pw[\overline{x}_{(i_0+k)2}\overline{z}] \pmod{p^3}$$

and then deduce that

$$\begin{aligned} \overline{x}_{(i_0+k+1)0} &= \overline{wx}_{(i_0+k)0} - \overline{wx}_{(i_0+k)1}\overline{z} \\ \overline{x}_{(i_0+k+1)1} &= \overline{wx}_{(i_0+k)2}\overline{z} + P((\overline{wx}_{(i_0+k)0})^{1/p}, (\overline{wx}_{(i_0+k)1}\overline{z})^{1/p}) \end{aligned}$$

for $P(x, y) = p^{-1}(x^p - y^p - (x-y)^p) \in \mathbb{Z}[x, y]$. From this we deduce

$$\begin{aligned} \alpha(\overline{x}_{(i_0+k+1)0}) &= \alpha(\overline{wx}_{(i_0+k)0}) \\ \alpha(\overline{x}_{(i_0+k+1)1}) &\leqslant \max\{\alpha(\overline{wx}_{(i_0+k)2}\overline{z}), \alpha(\overline{wx}_{(i_0+k)0})^{(p-1)/p}\alpha(\overline{wx}_{(i_0+k)1}\overline{z})^{1/p}\}. \end{aligned}$$

Since $\alpha(\overline{z}) = p^{-1}$, this yields the analogue of (3.3.9.3) with k replaced by $k+1$. □

3.4. Inverse perfection

We have already introduced one method for passing from an $\mathbb{F}_p$-algebra to a perfect $\mathbb{F}_p$-algebra, that of *direct perfection.* We now consider the dual operation of *inverse perfection*, which has the advantage of capturing useful information from characteristic 0.

Definition 3.4.1. — For any ring A, define the *inverse perfection* A^{frep} of A as the inverse limit of the system

$$\cdots \xrightarrow{\overline{\varphi}} A/pA \xrightarrow{\overline{\varphi}} A/pA.$$

This evidently gives a perfect $\mathbb{F}_p$-algebra. There is a natural projection $A^{\mathrm{frep}} \to A/pA$ by projection onto the last factor; this is surjective as long as $\overline{\varphi} : A/pA \to A/pA$ is surjective.

Lemma 3.4.2. — *Let A be a ring. For any ideal I of A satisfying $I^m \subseteq (p) \subseteq I$ for some positive integer m, the natural map $A^{\mathrm{frep}} \to (A/I)^{\mathrm{frep}}$ is an isomorphism.*

Proof. — We may assume $m = p^k$ for some positive integer k. Let $y = (\dots, y_1, y_0)$ be an element of A^{frep} whose image in $(A/I)^{\mathrm{frep}}$ is zero. For each nonnegative integer n, we then have $y_{n+k} \equiv 0 \pmod{I}$, and so

$$y_n \equiv y_{n+k}^{p^k} \equiv 0 \pmod{(p) + I^{p^k}}.$$

Hence $y_n \equiv 0 \pmod{p}$, and so $y = 0$ in A^{frep}.

Given $z = (\dots, z_1, z_0) \in (A/I)^{\mathrm{frep}}$, choose any lifts $\tilde{z}_n \in A$ of z_n. Put $y_n = \tilde{z}_{n+k}^{p^k}$; then the congruence $\tilde{z}_{n+k+1}^p \equiv \tilde{z}_{n+k} \pmod{I}$ implies $y_{n+1}^p \equiv y_n \pmod{(p) + I^{p^k}}$. Hence $y = (\dots, y_1, y_0)$ forms an element of A^{frep} lifting z. □

Definition 3.4.3. — Let A be a ring, and let $\widehat{A}$ denote the p-adic completion of A. From the projection $A^{\mathrm{frep}} \to A/pA$, we obtain first a multiplicative map $A^{\mathrm{frep}} \to \widehat{A}$ and then by Lemma 3.2.2 a homomorphism $\theta : W(A^{\mathrm{frep}}) \to \widehat{A}$. Note that θ is surjective if and only if $\overline{\varphi} : A/pA \to A/pA$ is surjective.

Remark 3.4.4. — Let A be a p-adically separated ring, let β be any submultiplicative (resp. power-multiplicative, multiplicative) seminorm on A bounded by the p-adic norm, and extend β to $\widehat{A}$ by continuity. Then $\alpha = \mu(\theta^*(\beta))$ is a submultiplicative (resp. power-multiplicative, multiplicative) seminorm on A^{frep} bounded by the trivial norm. In particular, the map θ is bounded for the seminorm $\lambda(\alpha)$ on $W(A^{\mathrm{frep}})$ and the seminorm β on $\widehat{A}$.

Lemma 3.4.5. — *In Remark 3.4.4, suppose that β is power-multiplicative (resp. multiplicative) norm and that A is complete under β. Then $\alpha = \mu(\theta^*(\beta))$ is a power-multiplicative (resp. multiplicative) norm under which A^{frep} is complete.*

Proof. — For $x = (\dots, \overline{x}_1, \overline{x}_0) \in A^{\mathrm{frep}}$ and any lifts $x_i \in A$ of $\overline{x}_i$, we have

$$\max\left\{\alpha(x), \beta(p)^i\right\} = \max\left\{\beta(x_i)^{p^i}, \beta(p)^i\right\}$$

for all i. In particular, if $\alpha(x) \neq 0$, then $x_i \neq 0$ for all sufficiently large i, so $x \neq 0$. Hence α is a norm.

Let $y_0, y_1, \dots$ be a sequence in A^{frep} which is Cauchy with respect to α. For each nonnegative integer i, the sequence $y_0^{p^{-i}}, y_1^{p^{-i}}, \dots$ is also Cauchy with respect to α, so $[y_0^{p^{-i}}], [y_1^{p^{-i}}], \dots$ is Cauchy with respect to $\lambda(\alpha)$. The images of $[y_0^{p^{-i}}], [y_1^{p^{-i}}], \dots$ in A then form a Cauchy sequence with respect to β, which by hypothesis has a limit $\tilde{z}_i \in A$. Let $z_i \in A/pA$ be the image of $\tilde{z}_i$; then $(\dots, z_1, z_0)$ forms an element of A^{frep} which is the limit of the y_n. Hence A^{frep} is complete. □

Remark 3.4.6. — For R a perfect $\mathbb{F}_p$-algebra, the natural map $W(R)^{\mathrm{frep}} \cong R$ is an isomorphism, as then is the map $\theta : W(W(R)^{\mathrm{frep}}) \to W(R)$.

Lemma 3.4.7. — *Let A_1 be a p-adically separated ring written as a union $\cup_{i \in I} A_{1,i}$ such that for each $i \in I$, $\overline{\varphi}$ is surjective on $A_{1,i}/pA_{1,i}$. Let A_2 be a p-adically separated ring equipped with a norm β_2 bounded by the p-adic norm. Let $\psi : A_1 \to A_2$ be a homomorphism. Put $\alpha_2 = \mu(\theta^*(\beta_2))$.*

(a) *Suppose that ψ has dense image. Then the induced map $\psi^{\mathrm{frep}} : \cup_{i\in I} A_{1,i}^{\mathrm{frep}} \to A_2^{\mathrm{frep}}$ has dense image.*

(b) *Suppose that the image of ψ has dense intersection with $\mathfrak{m}_{A_2}$. Then the image of ψ^{frep} has dense intersection with $\mathfrak{m}_{A_2^{\mathrm{frep}}}$.*

Proof. — Given $w = (\dots, w_1, w_0) \in A_2^{\mathrm{frep}}$, if we choose a nonnegative integer n, we can choose $i \in I$ and $\tilde{w}_n \in A_{1,i}$ so that $\alpha_2(\psi(\tilde{w}_n) - w_n) \leqslant p^{-1}$. Since $\overline{\varphi}$ is surjective on $A_{1,i}/pA_{1,i}$, we can find $x = (\dots, x_1, x_0) \in A_{1,i}^{\mathrm{frep}}$ with x_n equal to the image of $\tilde{w}_n$ in $A_{1,i}/pA_{1,i}$. Let $y = (\dots, y_1, y_0) \in A_2^{\mathrm{frep}}$ be the image of x under ψ^{frep}; then $\alpha_2(y_n - w_n) \leqslant p^{-1}$, so $\alpha_2(y - w) \leqslant p^{-p^n}$. Since this holds for any n (for some i, y depending on n), it follows that $\cup_{i\in I} A_{1,i}^{\mathrm{frep}}$ has dense image in A_2^{frep}. This proves (a); the proof of (b) is similar. □

Lemma 3.4.8. — *Let A be a p-adically separated p-torsion-free ring complete under a power-multiplicative norm β bounded by the p-adic norm, and put $\alpha = \mu(\theta^*(\beta))$. Suppose that there exists $z \in W(A^{\mathrm{frep}})$ primitive of degree 1 with $\theta(z) = 0$. Extend θ to a map $W(A^{\mathrm{frep}})[[\overline{z}]^{-1}] \to A[\theta([\overline{z}])^{-1}]$.*

(a) *The ideal $\ker(\theta) \subset W(A^{\mathrm{frep}})[[\overline{z}]^{-1}]$ is generated by z.*

(b) *The extended map θ is optimal.*

(c) *The map $\overline{\varphi} : A/(p) \to A/(p)$ is surjective if and only if θ has dense image in $A[\theta([\overline{z}])^{-1}]$.*

Proof. — Given $x \in W(A^{\mathrm{frep}})$ not divisible by z, choose $y = \sum_{i=0}^{\infty} p^i [\overline{y}_i]$ as in Lemma 3.3.6. Then $\theta(x) = \theta(y) = \theta([\overline{y}_0]) + \theta(y - [\overline{y}_0])$ and

$$\beta(\theta([\overline{y}_0])) = \alpha(\overline{y}_0) > \lambda(\alpha)(y - [\overline{y}_0]) \geqslant \beta(\theta(y - [\overline{y}_0])).$$

Consequently, $\beta(\theta(x)) = \alpha(\overline{y}_0) > 0$. This implies (a) and (b). To check (c), note that strictness of θ (from (b)) implies that θ is surjective if and only if it has dense image. □

Remark 3.4.9. — If $\overline{\varphi} : A/pA \to A/pA$ is surjective, then so is $\overline{\varphi} : A/I \to A/I$ for any ideal I for which $I^m \subseteq (p) \subseteq I$ for some positive integer m. The converse is not true: *e.g.*, take $A = \mathbb{Z}_p[\sqrt{p}]$ and $I = (\sqrt{p})$.

One correct partial converse is that if there exist $x, y \in A$ such that $\overline{\varphi} : A/(x,p) \to A/(x,p)$ is surjective, $x^m \in (p)$ for some positive integer m, and $y^p \equiv x \pmod{(x^2, p)}$, then $\overline{\varphi} : A/pA \to A/pA$ is surjective. To see this, we prove by induction that $\overline{\varphi} : A/(x^i, p) \to A/(x^i, p)$ is surjective for $i = 1, \dots, m$, the case $i = 1$ being given and the case $i = m$ being the desired result. Given the claim for some i, note first that $(y^p, p) = (x, p)$ and that $y^{pi} \equiv x^i \pmod{(x^{i+1}, p)}$. for any $z \in A$, we can find $w_0, z_1 \in A$ with $z - w_0^p - x^i z_1 \in pA$. We can then find $w_1 \in A$ with $z_1 - w_1^p \in (x^i, p)$; then $w = w_0 + y^i w_1$ satisfies $w^p \equiv w_0^p + y^{pi} w_1^p \equiv w_0^p + y^{pi} z_1 \equiv w_0^p + x^i z_1 \equiv z \pmod{(x^{i+1}, p)}$.

Remark 3.4.10. — If A is p-adically complete, then θ (surjective or not) induces an isomorphism of multiplicative monoids

$$A^{\mathrm{frep}} \cong \varprojlim_{x \mapsto x^p} A$$

whose inverse is reduction modulo p. This can be used to reformulate the perfectoid correspondence; see Proposition 3.6.25.

3.5. The perfectoid correspondence for analytic fields

In order to bring nonabelian Artin-Schreier-Witt theory to bear upon p-adic Hodge theory, one needs a link between étale covers of spaces of characteristic 0 and characteristic p. We first make this link at the level of analytic fields; this extends the *field of norms* correspondence introduced by Fontaine and Wintenberger [**49**], upon which usual p-adic Hodge theory is based. Similar results have been obtained by Scholze [**114**] using a slightly different method; see Remark 3.5.13. (See also [**89**] for a self-contained presentation of the correspondence following the approach taken here.) See §3.6 for an extension to more general Banach algebras.

Definition 3.5.1. — An analytic field F is *perfectoid* if F is of characteristic 0, κ_F is of characteristic p, F is not discretely valued, and $\overline{\varphi}$ is surjective on $\mathfrak{o}_F/(p)$. For example, any field F appearing in Example 3.3.8 is perfectoid; the converse is also true by Lemma 3.5.2 below.

Lemma 3.5.2. — *Let F be a perfectoid analytic field with norm β. Put $R = (\mathfrak{o}_F/(p))^{\mathrm{frep}}$, let $\theta : W(R) \to \mathfrak{o}_F$ be the surjective homomorphism from Definition 3.4.3, and define the multiplicative norm $\alpha = \mu(\theta^*(\beta))$ on R as in Remark 3.4.4.*

(a) *The ring $K = \mathrm{Frac}(R)$ is an analytic field under α which is perfect of characteristic p, and $R = \mathfrak{o}_K$.*

(b) *We have $\beta(F^\times) = \alpha(K^\times)$.*

(c) *For any $\overline{z} \in K$ with $\alpha(\overline{z}) = p^{-1}$ (which exists by (b)), there is a natural (in F) isomorphism $\mathfrak{o}_F/(p) \cong \mathfrak{o}_K/(\overline{z})$. In particular, we obtain a natural isomorphism $\kappa_F \cong \kappa_K$.*

(d) *There exists $z \in W(\mathfrak{o}_K)$ in $\ker(\theta)$ which is primitive of degree 1. Consequently (by Lemma 3.4.8), the kernel of $\theta : W(\mathfrak{o}_K)[[\overline{z}]^{-1}] \to F$ is generated by z.*

Proof. — Note that R is already complete under α by Lemma 3.4.5. Hence to prove (a), it suffices to check that for any nonzero $x, y \in R$ with $\alpha(x) \leqslant \alpha(y)$, x is divisible by y in R. Write $x = (\dots, \overline{x}_1, \overline{x}_0), y = (\dots, \overline{y}_1, \overline{y}_0)$ and lift $\overline{x}_n, \overline{y}_n$ to $x_n, y_n \in \mathfrak{o}_F$. Choose $n_0 \geqslant 0$ so that $\alpha(x), \alpha(y) > p^{-p^{n_0}}$. For $n \geqslant n_0$, as in the proof of Lemma 3.4.5 we have $\beta(x_n) = \alpha(x)^{p^{-n}}, \beta(y_n) = \alpha(y)^{p^{-n}}$, so $\beta(x_n) \leqslant \beta(y_n)$. Since $\mathfrak{o}_F$ is a valuation ring, $z_n = x_n/y_n$ belongs to $\mathfrak{o}_F$. Since $\alpha(x_{n+1}^p - x_n), \alpha(y_{n+1}^p - y_n) \leqslant p^{-1}$, we have $\alpha(z_{n+1}^p - z_n) \leqslant p^{-1}/\alpha(y_n)$. This last quantity is bounded away from 1 for $n \geqslant n_0$, so by Lemma 3.4.2, the z_n define an element $z \in R$ for which $x = yz$.

To establish (b), note that the group $\alpha(K^\times)$ is p-divisible and that $\alpha(K^\times) \cap (p^{-1}, 1) = \beta(F^\times) \cap (p^{-1}, 1)$ by (a). Since F is not discretely valued, we can choose $r \in \beta(F^\times) \cap (1, p^{1/p})$. For any such r, we have $r^{-1}, p^{-1}r^p \in \beta(F^\times) \cap (p^{-1}, 1) \subseteq \alpha(K^\times)$, so $p^{-1} \in \alpha(K^\times)$.

To establish (c), note that from the definition of the inverse perfection, we obtain a homomorphism $\mathfrak{o}_K \to \mathfrak{o}_F/(p)$. By comparing norms, we see that the kernel of this map is generated by $\overline{z}$.

To establish (d), keep notation as in (c). Note that $\theta([\overline{z}])$ is divisible by p in $\mathfrak{o}_F$ and that $\theta : W(\mathfrak{o}_K) \to \mathfrak{o}_F$ is surjective. We can thus find $z_1 \in W(\mathfrak{o}_K)^\times$ with $\theta(z_1) = \theta([\overline{z}])/p$; we then take $z = [\overline{z}] - pz_1$. □

***Theorem 3.5.3* (Perfectoid correspondence).** — *The constructions*

$$F \rightsquigarrow (\mathrm{Frac}((\mathfrak{o}_F/(p))^{\mathrm{frep}}), \ker(\theta)), \qquad (L, I) \rightsquigarrow \mathrm{Frac}(W(\mathfrak{o}_L)/I)$$

define a equivalence of categories between perfectoid analytic fields F and pairs (L, I) in which L is a perfect analytic field of characteristic p and I is a principal ideal of $W(\mathfrak{o}_L)$ admitting a generator which is primitive of degree 1.

Proof. — This follows immediately from Example 3.3.8 and Lemma 3.5.2. □

We next study the compatibility of this correspondence with finite extensions of fields. Moving from characteristic p to characteristic 0 turns out to be straightforward.

Lemma 3.5.4. — *Fix F and (L, I) corresponding as in Theorem 3.5.3. Then for any finite extension M of L, the pair $(M, IW(\mathfrak{o}_M))$ corresponds* via *Theorem 3.5.3 to a finite extension E of F with $[E : F] = [M : L]$.*

Proof. — Suppose first that M is Galois over L, and put $G = \mathrm{Gal}(M/L)$. Since I is a principal ideal, averaging over G induces a projection

$$E = \frac{W(\mathfrak{o}_M)[p^{-1}]}{IW(\mathfrak{o}_M)[p^{-1}]} \longrightarrow \frac{W(\mathfrak{o}_L)[p^{-1}]}{W(\mathfrak{o}_L)[p^{-1}] \cap IW(\mathfrak{o}_M)[p^{-1}]} = \frac{W(\mathfrak{o}_L)[p^{-1}]}{IW(\mathfrak{o}_L)[p^{-1}]} = F.$$

Consequently, $E^G = F$, so by Artin's lemma, E is a finite Galois extension of F and $[E : F] = \#G = [M : L]$. This proves the claim when M is Galois; the general case follows by Artin's lemma again. □

For the reverse direction, the crucial case is when the characteristic p field is algebraically closed.

Lemma 3.5.5. — *Fix F and (L, I) corresponding as in Theorem 3.5.3. If L is algebraically closed, then so is F.*

Proof. — Let β denote the norm on F. Let $P(T) \in \mathfrak{o}_F[T]$ be an arbitrary monic polynomial of degree $d \geqslant 1$; it suffices to check that $P(T)$ has a root in $\mathfrak{o}_F$. We will achieve this by exhibiting a sequence $x_0, x_1, \dots$ of elements of $\mathfrak{o}_F$ such that for all $n \geqslant 0$, $\beta(P(x_n)) \leqslant p^{-n}$ and $\beta(x_{n+1} - x_n) \leqslant p^{-n/d}$. This sequence will then have a limit $x \in \mathfrak{o}_F$ which is a root of P.

To begin, take $x_0 = 0$. Given $x_n \in \mathfrak{o}_F$ with $\beta(P(x_n)) \leqslant p^{-n}$, write $P(T + x_n) = \sum_i Q_i T^i$. If $Q_0 = 0$, we may take $x_{n+1} = x_n$, so assume hereafter that $Q_0 \neq 0$. Put

$$c = \min\{\beta(Q_0/Q_j)^{1/j} : j > 0, Q_j \neq 0\};$$

by taking $j = d$, we see that $c \leqslant \beta(Q_0)^{1/d}$. Also, $\beta(F^\times) = \alpha(L^\times)$ by Lemma 3.5.2, and the latter group is divisible because L is algebraically closed; we thus have $c = \beta(u)$ for some $u \in \mathfrak{o}_F$.

Apply Lemma 3.5.2 to construct $\overline{z} \in \mathfrak{o}_F$ with $\alpha(\overline{z}) = p^{-1}$. For each i, choose $\overline{R}_i \in \mathfrak{o}_L$ whose image in $\mathfrak{o}_L/(\overline{z}) \cong \mathfrak{o}_F/(p)$ is the same as that of $Q_i u^i/Q_0$. Define the polynomial $\overline{R}(T) = \sum_i \overline{R}_i T^i \in \mathfrak{o}_L[T]$. By construction, the largest slope in the Newton polygon of $\overline{R}$ is 0; by this observation plus the fact that L is algebraically closed, it follows that $\overline{R}(T)$ has a root $y' \in \mathfrak{o}_L^\times$. Choose $y \in \mathfrak{o}_F^\times$ whose image in $\mathfrak{o}_F/(p) \cong \mathfrak{o}_L/(\overline{z})$ is the same as that of y', and take $x_{n+1} = x_n + uy$. Then $\sum_i Q_i u^i y^i/Q_0 \equiv 0 \pmod p$, so $\beta(P(x_{n+1})) \leqslant p^{-1}\beta(Q_0) \leqslant p^{-n-1}$ and $\beta(x_{n+1} - x_n) = \beta(u) \leqslant \beta(Q_0)^{1/d} \leqslant p^{-n/d}$. We thus obtain the desired sequence, proving the claim. □

Theorem 3.5.6. — *For F and (L, I) corresponding as in Theorem 3.5.3, the correspondence described in Lemma 3.5.4 induces a tensor equivalence $\mathbf{F\acute{E}t}(F) \cong \mathbf{F\acute{E}t}(L)$. In particular, every finite extension of F is perfectoid, and the absolute Galois groups of F and L are homeomorphic.*

Proof. — Let M be the completion of an algebraic closure of L. *Via* Theorem 3.5.3, $(M, IW(\mathfrak{o}_M))$ corresponds to a perfectoid analytic field E, which by Lemma 3.5.5 is algebraically closed.

By Lemma 3.5.4, each finite Galois extension of L within M corresponds to a finite Galois extension of F within E which is perfectoid. The union of the latter is an algebraic extension of F whose closure is the algebraically closed field E; the union is thus forced to be separably closed by Krasner's lemma. Since F is of characteristic 0 and hence perfect, every finite extension of F is thus forced to lie within a finite Galois extension which is perfectoid; the rest follows from Theorem 3.5.3. □

Remark 3.5.7. — Using Theorem 3.5.6, it is not difficult to show that the functor $F \rightsquigarrow L$ induced by Theorem 3.5.3 by forgetting the ideal I is not fully faithful. For instance, as F varies over finite totally ramified extensions of the completion of $\mathbb{Q}_p(\mu_{p^\infty})$ of a fixed degree, the fields L are all isomorphic.

Theorem 3.5.6 implies that the perfectoid property moves up along finite extensions of analytic fields. It also moves in the opposite direction. (See Proposition 3.6.22 for a more general result.)

Lemma 3.5.8. — *Let K be a perfect analytic field of characteristic p, and let G be a finite group that acts faithfully on K by isometric automorphisms. Then $H^1(G, 1 + \mathfrak{m}_K) = 0$.*

Proof. — We start with an observation concerning additive Galois cohomology. Since K is an acyclic $K^G[G]$-module by the normal basis theorem, the complex

$$K \longrightarrow \operatorname{Hom}(G, K) \longrightarrow \operatorname{Hom}(G^2, K) \longrightarrow \cdots$$

computing Galois cohomology is exact. Using the inverse of Frobenius as in Remark 3.1.6, we see that this complex is in fact almost optimal exact for the supremum norm on each factor.

Now let $f : G \to 1 + \mathfrak{m}_K$ be a 1-cocycle, and put $\delta = \max\{|f(g) - 1| : g \in G\} < 1$. If we view f as an element of $\operatorname{Hom}(G, K)$, its image in $\operatorname{Hom}(G^2, K)$ has supremum norm at most δ^2. By the previous paragraph, we can modify f by an element of $1 + \mathfrak{m}_K$ of norm at most $\delta^{1/2}$ to get a new multiplicative cocycle f' such that $\max\{|f'(g) - 1| : g \in G\} \leqslant \delta^{3/2}$. By iterating the construction, we obtain the desired conclusion. □

Proposition 3.5.9. — *Let E/F be a finite extension of analytic fields such that E is perfectoid. Then F is also perfectoid.*

Proof. — By Theorem 3.5.6, we are free to enlarge E, so we may assume E/F is Galois with group G. Let (L, I) be the pair corresponding to E *via* Theorem 3.5.3; then G acts on both L and I.

We first check that I admits a G-invariant generator (this is immediate if F is already known to contain a perfectoid field, but not otherwise). Let $z \in I$ be any generator. Write z as $[\overline{z}] + pz_1$ with $z_1 \in W(\mathfrak{o}_E)^\times$; then $z_1^{-1}z$ is also a generator. Define the function $f : G \to W(\mathfrak{o}_E)^\times$ taking $g \in G$ to $g(z_1^{-1}z)/(z_1^{-1}z)$. The composition

$G \to W(\mathfrak{o}_E)^\times \to W(\kappa_E)^\times$ is identically 1, so we may apply Lemma 3.5.8 to trivialize the 1-cocycle; that is, there exists $y \in W(\mathfrak{o}_E)^\times$ with $f(g) = g(y)/y$ for all $g \in G$. Then $(yz_1)^{-1}z$ is a G-invariant generator of I.

Put $K = L^G$; by Artin's lemma, L is Galois over K of degree $\#G = [E : F]$. Since I admits a generator contained in $W(\mathfrak{o}_K)$ (which is then primitive of degree 1), we may apply Theorem 3.5.3 to the pair $(K, I \cap W(\mathfrak{o}_K))$ to obtain a perfectoid field F'. By Lemma 3.5.4, E is Galois over F' of degree $[L : K] = [E : F']$ with Galois group G; consequently, $F' = E^G = F$. This proves the claim. □

Definition 3.5.10. — An analytic field K is *deeply ramified* if for any finite extension L of K, $\Omega_{\mathfrak{o}_L/\mathfrak{o}_K} = 0$; that is, the morphism $\mathrm{Spec}(\mathfrak{o}_L) \to \mathrm{Spec}(\mathfrak{o}_K)$ is formally unramified. (Beware that this morphism is usually not of finite type if K is not discretely valued.)

Theorem 3.5.11. — *Any perfectoid analytic field is deeply ramified. (The converse is also true; see* [**52**, Proposition 6.6.6]*.)*

Proof. — Let F be a perfectoid field, and let E be a finite extension of F. Since E/F is separable, $\Omega_{E/F} = 0$; it follows easily that $\Omega_{\mathfrak{o}_E/\mathfrak{o}_F}$ is killed by some nonzero element of $\mathfrak{o}_F$. On the other hand, since E is perfectoid by Theorem 3.5.6, for any $x \in \mathfrak{o}_E$, we can find $y \in \mathfrak{o}_E$ for which $x \equiv y^p \pmod{p}$. Hence $\Omega_{\mathfrak{o}_E/\mathfrak{o}_F} = p\Omega_{\mathfrak{o}_E/\mathfrak{o}_F}$; it now follows that $\Omega_{\mathfrak{o}_E/\mathfrak{o}_F} = 0$. □

Remark 3.5.12. — Many cases of Theorem 3.5.6 in which F is the completion of an algebraic extension of $\mathbb{Q}_p$ arise from the *field of norms* construction of Fontaine and Wintenberger [**49, 127**]. For instance, one may take any *arithmetically profinite* extension of $\mathbb{Q}_p$ thanks to Sen's theory of ramification in p-adic Lie extensions [**117**]. The approach to Theorem 3.5.6 instead requires only checking the perfectoid condition for a single analytic field, as then it is transmitted along finite extensions. For example, for F the completion of $\mathbb{Q}_p(\mu_{p^\infty})$, the perfectoid condition is trivial to check.

Remark 3.5.13. — Theorems 3.5.6 and 3.5.11 have also been obtained by Scholze [**114**] using an analysis of valuation rings made by Gabber and Ramero [**52**, Chapter 6] in the language of *almost ring theory.* This generalizes the alternate proof of the Fontaine-Wintenberger theorem introduced by Faltings; see [**22**, Exercise 13.7.4]. Scholze uses the term *tilting* to refer to the relationship between F and L, as well as to the corresponding relationship between Banach algebras introduced in Theorem 3.6.5. (The term *perfectoid* is also due to Scholze.)

3.6. The perfectoid correspondence for adic Banach algebras

We next extend Theorem 3.5.3 to a correspondence of adic Banach algebras. A parallel development appears in the work of Scholze [**114**], but he fixes a pair of corresponding fields and works over these; our treatment does not require this, and gives rise to perfectoid algebras which need not be defined over a perfectoid field. Our treatment is much closer in spirit to that given in the Bourbaki seminar of

Fontaine [**48**]. The development in [**53**] takes a similar (albeit even more general) approach, but in common with Scholze's treatment it depends heavily on almost ring theory, which ours does not; we achieve similar effects by keeping track of norms. If one is interested in the statements in their almost-ring-theoretic form, these can be recovered after the fact (see for example §5.5).

The form of the following definition is taken from [**29**], where the perfectoid condition is studied from a purely ring-theoretic point of view.

Definition 3.6.1. — A uniform adic Banach $\mathbb{Q}_p$-algebra (A, A^+) is *perfectoid* if $\overline{\varphi} : A^+/(p) \to A^+/(p)$ is surjective and there exists $x \in A^+$ with $x^p \equiv p \pmod{p^2 A^+}$. A uniform Banach algebra A is *perfectoid* if (A, A°) is perfectoid. Note that the condition of (A, A^+) being perfectoid only depends on A (see Proposition 3.6.2).

It is worth pointing out some equivalent formulations of the perfectoid property.

Proposition 3.6.2. — *Let F be an analytic field containing $\mathbb{Q}_p$ and let (A, A^+) be a uniform adic Banach F-algebra.*

(a) *If $|F^\times| \neq |\mathbb{Q}_p^\times|$, then (A, A^+) is perfectoid if and only if $\overline{\varphi} : A^+/(p) \to A^+/(p)$ is surjective.*

(b) *The field F is perfectoid as a uniform Banach $\mathbb{Q}_p$-algebra as per Definition 3.6.1 if and only if it is perfectoid as an analytic field as per Definition 3.5.1.*

(c) *Equip A with the spectral norm. Then A is perfectoid if and only if there exists $c \in (0, 1)$ such that for every $x \in A$, there exists $y \in A$ with $|x - y^p| \leqslant c|x|$. Moreover, if this holds for some c, it holds for any $c \in (p^{-1}, 1)$.*

(d) *The ring A is perfectoid if and only if (A, A^+) is perfectoid.*

(e) *The ring A is perfectoid if and only if there exists a topologically nilpotent unit $\varpi \in A$ such that ϖ^p divides p in A^+ and $\overline{\varphi} : A^+/(\varpi) \to A^+/(\varpi^p)$ is surjective. (This criterion of Fontaine can be used to define perfectoid rings which are not $\mathbb{Q}_p$-algebras; see Remark 3.6.28.)*

Proof. — To check (a), assume that $\overline{\varphi} : A^+/(p) \to A^+/(p)$ is surjective. By hypothesis, there exists $a \in F$ with $p^{-1} < |a| < 1$, and there exist $b, c \in A^+$ with $b^p \equiv a \bmod pA^+$, $c^p \equiv (p/a) \pmod{pA^+}$. In particular, b^p/a and ac^p/p are elements of A^+ congruent to 1 modulo p/a and a, respectively, and so are units. We can thus find $d \in A^+$ with $d^p \equiv (b^p/a)^{-1}(ac^p/p)^{-1} \pmod{pA^+}$, and then $x = bcd$ has the property that $x^p \equiv p \pmod{p^2 A^+}$. Hence A is perfectoid. This yields (a), from which (b) follows by taking $(A, A^\circ) = (F, \mathfrak{o}_F)$.

To check (c), suppose first that the given condition holds for some $c \in (0, 1)$. We may then construct a sequence $x_1, x_2, \dots$ in A such that $|p - x_1^p| \leqslant cp^{-1}$ and $\left|x_n - x_{n+1}^p\right| \leqslant cp^{p^{-n}}$ for $n \geqslant 1$. For n sufficiently large, the conditions of Remark 3.4.9 are satisfied for $x = x_n, y = x_{n+1}$, and so $\overline{\varphi} : \mathfrak{o}_A/(p) \to \mathfrak{o}_A/(p)$ is surjective. We may then choose $y \in \mathfrak{o}_A$ with $y^p \equiv p/x_1^p \pmod{p\mathfrak{o}_A}$, and then $x = x_1 y$ satisfies $x^p \equiv p \pmod{p^2\mathfrak{o}_A}$. We conclude that A is perfectoid.

Conversely, suppose that A is perfectoid. Then for every nonnegative integer m, we can find $x_m \in A$ with $x_m^{p^m}/p \in \mathfrak{o}_A^\times$. Given $x \in A$ nonzero and $c \in (p^{-1}, 1)$, choose a nonnegative integer m and an integer t such that $p^{-1}/c < |x/x_m^{pt}| \leqslant 1$. Since A is perfectoid, we can find $w \in \mathfrak{o}_A$ with $w^p \equiv (x/x_m^{pt}) \pmod{p}$; then $y = x_m^t w$ satisfies $|x - y^p| \leqslant c|x|$.

To check (d), suppose first that (A, A^+) is perfectoid. To check that A is perfectoid, we check the criterion of (c) for any $c \in (p^{-1}, 1)$. Choose $x_1, x_2, \dots \in A^+$ with $x_1^p \equiv p \pmod{p^2 A^+}$ and $x_{n+1}^p \equiv x_n \pmod{pA^+}$ for $n \geqslant 1$. Given $x \in A$, we can find a positive integer n and some integer m such that $\left|x_{n+1}^{pm} x\right| \in (p^{-1}/c, 1)$. Then $x_{n+1}^{pm} x \in A^+$, so we can find $y \in A^+$ with $y^p \equiv x_{n+1}^{pm} x \pmod{pA^+}$. We then have $\left|(y/x_{n+1}^m)^p - x\right| \leqslant c\,|x|$, verifying the criterion.

Conversely, suppose that A is perfectoid. We then make the following observations.

(i) There exists $x_1 \in A^\circ$ with $x_1^p \equiv p \pmod{p^2 A^\circ}$. Since x_1 and (p/x_1) are topologically nilpotent, they belong to A^+.

(ii) For any $y \in A^+$, there exists $z \in A^\circ$ with $z^p \equiv y \pmod{pA^\circ}$. Since $z^p - y$ is topologically nilpotent, it belongs to A^+. Since A^+ is integrally closed, z belongs to A^+. Consequently, $\overline{\varphi} : A^+/(x_1, p) \to A^+/(x_1, p)$ is surjective.

(iii) By (ii), there exist $x_2, x_3 \in A^+$ with $x_2^p \equiv x_1 \pmod{pA^\circ}$, $x_3^p \equiv x_2 \pmod{pA^\circ}$. In particular, $x_3^p \equiv x_2 \pmod{A^+/(x_2, p)}$.

(iv) By Remark 3.4.9, $\overline{\varphi} : A^+/(p) \to A^+/(p)$ is surjective.

(v) By (iv), there exists $y \in A^+$ with $y^p \equiv x_1^p/p \pmod{pA^+}$. Then $x = x_1 y$ satisfies $x^p \equiv p \pmod{p^2 A^+}$. Hence (A, A^+) is perfectoid.

To check (e), note that if A is perfectoid, then the stated criterion holds for $\varpi = x$ for any $x \in A^+$ with $x^p \equiv p \pmod{p^2 A^+}$. Conversely, if the criterion holds, then $\overline{\varphi} : A^+/(p) \to A^+/(p)$ is surjective by Remark 3.4.9. We may thus choose $x_0 \in A^+$ with $x_0^p \equiv (p/\varpi^p) \pmod{pA^+}$, and then $x_1 = \varpi x_0$ satisfies $x_1^p \equiv p \pmod{p\varpi^p A^+}$. In particular, x_1^p/p is congruent to 1 modulo $\varpi^p A^+$ and hence is a unit in A^+. We may thus choose $x_2 \in A^+$ with $x_2^p \equiv (p/x_1^p) \pmod{pA^+}$, and then $x = x_1 x_2$ satisfies $x^p \equiv p \pmod{p^2 A^+}$. □

Lemma 3.6.3. — *Let (A, A^+) be a perfectoid uniform Banach $\mathbb{Q}_p$-algebra. Then the homomorphism $\theta : W(A^{+,\mathrm{frep}}) \to A^+$ is surjective, with kernel generated by an element z which is primitive of degree 1.*

Proof. — We may assume A carries its spectral norm. The map θ is surjective because $\overline{\varphi}$ is surjective on $A^+/(p)$. Choose $x \in A^+$ with $x^p \equiv p \pmod{p^2}$. Choose $\overline{z} = (\dots, \overline{z}_1, \overline{z}_0) \in A^{+,\mathrm{frep}}$ with $\overline{z}_1$ equal to the reduction of x modulo p; then $\theta([\overline{z}]) \equiv p \pmod{p^2 A^+}$. For $n \geqslant 1$, choose $z_n \in A^+$ lifting $\overline{z}_n$; then $z_n^{p^n} \equiv p \pmod{p^2 A^+}$, so $z_n \in A^\times$ and $|z_n y| = p^{p^{-n}} |y|$ for all $y \in A$.

Since $\theta([\overline{z}])$ is divisible by p in A^+, we can find $t \in W(A^+)^\times$ with $\theta(t) = \theta([\overline{z}])/p$. Then $z = [\overline{z}] - pt$ is primitive of degree 1 and belongs to the kernel of θ. By parts (a) and (b) of Lemma 3.4.8, z generates the kernel of θ on $W(\mathfrak{o}_A^{\mathrm{frep}})$. In particular, given $y \in \ker(\theta : W(A^{+,\mathrm{frep}}) \to A^+)$, there is a unique $x = \sum_{n=0}^{\infty} p^n[\overline{x}_n] \in W(\mathfrak{o}_A^{\mathrm{frep}})$ satisfying $xz = y$. But since $t \in W(A^+)^\times$, each $\overline{x}_n$ belongs to $A^{+,\mathrm{frep}} + p\mathfrak{m}_A^{\mathrm{frep}} = A^{+,\mathrm{frep}}$, so y is divisible by z. This proves the claim. $\square$

Definition 3.6.4. — Let (A, A^+) be a perfectoid uniform adic Banach $\mathbb{Q}_p$-algebra. By Lemma 3.6.3, the kernel of $\theta : W(A^{+,\mathrm{frep}}) \to A^+$ is generated by some element z which is primitive of degree 1. Let $\overline{z} \in A^{+,\mathrm{frep}}$ be the reduction of z, and define $R(A) = A^{+,\mathrm{frep}}[\overline{z}^{-1}]$ and $R^+(A^+) = A^{+\,\mathrm{frep}}$. Note that this construction does not depend on the choice of z and that $R(A)$ does not depend on A^+ (hence the notation). By Lemma 3.4.5, $(R(A), R^+(A^+))$ is a perfect uniform Banach $\mathbb{F}_p$-algebra. Also write $I(A, A^+) = \ker(\theta) = zW(A^{+,\mathrm{frep}})$; we also write $I(A)$ for $I(A, A^\circ)$.

For (R, R^+) a perfect uniform adic Banach $\mathbb{F}_p$-algebra and I an ideal of $W(R^+)$ generated by an element z which is primitive of degree 1, write $A(R, I) = (W(R^+)/I)[p^{-1}]$ and $A^+(R^+, I) = W(R^+)/I$. By Lemma 3.3.6, the surjective map $W(R^+)[[\overline{z}]^{-1}] \to A(R, I)$ is optimal, so $(A(R, I), A^+(R^+, I))$ is a perfectoid uniform adic Banach $\mathbb{Q}_p$-algebra; we sometimes denote this object by $(A(R), A^+(R^+))$ when the choice of I is to be understood.

Note that both of these constructions transfer idempotents to idempotents: if $e \in A$ is idempotent, then $e \in A^+$ because A^+ is integrally closed, and so $(\dots, e, e)$ is an idempotent of R^+; conversely, if $\overline{e} \in R$ is idempotent, then so is $\theta([\overline{e}])$. Consequently, for A and (R, I) corresponding as in Theorem 3.6.5, $\mathrm{Spec}(R)$ and $\mathrm{Spec}(A)$ below have the same closed-open subsets; however, they need not have the same irreducible components.

***Theorem 3.6.5* (Perfectoid correspondence).** — *The functors*

$$A \rightsquigarrow (R(A), I(A)), \qquad (R, I) \rightsquigarrow A(R, I)$$

define an equivalence of categories between perfectoid uniform Banach $\mathbb{Q}_p$-algebras A and pairs (R, I) in which R is a perfect uniform Banach $\mathbb{F}_p$-algebra and I is a principal ideal of $W(\mathfrak{o}_R)$ generated by an element which is primitive of degree 1. Similarly, the functors

$$(A, A^+) \rightsquigarrow ((R(A), R^+(A^+)), I(A, A^+)), \qquad ((R, R^+), I) \rightsquigarrow (A(R, I), A^+(R^+, I))$$

define an equivalence of categories between perfectoid uniform adic Banach $\mathbb{Q}_p$-algebras (A, A^+) and pairs $((R, R^+), I)$ in which (R, R^+) is a perfect uniform adic Banach $\mathbb{F}_p$-algebra and I is a principal ideal of $W(R^+)$ generated by an element which is primitive of degree 1.

Proof. — The proofs of the two assertions are similar, so we give only the first one. Given A carrying its spectral norm, the surjectivity of $\theta : W(\mathfrak{o}_A^{\mathrm{frep}}) \to \mathfrak{o}_A$ provides a natural isomorphism $A(R(A), I(A)) \cong A$. Conversely, given (R, I) with R carrying its spectral norm, note that $\mathfrak{o}_{A(R,I)} = W(\mathfrak{o}_R)/I$, so $\mathfrak{o}_{A(R,I)}/(p) = W(\mathfrak{o}_R)/(p, I) = \mathfrak{o}_R/(\overline{z})$ for any $z \in I$ which is primitive of degree 1. This yields a natural isomorphism $\mathfrak{o}_{A(R,I)}^{\mathrm{frep}} \cong \mathfrak{o}_R$, under which $I(A(R, I)) \subset W(\mathfrak{o}_{A(R,I)}^{\mathrm{frep}})$ corresponds to $I \subset W(\mathfrak{o}_R)$. $\square$

We introduce a key example: the perfectoid analogue of a Tate algebra.

Example 3.6.6. — Suppose that A and (R, I) correspond as in Theorem 3.6.5, *e.g.*, $A = F$ and $R = L$ where F and (L, I) correspond as in Theorem 3.5.3. For $r_1, \dots, r_n > 0$, let B, S be the completions of

$$A\{T_1/r_1, \dots, T_n/r_n\}[T_1^{p^{-\infty}}, \dots, T_n^{p^{-\infty}}], R\{T_1/r_1, \dots, T_n/r_n\}[T_1^{p^{-\infty}}, \dots, T_n^{p^{-\infty}}]$$

under the extension of the weighted Gauss norm. That is, the norm of $\sum_{i_1,\dots,i_n} a_{i_1,\dots,i_n} T_1^{i_1} \cdots T_n^{i_n}$ is the maximum of the norm of $a_{i_1,\dots,i_n}$ times $r_1^{i_1} \cdots r_n^{i_n}$ over all $i_1, \dots, i_n \geqslant 0$. Then B is perfectoid, S is perfect, and B corresponds to $(S, IW(\mathfrak{o}_S))$ *via* Theorem 3.5.3 with the element $(T_i, T_i^{1/p}, \dots)$ of $\mathfrak{o}_B^{\mathrm{frep}}$ corresponding to $T_i \in S$.

For some applications, it will be useful to have the following refinement of the criterion from Proposition 3.6.2 (c).

Corollary 3.6.7. — *Let A be a perfectoid uniform Banach $\mathbb{Q}_p$-algebra with spectral norm $|\cdot|$, and let R be the perfect uniform Banach $\mathbb{F}_p$-algebra corresponding to A* via *Theorem 3.6.5. Then for any $\epsilon > 0$, any nonnegative integer m, and any $x \in A$, there exists $y \in A$ of the form $\theta([\overline{y}])$ for some $\overline{y} \in R$, such that*

$$\beta(x - y^p) \leqslant \max\{p^{-1-p^{-1}-\cdots-p^{-m}} \beta(x), \epsilon\} \qquad (\forall \beta \in \mathcal{M}(A)). \tag{3.6.7.1}$$

In particular, we may choose y such that

$$|x - y^p| \leqslant p^{-1-p^{-1}-\cdots-p^{-m}} |x|. \tag{3.6.7.2}$$

Proof. — Lift x along θ to $\tilde{x} \in W(\mathfrak{o}_R)[[\overline{z}]^{-1}]$ and then apply Lemma 3.3.9 with $\tilde{x}$ playing the role of x. Let $\sum_{n=0}^{\infty} p^n [\overline{y}_n]$ be the resulting element of $W(\mathfrak{o}_R)[[\overline{z}]^{-1}]$; then $y = \theta([\overline{y}_0^{1/p}])$ satisfies (3.6.7.1). To obtain (3.6.7.2), take $y = 0$ if $x = 0$, and otherwise apply (3.6.7.1) with ϵ equal to the right side of (3.6.7.2). □

Remark 3.6.8. — The constant in (3.6.7.2) cannot be improved to $p^{-p/(p-1)}$. See [**29**, Example 5.9].

Using Theorem 3.6.5, we may replicate the conclusions of Remark 3.1.6 with perfect uniform Banach $\mathbb{F}_p$-algebras replaced by perfectoid algebras, in the process obtaining compatibility of the correspondence described in Theorem 3.6.5 with various natural operations on adic Banach rings. We begin with a correspondence between strict maps that includes an analogue of Remark 3.1.6 (a) in characteristic 0.

Proposition 3.6.9. — *Keep notation as in Theorem 3.6.5, and equip all uniform Banach rings with their spectral norms.*

(a) *Let $\overline{\psi} : R \to S$ be a strict (and hence almost optimal, by Remark 3.1.6) homomorphism of perfect uniform Banach $\mathbb{F}_p$-algebras, and apply the functor A to obtain $\psi : A \to B$. Then ψ is almost optimal (and surjective if ψ is).*

(b) *Let $\overline{\psi}_1, \overline{\psi}_2 : R \to S$ be homomorphisms of perfect uniform Banach $\mathbb{F}_p$-algebras, and apply the functor A to obtain $\psi_1, \psi_2 : A \to B$. If $\overline{\psi}_1 - \overline{\psi}_2$ is strict surjective (and hence almost optimal, by Remark 3.1.6), then $\psi_1 - \psi_2$ is almost optimal and surjective.*

(c) *Let $\psi : A \to B$ be a strict homomorphism of perfectoid uniform Banach $\mathbb{Q}_p$-algebras. Then ψ is almost optimal and $\psi(A)$ is perfectoid.*

(d) *With notation as in (c), apply the functor R to obtain $\overline{\psi} : R \to S$. Then $\overline{\psi}$ is also almost optimal (and surjective if ψ is).*

Proof. — We first check (a) in case $\overline{\psi}$ is strict surjective. By Remark 3.1.6, $\overline{\psi}$ is almost optimal; in particular, every element of S of norm strictly less than 1 lifts to an element of R of norm strictly less than 1. By Lemma 3.3.6, every element of B of norm strictly less than 1 lifts to an element of A of norm strictly less than 1. Consequently, ψ is almost optimal and surjective. A similar argument yields (b).

We now check (a) in the general case. We may factor ψ as a composition $R \to S_0 \to S$ with $R \to S_0$ strict surjective and $S_0 \to S$ an isometric injection (since R and S are uniform). For z a generator of I, we have $zW(\mathfrak{o}_S) \cap W(\mathfrak{o}_{S_0}) = zW(\mathfrak{o}_{S_0})$, so the map $S_0 \to S$ corresponds to an isometric injection $B_0 \to B$. We may thus deduce (a) from the previous paragraph.

To check (c), let α, β be the spectral norms on A, B. Choose a constant $c \geqslant 1$ such that every $b \in \text{image}(\psi)$ admits a lift $a \in A$ with $\alpha(a) \leqslant c\beta(b)$. We will then prove that the same conclusion holds with c replaced by $c^{1/p}$; this is enough to imply the desired result.

Suppose that $b_l \in \text{image}(\psi)$ for some nonnegative integer l. Lift b_l^p to $a_l \in A$ with $\alpha(a_l) \leqslant c\beta(b_l^p) = c\beta(b_l)^p$. Apply Lemma 3.3.9 (or [**87**, Lemma 5.16]) to find $\overline{x} \in R$ such that

$$\gamma(a_l - \theta([\overline{x}])) \leqslant p^{-1} \max\{\gamma(a_l), \beta(b_l)^p\} \qquad (\gamma \in \mathcal{M}(A)). \tag{3.6.9.1}$$

In particular,

$$\gamma(\theta([\overline{x}])) \leqslant \max\{\gamma(a_l), p^{-1}\gamma(a_l), p^{-1}\beta(b_l)^p\} \leqslant c\beta(b_l)^p.$$

Put $u_l = \theta([\overline{x}^{1/p}])$, $v_l = \psi(u_l)$, and $b_{l+1} = b_l - v_l$; note that $\alpha(u_l) \leqslant c^{1/p}\beta(b_l)$.

For each $\gamma \in \mathcal{M}(B)$, by applying (3.6.9.1) to the restriction of γ to A, we find that $\gamma(b_l^p - v_l^p) \leqslant p^{-1} \max\{\gamma(b_l)^p, \beta(b_l)^p\} = p^{-1}\beta(b_l)^p$. We now consider three cases.

(i) If $\gamma(b_l^p - v_l^p) > \gamma(b_l)^p$, then $\gamma(b_l^p - v_l^p) = \gamma(v_l)^p$, so $\gamma(b_{l+1}) = \gamma(v_l) > \gamma(b_l)$. It follows that $\gamma(b_{l+1}) = \gamma(b_l^p - v_l^p)^{1/p} \leqslant p^{-1/p}\beta(b_l)$.

(ii) If $p^{-1}\gamma(b_l)^p \leqslant \gamma(b_l^p - v_l^p) \leqslant \gamma(b_l)^p$, we may apply [**85**, Lemma 10.2.2] to deduce that $\gamma(b_{l+1}) \leqslant \gamma(b_l^p - v_l^p)^{1/p} \leqslant p^{-1/p}\beta(b_l)$.

(iii) If $\gamma(b_l^p - v_l^p) \leqslant p^{-1}\gamma(b_l)^p$, then by [**85**, Lemma 10.2.2] again, $\gamma(b_{l+1}) \leqslant p^{-1/p}\gamma(b_l) \leqslant p^{-1/p}\beta(b_l)$.

It follows that $\gamma(b_{l+1}) \leqslant p^{-1/p}\beta(b_l)$ for all $\gamma \in \mathcal{M}(B)$, and so $\beta(b_{l+1}) \leqslant p^{-1/p}\beta(b_l)$.

If we now start with $b = b_0 \in \operatorname{image}(\psi)$ and recursively define b_l, u_l, v_l as above, the b_l converge to 0, so the series $\sum_{l=0}^{\infty} v_l$ converges to b. Meanwhile, the series $\sum_{l=0}^{\infty} u_l$ converges to a limit $a \in A$ satisfying $\psi(a) = b$ and $\alpha(a) \leqslant c^{1/p}\beta(b)$. Hence ψ is almost optimal; by Proposition 3.6.2 (c), $\psi(A)$ is perfectoid. This proves (c).

To obtain (d), by noting that a strict injection of uniform Banach rings is isometric and invoking (c), we may reduce to the case where ψ is strict surjective. Let $\overline{\alpha}, \overline{\beta}$ be the norms on R, S. Given $\overline{y} \in S$, by (c) we may lift $\theta([\overline{y}]) \in B$ to some $a \in A$ with $\alpha(a) \leqslant p^{1/2}\overline{\beta}(\overline{y})$. By Lemma 3.3.9 again, we may find $\overline{x} \in R$ such that $\gamma(a - \theta([\overline{x}])) \leqslant p^{-1}\max\{\gamma(a), \overline{\beta}(\overline{y})\}$ for all $\gamma \in \mathcal{M}(A)$; in particular,

$$\gamma(\theta([\overline{x}])) \leqslant \max\{\gamma(a), p^{-1}\gamma(a), p^{-1}\overline{\beta}(\overline{y})\} \leqslant p^{1/2}\overline{\beta}(\overline{y}).$$

Let $\overline{z} \in S$ be the image of $\overline{x}$. For all $\gamma \in \mathcal{M}(B)$, we have $\gamma(\theta([\overline{z}])) \leqslant p^{1/2}\overline{\beta}(\overline{y})$ and

$$\gamma(\theta([\overline{y}] - [\overline{z}])) \leqslant p^{-1}\max\{\gamma(\theta([\overline{y}])), \overline{\beta}(\overline{y})\} = p^{-1}\overline{\beta}(\overline{y}).$$

Put $\overline{\gamma} = \mu(\theta^*(\gamma)) \in \mathcal{M}(S)$; then $\overline{\gamma}(\overline{z}) \leqslant p^{1/2}\overline{\beta}(\overline{y})$. If we expand $[\overline{y}] - [\overline{z}] = \sum_{i=0}^{\infty} p^i[\overline{w}_i]$, for $i > 0$ we have $\gamma(\theta(p^i[\overline{w}_i])) \leqslant p^{-1}\max\{\overline{\gamma}(\overline{y}), \overline{\gamma}(\overline{z})\} \leqslant p^{-1}\overline{\beta}(\overline{y})$. Since $\overline{w}_0 = \overline{y} - \overline{z}$, it follows that $\overline{\gamma}(\overline{y} - \overline{z}) = \gamma(\theta([\overline{y} - \overline{z}])) \leqslant p^{-1/2}\overline{\beta}(\overline{y})$. By iterating the construction as in the proof of (b), we see that every $\overline{y} \in S$ admits a lift $\overline{x} \in R$ for which $\overline{\alpha}(\overline{x}) \leqslant p^{1/2}\overline{\beta}(\overline{y})$. Hence $\overline{\psi}$ is strict, and hence almost optimal by Remark 3.1.6.

We thus may deduce (d) except for the fact that if ψ is surjective, then so is $\overline{\psi}$. This follows from (c) plus Lemma 3.4.7 (b). □

Remark 3.6.10. — Note that in Proposition 3.6.9 (a), strict surjectivity of $\overline{\psi}$ does not imply that $\mathfrak{o}_R$ surjects onto $\mathfrak{o}_S$. Similarly, in part (c), strict surjectivity of ψ does not imply that $\mathfrak{o}_A$ surjects onto $\mathfrak{o}_B$.

We next establish compatibility of the correspondence with completed tensor products, and obtain an analogue of Remark 3.1.6 (c).

Proposition 3.6.11. — *Let $A \to B, A \to C$ be morphisms of perfectoid uniform Banach $\mathbb{Q}_p$-algebras. Let (R, I) be the pair corresponding to A* via *Theorem 3.6.5, and put $S = R(B), T = R(C)$. Then the completed tensor product $B\widehat{\otimes}_A C$ with the tensor product norm is the perfectoid uniform Banach $\mathbb{Q}_p$-algebra corresponding to $S\widehat{\otimes}_R T$. (Note that this immediately implies the corresponding statement for adic Banach rings.)*

Proof. — Put $U = S\widehat{\otimes}_R T$, which is a perfect uniform Banach $\mathbb{F}_p$-algebra by Remark 3.1.6 (c). Put $D = A(U)$; then D is perfectoid. It remains to check that the natural map $B\widehat{\otimes}_A C \to D$ is an isometric isomorphism of Banach $\mathbb{Q}_p$-algebras. To see this, let $\alpha, \beta, \gamma, \delta, \overline{\alpha}, \overline{\beta}, \overline{\gamma}, \overline{\delta}$ denote the spectral norms on A, B, C, D, R, S, T, U, respectively. Choose any $x \in D$ and any $\epsilon > 1$, and apply Lemma 3.3.6 to find $y = \sum_{n=0}^{\infty} p^n[\overline{y}_n] \in W(\mathfrak{o}_U)[[\overline{z}]^{-1}]$ with $\theta(y) = x$ and $\overline{\delta}(\overline{y}_0) \geqslant \overline{\delta}(\overline{y}_n)$ for all $n > 0$. Then write each $\overline{y}_n$ as a convergent sum $\sum_{i=0}^{\infty} \overline{s}_{ni} \otimes \overline{t}_{ni}$ with $\overline{s}_{ni} \in S$, $\overline{t}_{ni} \in T$ and $\overline{\beta}(\overline{s}_{ni}), \overline{\gamma}(\overline{t}_{ni}) < (\epsilon\overline{\delta}(\overline{y}_n))^{1/2}$. (More precisely, by Remark 3.1.6 (c) we can ensure that $\overline{\beta}(\overline{s}_n i)\overline{\gamma}(\overline{t}_{ni}) < \epsilon\overline{\delta}(\overline{y}_n)$, but then we can enforce the desired inequality by transferring a suitable power of $\overline{z}$ between the two terms.) We can then write $[\overline{y}_n]$ as a convergent sum for the (p, z)-adic topology, each term of which is a power of p times the Teichmüller lift of an element of S times the Teichmüller lift of an element of T; moreover, each of these terms has norm at most $\epsilon\overline{\delta}(\overline{y}_n)$. It follows that x is the image of an element of $B\widehat{\otimes}_A C$ of norm at most $\epsilon\delta(y)$; since $\epsilon > 1$ was arbitrary, this yields the desired result. □

Remark 3.6.12. — At this point, we have analogues for perfectoid algebras of parts (a) and (c) of Remark 3.1.6. It would be useful to also have an analogue of part (b); that is, if $\psi_1, \psi_2 : A \to B$ are two homomorphisms of perfectoid uniform Banach $\mathbb{Q}_p$-algebras such that $\psi = \psi_1 - \psi_2$ is strict, one would expect that ψ is almost optimal. Unfortunately, the technique of proof of Proposition 3.6.9 (c) does not suffice to establish this, due to the fact that the image of ψ is not closed under taking p-th powers.

We next establish the compatibility of the perfectoid correspondence with rational localizations, starting with an explicit calculation in the special case of a simple Laurent covering. (See Remark 3.6.16 for a related observation.)

Lemma 3.6.13. — *Suppose that (A, A^+) and $((R, R^+), I)$ correspond as in Theorem 3.6.5. Choose $\overline{g} \in R$ and put $g = \theta([\overline{g}]) \in A$. Put*

$$B_- = A\{T\}/(T - g), \quad B_+ = A\{U\}/(Ug - 1),$$
$$S_- = R\{\overline{T}\}/(\overline{T} - \overline{g}), \quad S_+ = R\{\overline{U}\}/(\overline{U}\overline{g} - 1).$$

Then there are A-linear isomorphisms $A(S_-, IW(S_-^\circ)) \cong B_-$, $A(S_+, IW(S_+^\circ)) \cong B_+$ taking $[\overline{T}], [\overline{U}]$ to T, U.

Proof. — Equip A with the spectral norm. For $r \in \mathbb{Z}[p^{-1}]_{\geqslant 0}$, put $g^r = \theta([\overline{g}^r])$. By Lemma 2.8.8, for each nonnegative integer h,

$$B_- \cong A\{T^{p^{-h}}\}/(T^{p^{-h}} - g^{p^{-h}}),$$
$$B_+ \cong A\{U^{p^{-h}}\}/(U^{p^{-h}} g^{p^{-h}} - 1).$$

More precisely, for $y_- = \sum_i y_{-,i} \in A\{T^{p^{-h-1}}\}, y_+ = \sum_i y_{+,i} \in A\{U^{p^{-h-1}}\}$, put

$$z_- = \sum_i y_{-,i} g^{i - p^{-h}\lfloor p^h i\rfloor} T^{p^{-h}\lfloor p^h i\rfloor} \in A\{T^{p^{-h}}\},$$
$$z_+ = \sum_i y_{+,i} g^{p^{-h}\lceil p^h i\rceil - i} U^{p^{-h}\lceil p^h i\rceil} \in A\{U^{p^{-h}}\};$$

then y_* and z_* represent the same class in B_* and

$$|z_*| \leqslant \max\{1, |g|^{p^{-h}}\} |y_*|.$$

Write $|\bullet|_h$ for the quotient norms on B_-, B_+ induced from $A\{T^{p^{-h}}\}$, $A\{U^{p^{-h}}\}$; then

$$|x|_{h+1} \leqslant |x|_h \leqslant \max\{1, |g|\}^{p^{-h}} |x|_{h+1} \qquad (x \in B_*).$$

In particular,

$$|x|_h \leqslant |x|_0 \leqslant \max\{1, |g|\}^{p/(p-1)} |x|_h \qquad (x \in B_*; h = 0, 1, \dots). \tag{3.6.13.1}$$

By Proposition 3.1.7, S_* is perfect uniform. Let A'_-, A'_+ be the completions of $A\{T\}[T^{p^{-\infty}}]$, $A\{U\}[U^{p^{-\infty}}]$ for the Gauss norm. Let J_* be the closure in A'_* of the ideal of $A\{T\}[T^{p^{-\infty}}]$ generated by $T^{p^{-h}} - g^{p^{-h}}$ (if $* = -$) or the ideal of $A\{U\}[U^{p^{-\infty}}]$ generated by $U^{p^{-h}} g^{p^{-h}} - 1$ (if $* = +$) for $h = 0, 1, \ldots$. Note that J_* is itself an ideal of A'_*, so we may form the quotient $B'_* = A'_*/J_*$; the inclusions $A\{T\} \to A'_-, A\{U\} \to A'_+$ then induce maps $B_* \to B'_*$, which by (3.6.13.1) are isomorphisms.

From Example 3.6.6, we see that A'_-, A'_+ are perfectoid and we obtain identifications $R(A'_*) \cong S_*$ taking $(\ldots, T^{1/p}, T), (\ldots, U^{1/p}, U)$ to $\overline{T}, \overline{U}$. The resulting map $A'_* \to A(S_*, IW(S^\circ_*))$ is surjective; its kernel J'_* is the closure of the ideal generated by $\theta([\overline{T}^{p^{-h}} - \overline{g}^{p^{-h}}])$ (if $* = -$) or $\theta([\overline{U}^{p^{-h}} \overline{g}^{p^{-h}} - 1])$ (if $* = +$) for all h.

Under the map $A'_- \to A(S_-, IW(S^\circ_-))$, $T^{p^{-h}}$ and $g^{p^{-h}}$ both map to $[\overline{T}^{p^{-h}}] = [\overline{g}^{p^{-h}}]$; similarly, under the map $A'_+ \to A(S_+, IW(S^\circ_+))$, $U^{p^{-h}} g^{-p^{-h}}$ maps to $[\overline{U}^{p^{-h}} \overline{g}^{p^{-h}}] = [1] = 1$. This means that $J_* \subseteq J'_*$; we also have the reverse inclusion thanks to Remark 3.2.3. We conclude that the induced map $B'_* \to A(S_*, IW(S^\circ_*))$ is an isomorphism. □

With this calculation in hand, we may treat the general case.

Theorem 3.6.14. — *Suppose that (A, A^+) and $((R, R^+), I)$ correspond as in Theorem 3.6.5.*

(a) *The homeomorphism $\mathcal{M}(A) \cong \mathcal{M}(R)$ of Theorem 3.3.7 lifts to a functorial homeomorphism $\mathrm{Spa}(A, A^+) \cong \mathrm{Spa}(R, R^+)$.*

(b) *The homeomorphism in (a) identifies rational subspaces. More precisely, for $\overline{f}_1, \ldots, \overline{f}_n, \overline{g} \in R$, the rational subspace*

$$\{v \in \mathrm{Spa}(R, R^+) : v(\overline{f}_i) \leqslant v(\overline{g}) \quad (i = 1, \ldots, n)\}.$$

corresponds to

(3.6.14.1)
$$\{v \in \mathrm{Spa}(A, A^+) : v(f_i) \leqslant v(g) \quad (i = 1, \ldots, n)\} \quad (f_i = \theta([\overline{f}_i]), g = \theta([\overline{g}])),$$

and every rational subspace of $\mathrm{Spa}(A, A^+)$ can be written in the form.

(c) *Let $U \subseteq \mathrm{Spa}(A, A^+)$ and $V \subseteq \mathrm{Spa}(R, R^+)$ be rational subdomains corresponding as in (a). Let $(A, A^+) \to (B, B^+)$, $(R, R^+) \to (S, S^+)$ be the rational localizations representing U and V, respectively. Then (B, B^+) is again perfectoid, and there are natural identifications $(S, S^+) \cong (R(B), R^+(B^+))$, $(B, B^+) \cong (A(S, IW(S^+)), A^+(S^+, IW(S^+)))$.*

Proof. — To prove (a), we first define the map on points. For $\alpha \in \mathcal{M}(A)$ corresponding to $\beta \in \mathcal{M}(B)$, we have a canonical isomorphism $\mathfrak{o}_{\mathcal{H}(\alpha)}/(p) \cong \mathfrak{o}_{\mathcal{H}(\beta)}/(\overline{z})$ for any z as in Lemma 3.6.3. We may thus identify points of $\mathrm{Spa}(A, A^+)$ lifting α with valuation rings of $\mathcal{H}(\alpha)$ containing $\mathfrak{m}_{\mathcal{H}(\alpha)}$, then with valuation rings of $\kappa_{\mathcal{H}(\alpha)} \cong \kappa_{\mathcal{H}(\beta)}$, then with valuation rings of $\mathcal{H}(\beta)$ containing $\mathfrak{m}_{\mathcal{H}(\beta)}$, then with points of $\mathrm{Spa}(R, R^+)$ lifting β.

We now have a functorial bijection $\mathrm{Spa}(A, A^+) \to \mathrm{Spa}(R, R^+)$. To see that this is a homeomorphism, it suffices to prove (b), which we do by imitating the proof of Theorem 3.3.7. For $\overline{f}_1, \dots, \overline{f}_n, \overline{g} \in R$, if we put $f_i = \theta([\overline{f}_i]), g = \theta([\overline{g}])$, then $f_1, \dots, f_n, g$ generate the unit ideal in A if and only if $\overline{f}_1, \dots, \overline{f}_n, \overline{g}$ generate the unit ideal in R (by applying Corollary 2.3.6 in both A and R). This means that rational subspaces of $\mathrm{Spa}(R, R^+)$ correspond to rational subspaces of $\mathrm{Spa}(A, A^+)$ as described. Conversely, given a rational subspace U as in (2.4.3.2), pick $\epsilon > 0$ as in Remark 2.4.7, then apply Lemma 3.3.9 to find $\overline{f}_1, \dots, \overline{f}_n, \overline{g} \in R$ such that

$$\alpha(f_i - \theta([\overline{f}_i])) \leqslant p^{-1} \max\{\alpha(f_i), \epsilon\}, \quad \alpha(g - \theta([\overline{g}])) \leqslant p^{-1} \max\{\alpha(g), \epsilon\} \quad (\alpha \in U \cap \mathcal{M}(A)).$$

As in Remark 2.4.7, we see that the rational subspace defined by $\theta([\overline{f}_1]), \dots, \theta([\overline{f}_n])$, $\theta([\overline{g}])$ coincides with U. This proves (b).

To prove (c), by Proposition 2.4.24 we may assume that U is part of a simple Laurent covering. From the proof of (b), we may define this covering using a parameter of the form $g = \theta([\overline{g}])$ for some $\overline{g} \in R$. By Proposition 3.1.7, (S, S^+) is perfect uniform. By Lemma 3.6.13, B is perfectoid; by Proposition 3.6.2, (B, B^+) is also perfectoid. The other identifications now follow from Theorem 3.6.5 and the universal property of a rational localization. □

As a corollary, we obtain the Tate and Kiehl properties for perfectoid algebras.

Theorem 3.6.15. — *Any perfectoid adic Banach algebra is stably uniform and sheafy, and thus satisfies the Tate sheaf and Kiehl glueing properties (Definition 2.7.6).*

Proof. — The stably uniform property is immediate from Theorem 3.6.14. The sheafy property then follows from Theorem 2.8.10; alternatively, one may use Proposition 2.4.20 and Proposition 3.6.11 to reduce to the case of a simple Laurent covering of $\mathrm{Spa}(A, A^+)$, then argue as in Proposition 3.6.9 (a), (b) to derive the desired exact sequence from the corresponding exact sequence in characteristic p (Theorem 3.1.13). As usual, the Tate and Kiehl properties follow from the sheafy property *via* Theorem 2.7.7. □

We have the following analogue of Remark 3.1.8.

Remark 3.6.16. — Set notation as in Theorem 3.6.14, and equip A with the spectral norm. For $r \in \mathbb{Z}[p^{-1}]_{\geqslant 0}$ and $* \in \{f_1, \dots, f_n, g\}$, write $*^r$ for $\theta([\overline{*}^r])$. Choose $h_1, \dots, h_n, k \in A$ such that $h_1 f_1 + \cdots + h_n f_n + kg = 1$. Then

$$y = \sum_{i_1, \dots, i_n} y_{i_1, \dots, i_n} T_1^{i_1} \cdots T_n^{i_n} \in A\{T_1, \dots, T_n\}[T_1^{p^{-\infty}}, \dots, T_n^{p^{-\infty}}]$$

represents the same element of B as does

$$z = (k + h_1 T_1 + \cdots + h_n T_n)^n \times \\ \times \sum_{i_1, \dots, i_n} y_{i_1, \dots, i_n} f_1^{i_1 - \lfloor i_1 \rfloor} \cdots f_n^{i_n - \lfloor i_n \rfloor} g^{n - (i_1 - \lfloor i_1 \rfloor + \cdots + i_n - \lfloor i_n \rfloor)} T_1^{\lfloor i_1 \rfloor} \cdots T_n^{\lfloor i_n \rfloor},$$

which satisfies

$$|z| \leqslant c\,|y|, \qquad c = \max\{|h_1|, \dots, |h_n|, |k|\}^n \,|f_1| \cdots |f_n|\,|g|^n.$$

By replacing $\overline{f}_1, \dots, \overline{f}_n, \overline{g}$ by suitable p-power roots, for any given $c > 1$ we may obtain a strict surjection $A\{T_1, \dots, T_n\} \to B$ in which the quotient norm is at most c times the spectral norm.

We next establish compatibility with formation of quotients, and more generally with passage along homomorphisms with dense image.

Theorem 3.6.17. — *Suppose that A and (R, I) correspond as in Theorem 3.6.5.*

(a) *Let $\overline{\psi} : R \to S$ be a bounded homomorphism of uniform Banach $\mathbb{F}_p$-algebras with dense image. Then S is also perfect, and the corresponding homomorphism $\psi : A \to B$ of perfectoid uniform Banach $\mathbb{Q}_p$-algebras also has dense image.*

(b) *Let B be a uniform Banach $\mathbb{Q}_p$-algebra admitting a bounded homomorphism $\psi : A \to B$ with dense image. Then B is also perfectoid, and the corresponding homomorphism $\overline{\psi} : R \to S$ of perfect uniform Banach $\mathbb{F}_p$-algebras also has dense image.*

(c) *In (a) and (b), $\overline{\psi}$ is surjective if and only if ψ is.*

Proof. — In the setting of (a), the ring S is reduced and admits the dense perfect $\mathbb{F}_p$-subalgebra $\overline{\psi}(R)$, so S is also perfect. Let $\alpha, \beta, \overline{\alpha}, \overline{\beta}$ denote the spectral norms on A, B, R, S, respectively. By Lemma 3.3.6, for any $x \in B$ and $\epsilon > 0$, we can find a finite sum $\sum_{i=0}^n p^i[\overline{x}_i] \in W(S)$ such that $\beta(x - \theta(\sum_{i=0}^n p^i[\overline{x}_i])) < \epsilon$. For $i = 0, \dots, n$, if $\overline{x}_i = 0$, take $\overline{y}_i = 0$, otherwise choose $\overline{y}_i \in R$ with

$$\overline{\beta}(\overline{x}_i - \overline{\psi}(\overline{y}_i)) < \inf\{\epsilon^{p^j} p^{(i+j)p^j} \overline{\beta}(\overline{x}_i)^{1-p^j} : j = 0, 1, \dots\}.$$

(Note that the sequence whose infimum is sought tends to $+\infty$ as $j \to \infty$, since it is dominated by p^{jp^j}, so the infimum is positive.) Put $y = \sum_{i=0}^n p^i \theta([\overline{y}_i]) \in A$; then

$$\begin{aligned}
\beta\left(\sum_{i=0}^n p^i \theta([\overline{x}_i] - [\overline{\psi}(\overline{y}_i)])\right) & \\
&\leqslant \max\{p^{-i}\beta(\theta([\overline{x}_i] - [\overline{\psi}(\overline{y}_i)])) : i = 0, \dots, n\} \\
&\leqslant \max\{p^{-i-j}\overline{\beta}(\overline{x}_i)^{1-p^{-j}} \overline{\beta}(\overline{x}_i - \overline{\psi}(\overline{y}_i))^{p^{-j}} : i = 0, \dots, n; j = 0, 1, \dots\} \\
&< \epsilon.
\end{aligned}$$

It follows that $\beta(x - \psi(\sum_{i=0}^n p^i[\overline{y}_i])) < \epsilon$, yielding (a).

In the setting of (b), let α, β be the norms on A, B. Given $x \in \psi(A) \cap \mathfrak{o}_B$, choose $w \in \psi^{-1}(x)$. By Lemma 3.3.9, we can find $\overline{w} \in R$ such that

$$\gamma(w - \theta([\overline{w}])) \leqslant p^{-1} \max\{\gamma(w), \beta(x)\} \qquad (\gamma \in \mathcal{M}(A)).$$

Put $y = \psi(\theta([\overline{w}^{1/p}]))$; then $\gamma(x - y^p) \leqslant p^{-1}\beta(x)$ for all $\gamma \in \mathcal{M}(B)$, so $\beta(x - y^p) \leqslant p^{-1}\beta(x)$. Since $\psi(A)$ is dense in B, B is perfectoid. Given $\overline{x} \in S$, choose $w \in A$ with $\beta(\psi(w) - \theta([\overline{x}])) \leqslant p^{-1}\overline{\beta}(\overline{x})$, then apply Lemma 3.3.9 again to choose $\overline{w} \in R$ such that

$$\gamma(w - \theta([\overline{w}])) \leqslant p^{-1} \max\{\gamma(w), \overline{\beta}(\overline{x})\} \qquad (\gamma \in \mathcal{M}(A)).$$

Put $\overline{y} = \overline{\psi}(\overline{w})$; then $\gamma(\theta([\overline{x}] - [\overline{y}])) \leqslant p^{-1}\overline{\beta}(\overline{x})$ for $\gamma \in \mathcal{M}(B)$, so $\gamma(\overline{x} - \overline{y}) \leqslant p^{-1}\overline{\beta}(\overline{x})$. This yields (b).

In the setting of (c), if either $\overline{\psi}$ or ψ is surjective, then it is strict by the open mapping theorem (Theorem 2.2.8). Consequently, (c) follows from Proposition 3.6.9 (a), (d). □

Corollary 3.6.18. — *Let $\psi_1 : C \to A$, $\psi_2 : C \to B$ be bounded homomorphisms of uniform Banach $\mathbb{Q}_p$-algebras, and let $D = (A\widehat{\otimes}_C B)^u$ denote the uniform completion of $A \otimes_C B$. If A and B are perfectoid, then so is D.*

Proof. — Let α, β be the spectral norms on A, B. Let N be the multiplicative monoid of $R(B)$. Equip the monoid ring $A[N]$ with the weighted Gauss norm

$$\left|\sum_i a_i[n_i]\right| = \max_i\{\alpha(a_i)\mu(\beta)(n_i)\}.$$

Let E be the completion of $A[N]$; it is a perfectoid algebra (compare Example 3.6.6). The formula

$$\sum_i a_i[n_i] \longmapsto \sum_i a_i \otimes \theta([n_i])$$

defines a bounded homomorphism from E to the ordinary completion of $A \otimes_C B$ with dense image; consequently, the resulting homomorphism $E \to D$ also has dense image. By Theorem 3.6.17 (b) again, D is perfectoid. □

A related observation is that the perfectoid property is preserved under completions.

Proposition 3.6.19. — *Let A be a perfectoid uniform Banach $\mathbb{Q}_p$-algebra equipped with its spectral norm. Let J be a finitely generated ideal of $\mathfrak{o}_A$ which contains p. Equip each quotient $\mathfrak{o}_A/J^n$ with the quotient norm, equip the inverse limit R with the supremum norm, and put $B = R[p^{-1}]$. Then B is again a perfectoid uniform Banach $\mathbb{Q}_p$-algebra.*

Proof. — Choose generators $x_1, \dots, x_m$ of J; then R can also be written as the inverse limit of the quotients $\mathfrak{o}_A/(p^n, x_1^{pn}, \dots, x_m^{pn})$. Consequently, each element y of R can be written as an infinite series

$$\sum_{n=0}^{\infty} (a_{n0}p^n + a_{n1}x_1^{pn} + \cdots + a_{nm}x_m^{pn})$$

with all of the a_{ni} in $\mathfrak{o}_A$. Choose $b_{ni} \in \mathfrak{o}_A$ with $b_{ni}^p \equiv a_{ni} \pmod p$; then the series

$$b_{n0} + \sum_{n=0}^{\infty} (b_{n1}^p x_1^n + \cdots + b_{nm}^p x_m^n)$$

converges to an element z of R satisfying $z^p \equiv y \pmod{pR}$. From this, the claim follows at once. □

We finally establish compatibility with finite étale covers. As in the case of analytic fields, the first step is to lift from characteristic p.

Lemma 3.6.20. — *For A and (R, I) corresponding as in Theorem 3.6.5 and $S \in$ $\mathbf{F\acute{E}t}(R)$, the perfectoid Banach algebra B over $\mathbb{Q}_p$ corresponding to $(S, IW(\mathfrak{o}_S))$ via Theorem 3.6.5 belongs to $\mathbf{F\acute{E}t}(A)$ and its norm is equivalent to any norm given by Proposition 2.8.16.*

Proof. — We may assume that S is of constant rank $d > 0$ as an R-module. Let z be a generator of I which is primitive of degree 1. Since S is a finite R-module and $\overline{\varphi}$ is bijective on S, we can find $\overline{x}_1, \dots, \overline{x}_n \in \mathfrak{o}_S$ such that $\mathfrak{o}_S/(\overline{x}_1\mathfrak{o}_R + \cdots + \overline{x}_n\mathfrak{o}_R)$ is killed by $\overline{z}$. Using Remark 3.2.3, it follows that $W(\mathfrak{o}_S)/([\overline{x}_1]W(\mathfrak{o}_R) + \cdots + [\overline{x}_n]W(\mathfrak{o}_R))$ is killed by $[\overline{z}]$. Quotienting by z and then inverting p, we find that B is a finite A-module. By Proposition 2.8.4 and Lemma 3.5.4, B is locally free of constant rank d as an A-module.

We now know that B is a finite projective A-module. To check that $B \in \mathbf{F\acute{E}t}(A)$, it remains to check that the map $B \to \mathrm{Hom}_A(B, A)$ taking x to $y \mapsto \mathrm{Trace}_{B/A}(xy)$ is surjective. By Lemma 2.3.12, it suffices to check this pointwise; we may thus apply Lemma 3.5.4 to conclude.

To conclude, note that the equivalence between the norm on B and the one derived from Proposition 2.8.16 is a consequence of the open mapping theorem (Theorem 2.2.8). □

Theorem 3.6.21. — *For A and (R, I) corresponding as in Theorem 3.6.5, if we equip $B \in \mathbf{F\acute{E}t}(A)$ with a Banach norm provided by Proposition 2.8.16, then B is a perfectoid uniform Banach A-algebra. Moreover, the correspondence of Lemma 3.6.20 induces a tensor equivalence $\mathbf{F\acute{E}t}(A) \cong \mathbf{F\acute{E}t}(R)$. (As in Theorem 3.5.11, it follows that $\Omega_{\mathfrak{o}_B/\mathfrak{o}_A} = 0$.)*

Proof. — It suffices to check that any $B \in \mathbf{F\acute{E}t}(A)$ arises as in Lemma 3.6.20. Extend A to an adic Banach ring (A, A^+) corresponding to (R, R^+) *via* Theorem 3.6.5. Recall that by Theorem 3.3.7 and Theorem 3.6.14, there are compatible homeomorphisms $\mathcal{M}(A) \cong \mathcal{M}(R)$, $\mathrm{Spa}(A, A^+) \cong \mathrm{Spa}(R, R^+)$ matching up rational subdomains on both sides. For each $\delta \in \mathcal{M}(A)$, let $\gamma \in \mathcal{M}(R)$ be the corresponding point. Using Theorem 3.5.6, we may transfer $B \otimes_A \mathcal{H}(\delta) \in \mathbf{F\acute{E}t}(\mathcal{H}(\delta))$ to some $S(\gamma) \in \mathbf{F\acute{E}t}(\mathcal{H}(\gamma))$. By Lemma 2.2.3 (a) (and Theorem 1.2.8), there exists a rational localization $(R, R^+) \to (R_1, R_1^+)$ encircling γ such that $S(\gamma)$ extends to $S_1 \in \mathbf{F\acute{E}t}(R_1)$. Let $(A, A^+) \to (A_1, A_1^+)$ be the rational localization corresponding to $(R, R^+) \to (R_1, R_1^+)$ *via* Theorem 3.6.14. Applying Lemma 3.6.20, we may lift S_1 to $B_1 \in \mathbf{F\acute{E}t}(A_1)$. By Lemma 2.2.3 (a) again, by replacing $(A, A^+) \to (A_1, A_1^+)$ by another rational localization encircling δ, we can ensure that $B_1 \cong B \otimes_A A_1$. In particular, $B \otimes_A A_1$ is perfectoid.

By compactness, we obtain a strong rational covering $\{(A, A^+) \to (A_i, A_i^+)\}_i$ corresponding to a strong rational covering $\{(R, R^+) \to (R_i, R_i^+)\}$ as in Theorem 3.6.14, such that for each i, $B \otimes_A A_i$ corresponds to some $S_i \in \mathbf{F\acute{E}t}(R_i)$ as in Theorem 3.6.5. Put $(A_{ij}, A_{ij}^+) = (A_i, A_i^+)\widehat{\otimes}_{(A,A^+)}(A_j, A_j^+)$, so that $(A, A^+) \to (A_{ij}, A_{ij}^+)$ is the rational localization corresponding to $\mathrm{Spa}(A_i, A_i^+) \cap \mathrm{Spa}(A_j, A_j^+)$ by Proposition 3.6.11. Let $(R, R^+) \to (R_{ij}, R_{ij}^+)$ be the corresponding rational localization. Using Theorem 3.6.14, we may transfer the isomorphisms $(B \otimes_A A_i) \otimes_{A_i} A_{ij} \cong (B \otimes_A A_j) \otimes_{A_j} A_{ij}$ to obtain isomorphisms $S_i \otimes_{R_i} R_{ij} \cong S_j \otimes_{R_j} R_{ij}$ satisfying the cocycle condition. By Theorem 2.6.9, we may glue the S_i to obtain $S \in \mathbf{F\acute{E}t}(R)$.

Apply Lemma 3.6.20 to lift S to $C \in \mathbf{F\acute{E}t}(A)$. By Theorem 3.6.5, we have isomorphisms $B \otimes_A A_i \cong C \otimes_A A_i$ which again satisfy the cocycle condition. They thus glue to an isomorphism $B \cong C$ by Theorem 2.6.9 again, so B is perfectoid as desired. □

As for fields (see Proposition 3.5.9), we have the following converse result.

Proposition 3.6.22. — *Let $A \to B$ be a morphism of uniform Banach algebras over $\mathbb{Q}_p$ such that B is perfectoid.*

(a) *If $A \to B$ is faithfully finite étale, then A is perfectoid.*

(b) *Let $k \subseteq \ell$ be perfect fields of characteristic p and put $K = W(k)[p^{-1}]$, $L = W(\ell)[p^{-1}]$. Suppose that A is a Banach algebra over K and that $B = (A\widehat{\otimes}_K L)^u$. Then $A\widehat{\otimes}_K L = B$ and A is perfectoid.*

Proof. — Equip all uniform Banach rings in this argument with their spectral norms. To prove (a), we may assume that B is finite étale over A of constant degree $d > 0$. Let (S, J) be the pair corresponding to B *via* Theorem 3.6.5. Since the two natural morphisms $\iota_1, \iota_2 : B \to B \otimes_A B$ are both finite étale, by Theorem 3.6.21, on one hand $B \otimes_A B$ is perfectoid; on the other hand, if we put $T = R(B \otimes_A B)$, then ι_1, ι_2 correspond to two finite étale morphisms $\overline{\iota}_1, \overline{\iota}_2 : S \to T$.

Since $\mathrm{Spec}(B) \to \mathrm{Spec}(A)$ is étale, the surjection $B \otimes_{\mathbb{Z}} B \to B \otimes_A B$ defines a closed immersion $\mathrm{Spec}\, B \times_{\mathrm{Spec}\, A} \mathrm{Spec}\, B \to \mathrm{Spec}\, B \times_{\mathrm{Spec}\, \mathbb{Z}} \mathrm{Spec}\, B$ which is a finite étale equivalence relation on $\mathrm{Spec}(B)$ over $\mathrm{Spec}(\mathbb{Z})$. This transfers to a finite étale equivalence relation on $\mathrm{Spec}(S)$ over $\mathrm{Spec}(\mathbb{Z})$. However, by [**122**, Tag 07S5], any such equivalence relation on an affine scheme admits a quotient in the category of schemes; that is, for R the equalizer of $\overline{\iota}_1, \overline{\iota}_2$, the morphism $R \to S$ is finite étale of constant degree d and the induced map $S \otimes_R S \to T$ is an isomorphism.

Since R is the equalizer of $\overline{\iota}_1, \overline{\iota}_2 : R \to S$, $W(R)$ is the equalizer of $W(\overline{\iota}_1), W(\overline{\iota}_2) : W(R) \to W(S)$. Consequently, the image of $W(\mathfrak{o}_R)$ under θ belongs to the equalizer of ι_1, ι_2, which is A.

For $x \in \mathfrak{o}_A$, the image of x in $\mathfrak{o}_B$ belongs to the equalizer of ι_1 and ι_2. Consequently, the image of x in $\mathfrak{o}_S/(p)$ lifts to an element $\overline{x} \in \mathfrak{o}_S$ for which $|\overline{\iota}_1(x) - \overline{\iota}_2(x)| \leqslant p^{-1}$. By Remark 3.1.6, for any $\epsilon > 0$ there exists $\overline{y} \in \mathfrak{o}_R$ satisfying $|\overline{y} - \overline{x}| \leqslant p^{-1+\epsilon}$. If we put $y = \theta([\overline{y}^{1/p}])$, then $y \in \mathfrak{o}_A$ satisfies $|x - y^p| \leqslant p^{-1+\epsilon}$. It follows that $\overline{\varphi} : \mathfrak{o}_A/(p) \to \mathfrak{o}_A/(p)$ is surjective.

To prove that A is perfectoid, it now suffices to produce $x \in \mathfrak{o}_A$ with $x^p \equiv p \pmod{p^2\mathfrak{o}_A}$. To do this, choose an integer $n > d$ and choose integers a, b satisfying $pa + bd/p^n = 1$. Let $y = [\overline{y}] + py_1$ be a generator of J, and put

$$x_0 = p^a\theta([\mathrm{Norm}_{S/R}(-\overline{y}/\overline{y}_1)^{bp^{-n-1}}]);$$

then $x_0 \in A$ and $|x_0^p - p| \leqslant p^{-1-p^{-n}}$. By the previous paragraph, we can find $x_1 \in \mathfrak{o}_A$ such that $|x_1^p - x_0^p/p| \leqslant p^{-1}$; we may then take $x = x_0/x_1$. This proves (a).

To prove (b), we follow the proof of Lemma 2.8.6. Let S be a basis of ℓ over k containing 1; then each $y \in \mathfrak{o}_B$ has a unique convergent representation as $\sum_{s\in S} y_s \otimes [s]$ with $y_s \in A$, and $|y| = \max\{|y_s| : s \in S\}$. In particular, $B = A\widehat{\otimes}_K L$; more precisely,

$$\mathfrak{o}_B/p\mathfrak{o}_B \cong \mathfrak{o}_A/p\mathfrak{o}_A \otimes_{W(k)/pW(k)} W(\ell)/pW(\ell);$$

since $W(\ell)/pW(\ell) \cong \ell$ is flat over $W(k)/pW(k) \cong k$, the surjectivity of $\overline{\varphi}$ on $\mathfrak{o}_A/p\mathfrak{o}_A$ follows from the corresponding property on $\mathfrak{o}_B/p\mathfrak{o}_B$ by faithfully flat descent. To find $x \in \mathfrak{o}_A$ with $x^p \equiv p \pmod{p^2\mathfrak{o}_A}$, first choose $y \in \mathfrak{o}_B$ with $y^p \equiv p \pmod{p^2\mathfrak{o}_B}$. Then $|y_s| \leqslant p^{-1/p}$ for all $s \in S$, so

$$\sum_{s\in S} y_s^p \otimes [s^p] \equiv p \pmod{p^2\mathfrak{o}_B}.$$

Since k and ℓ are perfect, $\{s^p : s \in S\}$ is also a basis of ℓ over k, so we may take $x = y_1$ to deduce (b). □

Remark 3.6.23. — Scholze observes [**114**] that for A and (R, I) corresponding as in Theorem 3.6.5, Theorem 3.6.14 and Theorem 3.6.21 imply that the small étale sites of the adic spaces associated to R and A are naturally equivalent. This observation is the point of departure of his theory of *perfectoid spaces*, which casts the aforementioned results in more geometric terms; we will make contact with this construction in §8.

Besides the expected consequences for relative p-adic Hodge theory, as in the study of relative comparison isomorphisms between étale and de Rham cohomology [**115**] and in our own work in this paper, the perfectoid correspondence are some unexpected consequences. For instance, Scholze uses it to derive some new cases of the weight-monodromy conjecture in étale cohomology [**114**]. There may also be consequences in the direction of Hochster's direct summand conjecture in commutative algebra, as in the work of Bhatt [**16**] (see also the discussion in [**53**]). For further discussion, see [**113**].

Remark 3.6.24. — While the correspondence described above is sufficient for some applications, relative p-adic Hodge theory tends to requires somewhat more refined information. The most common approach to getting this extra information is through variants of the *almost purity theorem* of Faltings [**42, 43**]. We will instead take an alternative approach based on relative Robba rings, as introduced in §5. For the relationship between the two points of view, see §5.5.

We mention the characterization of the perfectoid correspondence used in [**114**], which avoids any reference to Witt vectors, as well as a related criterion not found in [**114**].

Proposition 3.6.25. — *The following statements are true.*

(a) *For A and (R, I) corresponding as in Theorem 3.6.5, there is an isomorphism*

$$R \longrightarrow \varprojlim_{x \mapsto x^p} A, \qquad \overline{x} \longmapsto (\theta([\overline{x}]), \theta([\overline{x}^{1/p}]), \dots).$$

(b) *For A perfectoid, $\varprojlim_{x \mapsto x^p} A$ generates a dense $\mathbb{Z}$-subalgebra of A.*

(c) *If A is a uniform Banach algebra over some perfectoid algebra B and $\varprojlim_{x \mapsto x^p} A$ generates a dense B-subalgebra of A, then A is perfectoid.*

Note that the converse of (b) would imply (c), but such a converse would require a different proof technique than that of (c).

Proof. — Put $S = \varprojlim_{x \mapsto x^p} A$. Let α be the norm on S which is defined to be the norm of the first element in the inverse limit. Part (a) is immediate from Remark 3.4.10. To prove (b), note that if A is perfectoid, then finite sums of images of S form a dense subring of A. To prove (c), note that the quotient of $B\{t_s/\alpha(s) : s \in S\}$ by the closure of the ideal $(t_s^p - t_{s^p} : s \in S)$ is a perfectoid algebra mapping to A with dense image. By Theorem 3.6.17, A is perfectoid. □

The following argument is a slight modification of an argument of Colmez [**115**, Proposition 4.8].

Lemma 3.6.26. — *Let A be a uniform Banach algebra over $\mathbb{Q}_p$. Let U be a subset of $\mathfrak{m}_A$ which generates a dense $\mathbb{Q}_p$-subalgebra of A. For each finite subset T of U and each nonnegative integer n, equip the finite étale A-algebra*

$$B_{T,n} = A[x_{t,n} : t \in T]/(x_{t,n}^{p^n} - 1 - t : t \in T) \otimes_{\mathbb{Q}_p} \mathbb{Q}_p(\mu_{p^n})$$

with the spectral seminorm. View these as a directed system running over T and n by identifying $x_{t,n}$ with $x_{t,n+1}^p$. Then the completed direct limit B of the $B_{T,n}$ is a perfectoid uniform Banach algebra over $\mathbb{Q}_p$ and the morphism $\mathcal{M}(B) \to \mathcal{M}(A)$ is surjective; consequently (by Remark 2.8.3), the map $A \to B$ is isometric.

Proof. — Let F be the completed direct limit of $\bigcup_{n=0}^{\infty} \mathbb{Q}_p(\mu_{p^n})$; this is a perfectoid field, as then is the completion C of $F[x_t^{p^{-n}} : t \in U]$ for the Gauss norm. However, C admits a bounded homomorphism to B taking x_t to $1+t$. Since this homomorphism has dense image, by Theorem 3.6.17 (b) B is perfectoid. The surjectivity of $\mathcal{M}(B) \to \mathcal{M}(A)$ follows by viewing $\mathcal{M}(B)$ as the inverse limit of the $\mathcal{M}(B_{T,n})$ and applying Lemma 2.3.14. □

Remark 3.6.27. — By analogy with Remark 3.1.17, one may ask whether an adic Banach algebra (A, A^+) admitting a rational covering by perfectoid algebras is itself perfectoid. For Banach algebras over a perfectoid field, this conjecture is made in [**114**, Conjecture 2.16], but one can exhibit a counterexample against this conjecture by adapting the construction of [**23**, Proposition 13].

By analogy with Proposition 3.1.16, one can ask whether one can salvage the conjecture by adding the assumption that (A, A^+) is uniform or even stably uniform. This is quite unclear: to analogize the proof of Proposition 3.1.16, one would need to show that an exact sequence as in Corollary 2.8.9 is not only strict (as would follow from the open mapping theorem), but strict with suitably small factors between the quotient and subspace norms. (In other words, one needs to control the cohomology of the integral structure sheaf.)

Remark 3.6.28. — In [**48**], Fontaine defines a *perfectoid algebra* to be a uniform f-adic ring A containing a topologically nilpotent unit ϖ such that ϖ^p divides p in A° (so in particular p is topologically nilpotent) and $\overline{\varphi} : A^\circ/(\varpi) \to A^\circ/(\varpi^p)$ is surjective. By Proposition 3.6.2 (e), a uniform Banach algebra over $\mathbb{Q}_p$ is perfectoid in our sense if and only if it is perfectoid in Fontaine's sense. However, Fontaine's definition includes perfect uniform Banach $\mathbb{F}_p$-algebras as a special case, as well as some rings which are not algebras over either $\mathbb{F}_p$ or $\mathbb{Q}_p$.

Although we will not do so in this paper, it is not difficult to extend the arguments used here to cover perfectoid algebras and spaces in the sense of Fontaine, modulo the formal adjustment of working with topological rings rather than Banach rings: dropping the requirement of working over $\mathbb{Q}_p$ means that not every continuous homomorphism is bounded (Remark 2.4.4). One important nonformal change is to replace Lemma 3.6.26 by an alternate construction using lifts of Artin-Schreier extensions instead of Kummer extensions; such a construction has recently been suggested by Scholze, but we will not include the details here.

3.7. Preperfectoid and relatively perfectoid algebras

We next consider some Banach algebras closely related to perfectoid algebras. Although these do not correspond directly to objects in characteristic p, they are close enough to perfectoid algebras to inherit some of their most useful properties.

Definition 3.7.1. — Consider the following properties of a uniform Banach algebra A over $\mathbb{Q}_p$.

(a) For some perfectoid field K, $A\widehat{\otimes}_{\mathbb{Q}_p}K$ is uniform and perfectoid.

(b) For every perfectoid field K, $A\widehat{\otimes}_{\mathbb{Q}_p}K$ is uniform and perfectoid.

(c) For every perfectoid algebra B, $A\widehat{\otimes}_{\mathbb{Q}_p}B$ is uniform and perfectoid.

In case (a), we say that A is *preperfectoid*; this definition and some examples are due to Scholze and Weinstein [**116**]. In case (b), we say that A is *strongly preperfectoid.*

In case (c), we say that C is *relatively perfectoid*. Note that perfectoid algebras are generally not preperfectoid; see Remark 2.8.5 for a typical example.

We say that a uniform adic Banach algebra (A, A^+) is *preperfectoid*, *strongly preperfectoid*, or *relatively perfectoid* if A has the corresponding property.

A typical example is the following analogue of Example 3.6.6.

Example 3.7.2. — For any $r_1, \dots, r_n > 0$, the completion of $\mathbb{Q}_p\{T_1/r_1, \dots, T_n/r_n\}[T_1^{p^{-\infty}}, \dots, T_n^{p^{-\infty}}]$ for the weighted Gauss norm is relatively perfectoid.

Although we cannot prove that a preperfectoid algebra is relatively perfectoid, we do have the following result.

Proposition 3.7.3. — *Let A be a preperfectoid Banach algebra. Then for any uniform Banach algebra B over $\mathbb{Q}_p$, $A\widehat{\otimes}_{\mathbb{Q}_p}B$ is uniform.*

Proof. — Choose a perfectoid field K such that $A\widehat{\otimes}_{\mathbb{Q}_p}K$ is perfectoid. Apply Lemma 3.6.26 to construct an isometric morphism $B \to C$ of uniform Banach algebras with C perfectoid. Put $D = (K\widehat{\otimes}_{\mathbb{Q}_p}C)^u$; it is perfectoid by Corollary 3.6.18. By Proposition 3.6.11, $A\widehat{\otimes}_{\mathbb{Q}_p}D = (A\widehat{\otimes}_{\mathbb{Q}_p}K)\widehat{\otimes}_K D$ is perfectoid. Since C and D are uniform and $\mathcal{M}(D) \to \mathcal{M}(C)$ is surjective by Lemma 2.3.13, by Remark 2.8.3 the bounded homomorphism $C \to D$ is isometric. Therefore the composition $B \to C \to D$ is isometric; by Lemma 2.2.9, it follows that the map $A\widehat{\otimes}_{\mathbb{Q}_p}B \to A\widehat{\otimes}_{\mathbb{Q}_p}D$ is a strict injection. This yields the claim. □

Theorem 3.7.4. — *Let (A, A^+) be a preperfectoid adic Banach algebra. Then any rational localization of (A, A^+) is again preperfectoid. In particular, (A, A^+) is stably uniform, sheafy (by Theorem 2.8.10), and satisfies the Tate sheaf and Kiehl glueing properties (by Theorem 2.7.7).*

Proof. — Choose a perfectoid field K such that $A_K = A\widehat{\otimes}_{\mathbb{Q}_p}K$ is perfectoid. By Theorem 3.6.14, A_K is stably uniform; by Remark 2.8.12, so is A. □

Proposition 3.7.5. — *Let A be a preperfectoid Banach algebra, and view $B \in$ **FÉt**(A) as a Banach algebra as per Proposition 2.8.16. Then B is again preperfectoid.*

Proof. — Choose a perfectoid field K such that $A_K = A\widehat{\otimes}_{\mathbb{Q}_p}K$ is perfectoid. By the open mapping theorem (Theorem 2.2.8), the tensor product norm on $B_K = B\widehat{\otimes}_{\mathbb{Q}_p}K$ is equivalent to the norm obtained by viewing B_K as an object of **FÉt**(A_K) and applying Proposition 2.8.16. We may thus conclude by applying Theorem 3.6.21 to deduce that B_K is perfectoid. □

Proposition 3.7.6. — *Let $(A_1, A_1^+) \to (A_2, A_2^+)$, $(A_1, A_1^+) \to (A_3, A_3^+)$ be morphisms of strongly preperfectoid (resp. relatively perfectoid) adic Banach algebras over $\mathbb{Q}_p$. Then the tensor product $(A_2, A_2^+)\widehat{\otimes}_{(A_1, A_1^+)}(A_3, A_3^+)$ is strongly preperfectoid (resp. relatively perfectoid).*

Proof. — This is immediate from Proposition 3.6.11. Note that the argument does not work in the preperfectoid case because we need to know that the same perfectoid field can be used for all three rings. □

CHAPTER 4

ROBBA RINGS AND φ-MODULES

An important feature of the approach to p-adic Hodge theory used in this series of papers (and in the work of Berger and others) is the theory of slopes of Frobenius actions on modules over certain rings. This bears some resemblance to the theory of slopes for vector bundles on Riemann surfaces, including the relationship of those slopes to unitary representations of fundamental groups. (A more explicit link to vector bundles appears in the work of Fargues and Fontaine [**45**]; see §6.3.) We review here some of the principal results of the first author which are pertinent to p-adic Hodge theory, mostly omitting proofs. Besides serving as a model, some of these results provide key inputs into our work on relative p-adic Hodge theory.

4.1. Slope theory over the Robba ring

We begin by introducing several key rings used in p-adic Hodge theory, and the basic theory of slopes of Frobenius actions. Our description here is rather minimal; see [**84**] for a more detailed discussion.

Hypothesis 4.1.1. — Throughout §4.1, put $K = \mathrm{Frac}(W(k))$ for some perfect field k of characteristic p. Equip K with the p-adic norm and the Frobenius lift φ_K induced by Witt vector functoriality. Fix also a choice of $\omega \in (0,1)$.

Definition 4.1.2. — For $r > 0$, put

$$\mathcal{R}_K^r = \left\{ \sum_{i\in\mathbb{Z}} c_i T^i : c_i \in K, \lim_{i\to\pm\infty} |c_i|\rho^i = 0 \quad (\rho \in [\omega^r, 1)) \right\}.$$

In other words, $\mathcal{R}_K^r$ consists of formal sums $\sum_{i\in\mathbb{Z}} c_i T^i$ in the indeterminate T with coefficients in K which converge on the annulus $\omega^r \leqslant |T| < 1$. The set $\mathcal{R}_K^r$ forms a ring under formal series addition and multiplication; let $\mathcal{R}_K^{\mathrm{int},r}$ be the subring of $\mathcal{R}_K^r$ consisting of series whose coefficients have norm at most 1, and put $\mathcal{R}_K^{\mathrm{bd},r} = \mathcal{R}_K^{\mathrm{int},r}[p^{-1}]$. Put

$$\mathcal{R}_K^{\mathrm{int}} = \bigcup_{r>0} \mathcal{R}_K^{\mathrm{int},r}, \quad \mathcal{R}_K^{\mathrm{bd}} = \bigcup_{r>0} \mathcal{R}_K^{\mathrm{bd},r}, \quad \mathcal{R}_K = \bigcup_{r>0} \mathcal{R}_K^{r}.$$

The ring $\mathcal{R}_K^{\mathrm{int}}$ is a local ring with residue field $k((\overline{T}))$ which is not complete but is henselian. (See for instance [**82**, Lemma 3.9]. For a similar argument, see Proposition 5.5.3.) The completion of the field $\mathcal{R}_K^{\mathrm{bd}}$ is the field

$$\mathcal{E}_K = \left\{ \sum_{i \in \mathbb{Z}} c_i T^i : c_i \in K, \sup_i \{|c_i|\} < +\infty, \lim_{i \to -\infty} |c_i| = 0 \right\}.$$

The units of $\mathcal{R}_K$ are precisely the nonzero elements of $\mathcal{R}_K^{\mathrm{bd}}$, as may be seen by considering Newton polygons [**82**, Corollary 3.23].

Definition 4.1.3. — We will need to consider several different topologies on the rings described above.

(a) Those rings contained in $\mathcal{E}_K$ carry both a *p-adic topology* (the metric topology defined by the Gauss norm) and a *weak topology* (in which a sequence converges if it is bounded for the Gauss norm and converges T-adically modulo any fixed power of p). For both topologies, $\mathcal{E}_K$ is complete.

(b) Those rings contained in $\mathcal{R}_K^r$ carry a *Fréchet topology*, in which a sequence converges if and only if it converges under the ω^s-Gauss norm for all $s \in (0, r]$. For this topology, $\mathcal{R}_K^r$ is complete.

(c) Those rings contained in $\mathcal{R}_K$ carry a *limit-of-Fréchet topology*, or *LF topology*. This topology is defined on $\mathcal{R}_K$ by taking the locally convex direct limit (in the sense of [**21**, §II.4]) of the $\mathcal{R}_K^r$ (each equipped with the Fréchet topology). In particular, a sequence converges in $\mathcal{R}_K$ if and only if it is a convergent sequence in $\mathcal{R}_K^r$ for some $r > 0$.

Remark 4.1.4. — The convergence of the formal expression $x = \sum_i c_i T^i$ for various of the topologies described in Definition 4.1.3 is useful for defining operations such as Frobenius lifts (see Definition 4.1.6 below). In $\mathcal{E}_K$, the formal sum converges for the weak topology but not the p-adic topology. In $\mathcal{R}_K^r$, the sum converges for the Fréchet topology. In $\mathcal{R}_K$, the sum converges for the LF topology.

Remark 4.1.5. — Note that a sequence of elements of $\mathcal{R}_K^{\mathrm{bd},r}$ which is p-adically bounded and convergent under the ω^r-Gauss norm also converges in the weak topology.

Definition 4.1.6. — A *Frobenius lift* φ on $\mathcal{R}_K$ is an endomorphism defined by the formula

$$\varphi\left(\sum_{i \in \mathbb{Z}} c_i T^i\right) = \sum_{i \in \mathbb{Z}} \varphi_K(c_i) u^i$$

for some $u \in \mathcal{R}_K^{\mathrm{int}}$ with $|u - T^p| < 1$, where the right side may be interpreted as a convergent sum using Remark 4.1.4. Such an endomorphism also acts on $\mathcal{R}_K^{\mathrm{int}}, \mathcal{R}_K^{\mathrm{bd}}, \mathcal{E}_K$, but not on $\mathcal{R}_K^r$ for any individual $r > 0$; rather, for $r > 0$ sufficiently small, φ carries $\mathcal{R}_K^r$ into $\mathcal{R}_K^{r/p}$. The action of φ is continuous for each of the topologies described in Definition 4.1.3.

Choose a Frobenius lift φ on $\mathcal{R}_K$. For $R \in \{\mathcal{R}_K^{\mathrm{int}}, \mathcal{R}_K^{\mathrm{bd}}, \mathcal{R}_K, \mathfrak{o}_{\mathcal{E}_K}, \mathcal{E}_K\}$, a *$\varphi$-module* over R is a finite free R-module M equipped with a semilinear φ-action (*i.e.*, an R-linear isomorphism $\varphi^* M \to M$). Since the action of φ takes any basis of M to another basis, the p-adic valuation of the matrix *via* which φ acts on a basis of M is both finite and independent of the choice of the basis. We call the negative of this quantity the *degree* of M, denoted $\deg(M)$. For M nonzero, we define the *slope* of M to be $\mu(M) = \deg(M)/\operatorname{rank}(M)$.

For $s \in \mathbb{Q}$, we say M is *pure of slope s* if for some (hence any) $c, d \in \mathbb{Z}$ with $d > 0$ and $c/d = s$, $p^c\varphi^d$ acts on M *via* a matrix U such that the entries of U and U^{-1} all have Gauss norm at most 1. This evidently implies $\mu(M) = s$. If $s = 0$, we also say M is *étale*. (Note that our definitions force any nonzero φ-module over $\mathcal{R}_K^{\mathrm{int}}$ or $\mathfrak{o}_{\mathcal{E}_K}$ to be étale.)

Remark 4.1.7. — For M a φ-module over a ring R, view the dual module $M^\vee = \operatorname{Hom}_R(M, R)$ as a φ-module as in Remark 1.5.6. If M is pure of slope s, then $M^\vee$ is pure of slope $-s$.

Proposition 4.1.8. — *For any $s \in \mathbb{Q}$, we have the following.*

(a) *The functor $M \rightsquigarrow M \otimes_{\mathcal{R}_K^{\mathrm{bd}}} \mathcal{E}_K$ gives a fully faithful functor from φ-modules over $\mathcal{R}_K^{\mathrm{bd}}$ which are pure of slope s to φ-modules over $\mathcal{E}_K$ which are pure of slope s.*

(b) *The functor $M \rightsquigarrow M \otimes_{\mathcal{R}_K^{\mathrm{bd}}} \mathcal{R}_K$ gives an equivalence of categories between φ-modules over $\mathcal{R}_K^{\mathrm{bd}}$ which are pure of slope s and φ-modules over $\mathcal{R}_K$ which are pure of slope s.*

Proof. — See [**84**, Proposition 1.2.7] for (a) and [**84**, Theorem 1.6.5] for (b). For (b), see also Remark 4.3.4. □

The main theorem about slopes of φ-modules over $\mathcal{R}_K$ can be formulated in several ways. This formulation asserts that M is pure if and only if M is semistable with respect to slope in the sense of geometric invariant theory [**101**].

Theorem 4.1.9. — *Let M be a nonzero φ-module over $\mathcal{R}_K$ with $\mu(M) = s$. Then M is pure of slope s if and only if there exists no nonzero proper φ-submodule of M of slope greater than s.*

Proof. — See [**84**, Theorem 1.7.1]. □

An essentially equivalent formulation, incorporating an analogue of the Harder-Narasimhan filtration for vector bundles, is the following. (See Remark 4.2.18 for further discussion of this analogy.)

Theorem 4.1.10. — *Let M be a nonzero φ-module over $\mathcal{R}_K$. Then there exists a unique filtration $0 = M_0 \subset \cdots \subset M_l = M$ by saturated φ-submodules, such that $M_1/M_0, \dots, M_l/M_{l-1}$ are pure φ-modules and $\mu(M_1/M_0) > \cdots > \mu(M_l/M_{l-1})$.*

Proof. — See again [**84**, Theorem 1.7.1]. □

Remark 4.1.11. — Beware that Proposition 4.1.8 does not imply anything about maps between φ-modules which are pure of different slopes. For instance, it is common in p-adic Hodge theory (in the study of trianguline representations; see for instance [**26**]) to encounter short exact sequences of the form $0 \to M_1 \to M \to M_2 \to 0$ of φ-modules over $\mathcal{R}_K$ in which for some positive integer m, M_1 has rank 1 and slope $-m$, M_2 has rank 1 and slope m, and M is étale. While each term individually descends uniquely to $\mathcal{R}_K^{\mathrm{bd}}$ by Proposition 4.1.8 (b), the sequence cannot descend because there are no nonzero maps between pure φ-modules over $\mathcal{R}_K^{\mathrm{bd}}$ of different slopes. (Note that $\mu(M_1) < \mu(M)$, so there is no contradiction with Theorem 4.1.9.)

Remark 4.1.12. — The sign convention used here for degrees of φ-modules is opposite to that used in previous work of the first author [**82, 83, 84, 85**]. We have changed it in order to match the sign convention used in geometric invariant theory [**101**], in which the ample line bundle $\mathcal{O}(1)$ on any projective space has degree $+1$. This choice of sign also creates agreement with the work of Hartl and Pink [**67**] and of Fargues and Fontaine [**45**].

Convention 4.1.13. — It will be convenient at several points to speak also of φ^d-modules for d an arbitrary positive integer. We adopt the convention that the degree of a φ^d-module is defined by computing the p-adic valuation of the determinant of the matrix on which φ^d acts on some basis, then dividing by d. This has the advantage that the degree is preserved upon restriction of a φ^d-module to a φ^{de}-module. (One can even replace φ with a map on $\tilde{\mathcal{R}}_K$ lifting the p^d-power absolute Frobenius on $k((\overline{T}))$, not necessarily given by raising a p-power Frobenius lift to the d-th power, by modifying Definition 4.1.6 in the obvious way. We will not need this extra generality.)

4.2. Slope theory and Witt vectors

One can generalize the slope theory for Frobenius modules over the Robba ring by first making explicit the role of the $\overline{T}$-adic norm on the residue field of $\mathcal{R}_K^{\mathrm{bd}}$, then replacing this norm with something more general. This second step turns out to be a bit subtle unless we first perfect the residue field of $\mathcal{R}_K^{\mathrm{bd}}$ and pass to Witt vectors (as in Definition 3.2.10); this gives a slope theory introduced in [**83**] and reviewed here. In fact, this study is integral to the slope theory over the Robba ring itself; see Remark 4.3.5. We take up the relative version of this story starting in §5.

Hypothesis 4.2.1. — Throughout §4.2, let L be a perfect analytic field of characteristic p with norm α.

Definition 4.2.2. — For $r > 0$, let $\tilde{\mathcal{R}}_L^{\mathrm{int},r}$ be the set of $x = \sum_{i=0}^{\infty} p^i[\overline{x}_i] \in W(L)$ for which $\lim_{i\to\infty} p^{-i}\alpha(\overline{x}_i)^r = 0$. Thanks to the homogeneity property of Witt vector addition (Remark 3.2.3), this set forms a ring on which the formula

$$\lambda(\alpha^s)(x) = \max_i \{p^{-i}\alpha(\overline{x}_i)^s\}$$

defines a multiplicative norm $\lambda(\alpha^s)$ on $\tilde{\mathcal{R}}_L^{\mathrm{int},r}$ for each $s \in [0, r]$. (For an explicit argument, see Proposition 5.1.2.)

Put $\tilde{\mathcal{R}}_L^{\mathrm{bd},r} = \tilde{\mathcal{R}}_L^{\mathrm{int},r}[p^{-1}]$. Let $\tilde{\mathcal{R}}_L^r$ be the Fréchet completion of $\tilde{\mathcal{R}}_L^{\mathrm{bd},r}$ under the norms $\lambda(\alpha^s)$ for $s \in (0, r]$. Put

$$\tilde{\mathcal{R}}_L^{\mathrm{int}} = \cup_{r>0}\tilde{\mathcal{R}}_L^{\mathrm{int},r}, \quad \tilde{\mathcal{R}}_L^{\mathrm{bd}} = \cup_{r>0}\tilde{\mathcal{R}}_L^{\mathrm{bd},r}, \quad \tilde{\mathcal{R}}_L = \cup_{r>0}\tilde{\mathcal{R}}_L^r.$$

Again, $\tilde{\mathcal{R}}_L^{\mathrm{int}}$ is an incomplete but henselian (see [**83**, Lemma 2.1.12] or Proposition 5.5.3) local ring with residue field L; the completion of the field $\tilde{\mathcal{R}}_L^{\mathrm{bd}}$ is simply $\tilde{\mathcal{E}}_L = W(L)[p^{-1}]$. (For parallelism, we write $\tilde{\mathcal{E}}_L^{\mathrm{int}}$ as another notation for $W(L)$.) One can again identify the units of $\tilde{\mathcal{R}}_L$ as the nonzero elements of $\tilde{\mathcal{R}}_L^{\mathrm{bd}}$; see Corollary 4.2.5.

We call $\tilde{\mathcal{R}}_L$ the *extended Robba ring* with residue field L. This terminology is not used in [**83**], but is suggested in [**84**].

Lemma 4.2.3. — *For $x \in \tilde{\mathcal{R}}_L^r$, the function $s \mapsto \log\lambda(\alpha^s)(x)$ is continuous and convex on $(0, r]$.*

Proof. — The function is affine in case $x = p^i[\overline{x}]$ for some $i \in \mathbb{Z}, \overline{x} \in L$. In case x is a finite sum of such terms, the function is the maximum of finitely many affine functions, and hence is convex. Since such finite sums are dense in $\tilde{\mathcal{R}}_L^r$, the general case follows. □

Lemma 4.2.4. — *For $x \in \tilde{\mathcal{R}}_L$, we have $x \in \tilde{\mathcal{R}}_L^{\mathrm{bd}}$ if and only if for some $r > 0$, $\lambda(\alpha^s)(x)$ is bounded for $s \in (0, r]$.*

Proof. — If $x \in \tilde{\mathcal{R}}_L^{\mathrm{bd}}$, then as s tends to 0, $\lambda(\alpha^s)(x)$ tends to the p-adic norm of x. Hence for some $r > 0$, $\lambda(\alpha^s)(x)$ is bounded for $s \in (0, r]$. Conversely, suppose that for some $r > 0$, $\lambda(\alpha^s)(x)$ is bounded for $s \in (0, r]$. To prove that $x \in \tilde{\mathcal{R}}_L^{\mathrm{bd}}$, we may first multiply by a power of p; we may thus ensure that for some $r > 0$, $x \in \tilde{\mathcal{R}}_L^r$ and $\lambda(\alpha^s)(x) \leqslant 1$ for $s \in (0, r]$. We will show in this case that $x \in \tilde{\mathcal{R}}_L^{\mathrm{int},r}$.

Write x as the limit of a sequence $x_0, x_1, \dots$ with $x_i \in \tilde{\mathcal{R}}_L^{\mathrm{bd},r}$. For each positive integer j, we can find $N_j > 0$ such that

$$\lambda(\alpha^s)(x_i - x) \leqslant p^{-j} \qquad (i \geqslant N_j,\ s \in [p^{-j}r, r]).$$

Write $x_i = \sum_{l=m(i)}^{\infty} p^l[\overline{x}_{il}]$, and put $y_i = \sum_{l=0}^{\infty} p^l[\overline{x}_{il}] \in \tilde{\mathcal{R}}_L^{\mathrm{int},r}$. For $i \geqslant N_j$, we have $\lambda(\alpha^{p^{-j}r})(x_i) \leqslant 1$ and so

$$\alpha(\overline{x}_{il}) \leqslant p^{lp^j/r} \qquad (i \geqslant N_j,\ l < 0).$$

Since $p^{-l}p^{lp^j} \leqslant p^{1-p^j}$ for $l \leqslant -1$, we deduce that $\lambda(\alpha^r)(x_i - y_i) \leqslant p^{1-p^j}$. Consequently, the sequence $y_0, y_1, \dots$ converges to x under $\lambda(\alpha^r)$, and hence under $\lambda(\alpha^s)$

for $s \in (0, r]$ by Lemma 4.2.3; it follows that $x \in \tilde{\mathcal{R}}_L^{\mathrm{int},r}$ as desired. (See also [**83**, Corollary 2.5.6] for a slightly different argument.) □

Corollary 4.2.5. — *The units in $\tilde{\mathcal{R}}_L$ are precisely the nonzero elements of $\tilde{\mathcal{R}}_L^{\mathrm{bd}}$.*

Proof. — Suppose $x \in \tilde{\mathcal{R}}_L$ is a unit with inverse y. Choose $r > 0$ so that $x, y \in \tilde{\mathcal{R}}_L^r$. Then the functions $\log \lambda(\alpha^s)(x), \log \lambda(\alpha^s)(y)$ are convex by Lemma 4.2.3, but their sum is the constant function 0. Hence both functions are affine in s; in particular, $\lambda(\alpha^s)(x)$ is bounded for $s \in (0, r]$. By Lemma 4.2.4, this forces $x \in \tilde{\mathcal{R}}_L^{\mathrm{bd}}$. (Compare [**83**, Lemma 2.4.7].) □

Lemma 4.2.6. — *The rings $\tilde{\mathcal{R}}_L^r$ and $\tilde{\mathcal{R}}_L$ are Bézout domains,* i.e., *integral domains in which every finitely generated ideal is principal.*

Proof. — See [**83**, Theorem 2.9.6]. □

Remark 4.2.7. — A fact closely related to Lemma 4.2.6 is that for $0 < s \leqslant r$, the completion of $\tilde{\mathcal{R}}_L^r$ with respect to the norm $\max\{\lambda(\alpha^r), \lambda(\alpha^s)\}$ is a principal ideal domain, and even a Euclidean domain. See [**83**, Proposition 2.6.8]. (This ring will later be denoted $\tilde{\mathcal{R}}_L^{[s,r]}$; see Definition 5.1.1.)

Definition 4.2.8. — To each nonzero $x \in \tilde{\mathcal{R}}_L^r$ is associated its *Newton polygon, i.e.*, the convex dual of the function $f_x(t) = \log \lambda(\alpha^t)(x)$ (which is convex by Lemma 4.2.3). The slopes of the Newton polygon of x are the values t where f_x changes slope; for short, we call these the *slopes of* x. The *multiplicity* of a slope s is the width of the corresponding segment of the Newton polygon. Note that r fails to receive a multiplicity under this definition; to correct this, choose any $s > r$ and any $y \in \tilde{\mathcal{R}}_L^s$ with $\lambda(\alpha^r)(x - y) < \lambda(\alpha^r)(x)$, then define the multiplicity of r as a slope of x to be its multiplicity as a slope of y. This does not depend on the choice of y. (See [**83**, Definition 2.4.4] for an alternate definition.)

For our present purposes, the most important properties of slopes are the following.

(a) If $x = yz$, then the slopes of x are precisely the slopes of y and z. More precisely, the multiplicity of any $s \in (0, r]$ as a slope of x equals the sum of its multiplicities as a slope of y and of z.

(b) If $x \in \tilde{\mathcal{R}}_L^r$ is a unit, then by (a) it has no slopes. The converse is also true, by the following argument. If x has no slopes, then $x \in \tilde{\mathcal{R}}_L^{\mathrm{bd},r}$ by Lemma 4.2.4. Write $x = p^m y$ with $m \in \mathbb{Z}$ and $y \in \tilde{\mathcal{R}}_L^{\mathrm{int},r}$ not divisible by p; we must then have $\lambda(\alpha^r)(y - [\overline{y}]) < \lambda(\alpha^r)(y)$ by definition of the multiplicity of r. It follows that x is a unit in $\tilde{\mathcal{R}}_L^r$.

Definition 4.2.9. — The Frobenius lift φ on $W(L)$ acts on $\tilde{\mathcal{R}}_L^{\mathrm{int}}$, and extends by continuity to $\tilde{\mathcal{R}}_L^{\mathrm{bd}}, \tilde{\mathcal{R}}_L, \tilde{\mathcal{E}}_L$. Note the following useful identity:

$$\lambda(\alpha^s)(\varphi(x)) = \lambda(\alpha^{ps})(x) \qquad (x \in \tilde{\mathcal{R}}_L,\ s > 0). \tag{4.2.9.1}$$

We define φ-modules over these rings, degrees, slopes, and the pure and étale conditions as in Definition 4.1.6. (We also consider φ^d-modules for d a positive integer, keeping in mind Convention 4.1.13.)

Lemma 4.2.10. — *We have $\tilde{\mathcal{E}}_L^{\varphi} = \mathbb{Q}_p$ and $\tilde{\mathcal{R}}_L^{\varphi} = \mathbb{Q}_p$. More generally, for any positive integer a, the elements of $\tilde{\mathcal{E}}_L$ and $\tilde{\mathcal{R}}_L$ fixed by φ^a constitute the unramified extension of $\mathbb{Q}_p$ with residue field $\mathbb{F}_{p^a}$.*

Proof. — The first equality holds because $\tilde{\mathcal{E}}_L = W(L)[p^{-1}]$, so $\tilde{\mathcal{E}}_L^{\varphi} = W(L^{\varphi})[p^{-1}] = \mathbb{Q}_p$. For the second equality, suppose $x \in \tilde{\mathcal{R}}_L^{\varphi}$ is nonzero; then by (4.2.9.1), $\lambda(\alpha^s)(x) = \lambda(\alpha^{ps})(x)$ for all $s > 0$. It follows that $\lambda(\alpha^r)(x)$ is bounded over all $r > 0$, and so by Lemma 4.2.4, $x \in (\tilde{\mathcal{R}}_L^{\mathrm{bd}})^{\varphi} \subseteq \tilde{\mathcal{E}}_L^{\varphi} = \mathbb{Q}_p$. The final assertion is proved similarly. □

We have the following analogue of Proposition 4.1.8. One key difference is that the functor in part (a) can be shown to be essentially surjective; see Theorem 8.5.3.

Proposition 4.2.11. — *For any $s \in \mathbb{Q}$, we have the following.*

(a) *The functor $M \rightsquigarrow M \otimes_{\tilde{\mathcal{R}}_L^{\mathrm{bd}}} \tilde{\mathcal{E}}_L$ gives a fully faithful functor from φ-modules over $\tilde{\mathcal{R}}_L^{\mathrm{bd}}$ which are pure of slope s to φ-modules over $\tilde{\mathcal{E}}_L$ which are pure of slope s.*

(b) *The functor $M \rightsquigarrow M \otimes_{\tilde{\mathcal{R}}_L^{\mathrm{bd}}} \tilde{\mathcal{R}}_L$ gives an equivalence of categories between φ-modules over $\tilde{\mathcal{R}}_L^{\mathrm{bd}}$ which are pure of slope s and φ-modules over $\tilde{\mathcal{R}}_L$ which are pure of slope s.*

Proof. — See [**83**, Theorem 6.3.3(a,b)] or Remark 4.3.4. □

We also have analogues of Theorem 4.1.9 and Theorem 4.1.10.

Theorem 4.2.12. — *Let M be a nonzero φ-module over $\tilde{\mathcal{R}}_L$ with $\mu(M) = s$. Then M is pure of slope s if and only if there exists no nonzero proper φ-submodule of M of slope greater than s.*

Proof. — See [**83**, Proposition 6.3.5, Corollary 6.4.3]. □

Theorem 4.2.13. — *Let M be a nonzero φ-module over $\tilde{\mathcal{R}}_L$. Then there exists a unique filtration $0 = M_0 \subset \cdots \subset M_l = M$ by saturated φ-submodules, such that $M_1/M_0, \dots, M_l/M_{l-1}$ are pure and $\mu(M_1/M_0) > \cdots > \mu(M_l/M_{l-1})$.*

Proof. — See [**83**, Theorem 6.4.1]. □

Corollary 4.2.14. — *Let M be a nonzero φ-module over $\tilde{\mathcal{R}}_L$ with $\mu(M) = s$. Let L' be a perfect analytic field containing L with compatible norms. Then M is pure of slope s if and only if $M \otimes_{\tilde{\mathcal{R}}_L} \tilde{\mathcal{R}}_{L'}$ is pure of slope s.*

Proof. — From the definition of purity, it is clear that if M is pure of slope s, then so is $M \otimes_{\tilde{\mathcal{R}}_L} \tilde{\mathcal{R}}_{L'}$. On the other hand, from the alternate criterion for purity given by Theorem 4.2.12, it is also clear that if M fails to be pure of slope s, then so does $M \otimes_{\tilde{\mathcal{R}}_L} \tilde{\mathcal{R}}_{L'}$. □

In addition, in case L is algebraically closed, we get an analogue of Manin's classification of rational Dieudonné modules.

Proposition 4.2.15. — *Suppose that L is algebraically closed. Let M be a φ-module over $\tilde{\mathcal{E}}_L$ (resp. $\tilde{\mathcal{R}}_L^{\mathrm{bd}}$) which is pure of slope s. Then for any $c, d \in \mathbb{Z}$ with $d > 0$ and $c/d = s$, and any basis of M on which $p^c\varphi^d$ acts* via *an invertible matrix over $W(L)$ (resp. $\tilde{\mathcal{R}}_L^{\mathrm{int}}$), the $W(L)$-span (resp. $\tilde{\mathcal{R}}_L^{\mathrm{int}}$-span) of this basis admits another basis fixed by $p^c\varphi^d$.*

Proof. — Both assertions reduce easily to the case $s = 0$ provided that we allow φ to be replaced by a power (which does not affect the argument). The assertion about $\tilde{\mathcal{E}}_L$ is fairly standard; see for instance [**47**, Proposition A1.2.6]. The assertion about $\tilde{\mathcal{R}}_L^{\mathrm{bd}}$ follows from the assertion about $\tilde{\mathcal{E}}_L$ as in [**84**, Proposition 2.5.8]. See also Proposition 7.3.6 for a stronger statement. □

One has an analogue of Proposition 4.2.15 for φ-modules over $\tilde{\mathcal{R}}_L$ which need not be pure.

Proposition 4.2.16. — *Suppose that L is algebraically closed. Let M be a φ-module over $\tilde{\mathcal{R}}_L$. Then for some positive integer d, there exists a basis of M on which φ^d acts* via *a diagonal matrix with diagonal entries in $p^{\mathbb{Z}}$.*

Proof. — Using Theorem 4.2.13 and Proposition 4.2.15, this reduces to the assertion that for any positive integers c, d, the map $x \mapsto x^{\varphi^d} - p^c x$ on $\tilde{\mathcal{R}}_L$ is surjective. For this, see [**83**, Proposition 3.3.7(c)]. See also [**83**, Proposition 4.5.3] for a detailed proof of the original statement. □

Remark 4.2.17. — The use of growth conditions to cut out subrings of the rings of Witt vectors also appears in the work of Fargues and Fontaine [**45**] with which we make contact later (§6.3), as well as in the approach to constructing p-adic cohomology *via* the *overconvergent de Rham-Witt complex* of Davis, Langer, and Zink [**31, 30**].

Remark 4.2.18. — Theorems 4.2.12 and 4.2.13 together mean that the *slope filtration* of a φ-module M, as described in Theorem 4.2.13, coincides with the *Harder-Narasimhan filtration* of M in the category of φ-modules for the degree function $M \mapsto \mu(\wedge^{\mathrm{rank}(M)} M)$. That is to say, M_1 is the maximal nonzero φ-submodule of M of maximal slope, M_2/M_1 is the maximal nonzero φ-submodule of M of maximal slope, and so on. This equality implies among other things that the Harder-Narasimhan filtration is multiplicative (because the tensor product of pure φ-modules of slopes s_1, s_2 is pure of slope $s_1 + s_2$).

The analogy with stability of vector bundles will become even more explicit when we relate φ-modules to vector bundles on the relative Fargues-Fontaine curve. See §6.3 and §8.7.

Remark 4.2.19. — We take this opportunity to record some corrections to [**83**, §2.5-2.6] not included in the printed erratum. Thanks to Max Bender for reporting these.

- Lemma 2.5.3: in (a), $i \in \mathbb{Z}$ should be $i \geqslant 0$. In the last line of the proof of (a), both instances of n should be i. In the last line of the proof of (c), $n \to \infty$ should be $i \to \infty$.
- Lemma 2.5.4: it should be assumed that $r > 0$.
- Corollary 2.5.6: $v_{j,n}$ should be $v_{j,r}$.
- Lemma 2.5.11: it should also be assumed that $r \in I$. The statement is also correct when $r \notin I$ provided that the right side of the inequality is finite, but this is not used anywhere.
- Lemma 2.6.3: The first displayed equation should read
$$v_{n,r}(x - x') \geqslant w_r(x) + (1 - r/r_0) \qquad (n \geqslant m).$$
In the second displayed equation, x'/x should be x/x'. In the following line, the inequality
$$w_r(z_l \pi^m (1 - x'/x)) \geqslant w_r(y_l) + (1 - r/r_0)$$
should instead assert that
$$v_{n,r}(z_l \pi^m (1 - x/x')) \geqslant \min\{w_r(y_l) + (1 - r/r_0), \min_{n'>n}\{v_{n',r}(y_l)\}\} \text{ for } n \geqslant m.$$
Similarly, after the third displayed equation, the inequality
$$v_{n,r}(y_{l+1}) \geqslant w_r(y_l) + (1 - r/r_0)$$
should read
$$v_{n,r}(y_{l+1}) \geqslant \min\{w_r(y_l) + (1 - r/r_0), \min_{n'>n}\{v_{n',r}(y_l)\}\}.$$
The first two sentences of the last paragraph (from "It follows that...") must be replaced by the following:

  "It follows that, for $n \geqslant m$, we have
$$v_{n,r}(y_{l+1}) \geqslant \min\{w_r(y_l) + (1 - r/r_0), \min_{n'>n}\{v_{n',r}(y_l)\}\}$$
  We may assume that $y_{l+1}, y_{l+2}, \ldots$ also have height at least m, in which case $w_r(y_{l+h}) \geqslant w_r(y_l)$ for all $h > 0$ and $w_r(y_{l+h}) \geqslant w_r(y_l) + (1 - r/r_0)$ for some $h > 0$ (because the maximum value of n for which $v_{n,r}(y_{l+h}) < w_r(y_l) + (1 - r/r_0)$ decreases as h increases)."
- Remark 2.6.4: "[discreteness of the valuation] on K" should be "[...] on $\mathcal{O}$".
- Lemma 2.6.7: The sentence starting "Moreover, if it is ever less than $\min_{n<0}\{v_{n,r'}(u_l x)\} + c$," should continue "then the smallest value of n for which $v_{n,r'}(u_{l+1} x) \leqslant \min_{n<0}\{v_{n,r'}(u_l x)\} + c$ is strictly greater than the smallest value of n for which $v_{n,r'}(u_l x) \leqslant \min_{n<0}\{v_{n,r'}(u_l x)\} + c$."
- Proposition 2.6.8: the reference to Proposition 2.6.8 in the proof should be to Proposition 2.6.5.

4.3. Comparison of slope theories

The slope theories for φ-modules over $\mathcal{R}_K$ and $\tilde{\mathcal{R}}_L$ can be related as follows. Throughout §4.3, retain Hypothesis 4.1.1.

Definition 4.3.1. — Equip the field $k((\overline{T}))$ with the $\overline{T}$-adic norm α for the normalization $\alpha(\overline{T}) = \omega$. Let L be the completed perfection of $k((\overline{T}))$ for the unique multiplicative extension of α. Proceeding as in Definition 3.2.10, we obtain a map $s_\varphi : \mathcal{E}_K \to \tilde{\mathcal{E}}_L$; more precisely, $\tilde{\mathcal{E}}_L$ is the completion of the direct perfection of $\mathcal{E}_K$ for the weak topology. For $r > 0$ small enough that the $\omega^{r/p}$-Gauss norm of $\varphi(T)/T^p - 1$ is less than 1, s_φ takes $\mathcal{R}_K^{\mathrm{int},r}$ into $\tilde{\mathcal{R}}_L^{\mathrm{int},r}$. In fact, this map is isometric for the ω^r-Gauss norm on the source and the norm $\lambda(\alpha^r)$ on the target [**83**, Lemma 2.3.5]. We thus obtain a φ-equivariant homomorphism $\mathcal{R}_K \to \tilde{\mathcal{R}}_L$.

Example 4.3.2. — For the Frobenius lift $\varphi(T) = (T+1)^p - 1$ and $\omega = p^{-p/(p-1)}$, s_φ is isometric for the ω^r-Gauss norm for $r \in (0,1)$.

We can use the extended rings to trivialize φ-modules over the smaller rings, as follows.

Proposition 4.3.3. — *Let L' be the completed direct perfection of $\mathfrak{o}_{\widehat{\mathcal{E}_K^{\mathrm{unr}}}}/(p)$ (which is algebraically closed). Identify the completion of the maximal unramified extension $\widehat{\mathcal{E}_K^{\mathrm{unr}}}$ of $\mathcal{E}_K$ with a subring of $\tilde{\mathcal{E}}_{L'}$.*

(a) *Let M be an étale φ-module over $\mathfrak{o}_{\mathcal{E}_K}$. Then the $\mathbb{Z}_p$-module*
$$V = (M \otimes_{\mathfrak{o}_{\mathcal{E}_K}} \mathfrak{o}_{\widehat{\mathcal{E}_K^{\mathrm{unr}}}})^\varphi$$
has the property that the natural map
$$V \otimes_{\mathbb{Z}_p} \mathfrak{o}_{\widehat{\mathcal{E}_K^{\mathrm{unr}}}} \longrightarrow M \otimes_{\mathfrak{o}_{\mathcal{E}_K}} \mathfrak{o}_{\widehat{\mathcal{E}_K^{\mathrm{unr}}}}$$
is an isomorphism.

(b) *Let M be an étale φ-module over $\mathcal{R}_K^{\mathrm{int}}$. Then the $\mathbb{Z}_p$-module*
$$V = (M \otimes_{\mathcal{R}_K^{\mathrm{int}}} (\mathfrak{o}_{\widehat{\mathcal{E}_K^{\mathrm{unr}}}} \cap \tilde{\mathcal{R}}_{L'}^{\mathrm{int}}))^\varphi$$
has the property that the natural map
$$V \otimes_{\mathbb{Z}_p} (\mathfrak{o}_{\widehat{\mathcal{E}_K^{\mathrm{unr}}}} \cap \tilde{\mathcal{R}}_{L'}^{\mathrm{int}}) \longrightarrow M \otimes_{\mathcal{R}_K^{\mathrm{int}}} (\mathfrak{o}_{\widehat{\mathcal{E}_K^{\mathrm{unr}}}} \cap \tilde{\mathcal{R}}_{L'}^{\mathrm{int}})$$
is an isomorphism. Moreover,
$$V \otimes_{\mathbb{Z}_p} \mathbb{Q}_p = (M \otimes_{\mathcal{R}_K^{\mathrm{int}}} \tilde{\mathcal{R}}_{L'})^\varphi = (M \otimes_{\mathcal{R}_K^{\mathrm{int}}} \tilde{\mathcal{E}}_{L'})^\varphi.$$

Proof. — Both parts follow from Proposition 4.2.15 plus Lemma 4.2.10. □

Remark 4.3.4. — For any φ-modules M_1, M_2 over a ring R, there is a natural identification
$$\mathrm{Hom}_R(M_1, M_2) = M_1^\vee \otimes_R M_2.$$

If M_1, M_2 are both pure of slope s, then $M_1^\vee \otimes_R M_2$ is étale. By this reasoning, Proposition 4.1.8 reduces to Proposition 4.3.3, while Proposition 4.2.11 reduces to a similar consequence of Proposition 4.2.15 (derived using Lemma 4.2.10).

Remark 4.3.5. — Note that any pure φ-module over $\mathcal{R}_K$ remains pure upon base extension to $\tilde{\mathcal{R}}_L$, while any φ-module over $\mathcal{R}_K$ whose base extension to $\tilde{\mathcal{R}}_L$ is *semistable*, *i.e.*, which does not have any nonzero proper φ-submodule of larger slope, is also itself semistable. The reverse implications also hold, and in fact form part of the proof of Theorem 4.1.9 (in the form of a reduction to the somewhat more tractable Theorem 4.2.12). One approach to the reverse implications is to make somewhat careful calculations, as in [**83**]; a simpler approach is to use faithfully flat descent, as in [**84**, §3].

Remark 4.3.6. — On the topic of descent, we record some minor inaccuracies in the statement and proof of [**84**, Proposition 3.3.2].

(a) The module M should not be assumed to be a φ-module over R, but only an R-module equipped with an isomorphism $\varphi^* M \cong M$. That is, we should not assume M is finite free over R. That is because in the proof of [**84**, Theorem 3.1.3], we need to take $R = \mathcal{R}^{\mathrm{bd}}$ and $S = \tilde{\mathcal{R}}_L^{\mathrm{bd}}$, and to take M to be the restriction of scalars of a φ-module over $\mathcal{R}$.

(b) The conclusion should not state that N is a φ-module over R, only a finite *locally free* R-module equipped with an isomorphism $\varphi^* N \cong N$. The proof of [**84**, Proposition 3.2.2] invokes [**62**, Exposé VIII, Corollaire 1.3], which only guarantees the existence and uniqueness of the module N. It should instead invoke [**62**, Exposé XIII, Théorème 1.1] (*i.e.*, Theorem 1.3.4 (a)) to recover both N and the isomorphism $\varphi^* N \cong N$, plus [**62**, Exposé VIII, Proposition 1.10] (*i.e.*, Theorem 1.3.5) to deduce that N is finite locally free over R.

Note that the modified statement suffices for the applications to [**84**, Theorems 3.1.2 and 3.1.3] because in those cases R is a Bézout domain (Lemma 4.2.6), over which any finite locally free module is free [**84**, Remark 1.1.2].

CHAPTER 5

RELATIVE ROBBA RINGS

We now begin in earnest to consider the relative setting. Although for some applications it is necessary to consider analogues of the Robba ring itself, these are not so straightforward to construct, and we leave them to a subsequent paper. Here, we treat only the analogue of the extended Robba ring in which the field of positive characteristic (over which we define Witt vectors) is replaced by a more general ring. As noted in the introduction, this pertains to a "geometric" relativization of slope theory, which is rather different from an "arithmetic" relativization in which one works with power series over a more general ring. (See Remark 7.4.13 for some discussion of the latter.)

Hypothesis 5.0.1. — For the remainder of the paper, let (R, R^+) be a perfect uniform adic Banach algebra over $\mathbb{F}_p$ with spectral norm α, such that R is a Banach algebra over some analytic field. (The condition that R be defined over an analytic field is no restriction at all if we are willing to modify the norm on R without changing the norm topology; see Remark 2.8.18.) When R has been assumed to be an analytic field, we conventionally change its name from R to L, but this change is pointed out explicitly in each instance.

It is possible to further weaken the hypothesis on R in some of the results; see for example Remark 6.2.6. However, this extra generality is of no use to us: our ultimate goal is to consider perfectoid algebras, which always give rise to perfect uniform Banach algebras over analytic fields (see Definition 3.3.4).

Remark 5.0.2. — For $x = \sum_{i=m}^{\infty} p^i[\overline{x}_i] \in W(R)[p^{-1}]$, for each $h \in \mathbb{Z}$, the set

$$\{\beta \in \mathcal{M}(R) : \beta(\overline{x}_i) = 0 \text{ for all } i \leqslant h\}$$

is closed in $\mathcal{M}(R)$. Consequently, the p-adic absolute value of the image of x in $W(\mathcal{H}(\beta))[p^{-1}]$ is a lower semicontinuous function of $\beta \in \mathcal{M}(R)$. When x is a unit, this function is seen to be continuous by applying the same argument to x^{-1}.

5.1. Relative extended Robba rings

We start by generalizing the definition of the extended Robba rings and their subrings.

Definition 5.1.1. — For $* \in \{R, R^+\}$, define the rings $\tilde{\mathcal{E}}^{\mathrm{int}}_*, \tilde{\mathcal{E}}_*, \tilde{\mathcal{R}}^{\mathrm{int},r}_*, \tilde{\mathcal{R}}^{\mathrm{int}}_*, \tilde{\mathcal{R}}^{\mathrm{bd},r}_*$, $\tilde{\mathcal{R}}^{\mathrm{bd}}_*, \tilde{\mathcal{R}}^r_*, \tilde{\mathcal{R}}_*$ by changing L to $*$ in Definition 4.2.2. That is, for $r > 0$, put $\tilde{\mathcal{E}}^{\mathrm{int}}_* = W(*)$ and $\tilde{\mathcal{E}}_* = W(*)[p^{-1}]$, let $\tilde{\mathcal{R}}^{\mathrm{int},r}_*$ be the ring of $x = \sum_{i=0}^{\infty} p^i[\overline{x}_i] \in W(*)$ for which $\lim_{i\to\infty} p^{-i}\alpha(\overline{x}_i)^r = 0$, and extend $\lambda(\alpha^s)$ to a power-multiplicative norm on $\tilde{\mathcal{R}}^{\mathrm{int},r}_*$ for $s \in (0, r]$ by putting

$$\lambda(\alpha^s)\left(\sum_{i=0}^{\infty} p^i[\overline{x}_i]\right) = \max_i\{p^{-i}\alpha(\overline{x}_i)^s\}.$$

(See Proposition 5.1.2 (a) for more details about the case $* = R$.) Put $\tilde{\mathcal{R}}^{\mathrm{bd},r}_* = \tilde{\mathcal{R}}^{\mathrm{int},r}_*[p^{-1}]$, let $\tilde{\mathcal{R}}^r_*$ be the Fréchet completion of $\tilde{\mathcal{R}}^{\mathrm{bd},r}_*$ under $\lambda(\alpha^s)$ for $s \in (0, r]$, and drop r from the superscript to indicate the union over all $r > 0$.

For $0 < s \leqslant r$, let $\tilde{\mathcal{R}}^{[s,r]}_*$ be the Fréchet completion of $\tilde{\mathcal{R}}^{\mathrm{bd},r}_*$ under the norms $\lambda(\alpha^t)$ for $t \in [s, r]$; it will follow from Lemma 5.2.1 below that $\tilde{\mathcal{R}}^{[s,r]}_*$ is also complete under $\max\{\lambda(\alpha^r), \lambda(\alpha^s)\}$, and so is a Banach ring. Note that the rings $\tilde{\mathcal{E}}^{\mathrm{int}}_{R^+}, \tilde{\mathcal{R}}^{\mathrm{int},r}_{R^+}, \tilde{\mathcal{R}}^{\mathrm{int}}_{R^+}$ (resp. $\tilde{\mathcal{E}}^{\mathrm{bd}}_{R^+}, \tilde{\mathcal{R}}^{\mathrm{bd},r}_{R^+}, \tilde{\mathcal{R}}^{\mathrm{bd}}_{R^+}$) are all equal to $W(R^+)$ (resp. $W(R^+)[1/p]$); we also denote them by $\tilde{\mathcal{R}}^{\mathrm{int},+}_R$ (resp. $\tilde{\mathcal{R}}^{\mathrm{bd},+}_R$) later on. Let $\tilde{\mathcal{R}}^+_R$ be the Fréchet completion of $\tilde{\mathcal{R}}^{\mathrm{bd},+}_R$ under $\lambda(\alpha^s)$ for all $s > 0$. Note that ring is in general properly contained in $\tilde{\mathcal{R}}^{\infty}_R = \cap_{r>0}\tilde{\mathcal{R}}^r_R$.

We need the following mild extension of the basic constructions of [**87**, §4]. For more discussion of topological aspects (*e.g.*, continuity of λ and μ), see §5.4.

Proposition 5.1.2. — *Choose* $0 < s \leqslant r$.

(a) *The set* $\tilde{\mathcal{R}}^{\mathrm{int},r}_R$ *is a ring on which* $\lambda(\alpha^s)$ *is a power-multiplicative norm. Moreover,* $\lambda(\alpha^s)$ *is multiplicative in case* α *is.*

(b) *For* β *a submultiplicative (resp. power-multiplicative, multiplicative) (semi)norm on* R *dominated by* $\max\{\alpha^s, \alpha^r\}$, *the formula*

$$\lambda(\beta)\left(\sum_{i=0}^{\infty} p^i[\overline{x}_i]\right) = \max_i\{p^{-i}\beta(\overline{x}_i)\}$$

defines a submultiplicative (resp. power-multiplicative, multiplicative) (semi-) norm on $\tilde{\mathcal{R}}^{\mathrm{int},r}_R$ *dominated by* $\max\{\lambda(\alpha^s), \lambda(\alpha^r)\}$.

(c) *In (b), if* β *is power-multiplicative (resp. multiplicative), then* $\lambda(\beta)$ *extends to a power-multiplicative (resp. multiplicative) (semi)norm on* $\tilde{\mathcal{R}}^{\mathrm{bd},r}_R$, *and then extends further by continuity to* $\tilde{\mathcal{R}}^{[s,r]}_R$.

(d) *For γ a power-multiplicative (resp. multiplicative) (semi)norm on $\tilde{\mathcal{R}}_R^{\mathrm{int},r}$ dominated by* $\max\{\lambda(\alpha^s), \lambda(\alpha^r)\}$, *the formula*

$$\mu(\gamma)(\overline{x}) = \gamma([\overline{x}])$$

defines a power-multiplicative (resp. multiplicative) (semi)norm on R dominated by $\max\{\alpha^s, \alpha^r\}$. *Moreover, γ is dominated by $\lambda(\mu(\gamma))$.*

Proof. — To check (a), we follow the argument of [**87**, Lemma 4.1], omitting those details which remain unchanged. Closure of $\tilde{\mathcal{R}}_R^{\mathrm{int},r}$ under addition and the inequality $\lambda(\alpha)(x+y) \leqslant \max\{\lambda(\alpha)(x), \lambda(\alpha)(y)\}$ follow from the homogeneity of the Witt vector addition formula [**87**, Remark 3.7] (see also Remark 3.2.3). This easily implies that $\tilde{\mathcal{R}}_R^{\mathrm{int},r}$ is closed under multiplication and that $\lambda(\alpha)$ is a submultiplicative norm, as in [**87**, Lemma 4.1]. To check that $\lambda(\alpha)$ is multiplicative whenever α is, it is enough to check that $\lambda(\alpha)(xy) \geqslant \lambda(\alpha)(x)\lambda(\alpha)(y)$ in case the right side of this inequality is positive. Write $x = \sum_{i=0}^{\infty} p^i[\overline{x}_i]$, $y = \sum_{i=0}^{\infty} p^i[\overline{y}_i]$. Let j, k be the largest indices maximizing $p^{-j}\alpha(\overline{x}_j)$, $p^{-k}\alpha(\overline{y}_k)$. As in [**87**, Lemma 4.1], we use the fact that $\lambda(\alpha)$ is a submultiplicative norm to reduce to the case where $\overline{x}_i = 0$ for $i < j$ and $\overline{y}_i = 0$ for $i < k$. Then $xy = \sum_{i=j+k}^{\infty} p^i[\overline{z}_i]$ with $\overline{z}_{j+k} = \overline{x}_j\overline{y}_k$, proving the desired inequality. To check that $\lambda(\alpha)$ is power-multiplicative whenever α is, one makes the same argument with $y = x$. This yields (a); we may check (b) by imitating the proof of (a), and (c) is clear.

To check (d), we introduce an alternate proof of [**87**, Lemma 4.4]. Again from [**87**, Remark 3.7], we deduce that for $\overline{x}, \overline{y} \in R$, $\gamma([\overline{x}+\overline{y}]) \leqslant \max\{\gamma([\overline{x}]), \gamma([\overline{y}])\}$. (Note that this requires at least power-multiplicativity, not just submultiplicativity.) By rewriting this inequality as $\mu(\gamma)(\overline{x}+\overline{y}) \leqslant \max\{\mu(\gamma)(\overline{x}), \mu(\gamma)(\overline{y})\}$, we see that $\mu(\gamma)$ is a (semi)norm. The power-multiplicativity or multiplicativity of $\mu(\gamma)$ follows from the corresponding property of γ. The fact that γ is dominated by $\lambda(\mu(\gamma))$ follows as in [**87**, Theorem 4.5]. □

Definition 5.1.3. — As in Definition 4.1.3, we impose topologies on the aforementioned rings as follows.

(a) Those rings contained in $\tilde{\mathcal{E}}_R$ carry both a *p-adic topology* (the metric topology defined by the Gauss norm) and a *weak topology* (in which a sequence converges if it is bounded for the Gauss norm and converges under $\lambda(\alpha)$ modulo any fixed power of p). For both topologies, $\tilde{\mathcal{E}}_R$ is complete.

(b) Those rings contained in $\tilde{\mathcal{R}}_R^r$ carry a *Fréchet topology*, in which a sequence converges if and only if it converges under $\lambda(\alpha^s)$ for all $s \in (0, r]$. For this topology, $\tilde{\mathcal{R}}_R^r$ is complete.

(c) Those rings contained in $\tilde{\mathcal{R}}_R$ carry a *limit-of-Fréchet topology*, or *LF topology*. This topology is defined by taking the locally convex direct limit of the $\tilde{\mathcal{R}}_R^r$ (each equipped with the Fréchet topology).

The analogue of Remark 4.1.5 is true: a sequence in $\tilde{\mathcal{R}}_R^{\mathrm{bd},r}$ which is p-adically bounded and convergent under $\lambda(\alpha^r)$ also converges in the weak topology.

Remark 5.1.4. — If one extends the definitions of $\tilde{\mathcal{E}}_R, \tilde{\mathcal{R}}_R^{\mathrm{bd}}, \tilde{\mathcal{R}}_R$ to the case where α is the trivial norm, then in this case these rings all coincide. This makes it possible to abbreviate some arguments.

Remark 5.1.5. — Recall that $\overline{\varphi}$ acts as the identity map on R if and only if R is generated over $\mathbb{F}_p$ by idempotent elements (Lemma 3.1.3). In this case, the power-multiplicative norm α on R must be trivial, so by Remark 5.1.4, all of the topologies in Definition 5.1.3 coincide with the p-adic topology.

Remark 5.1.6. — All of the constructions in Definition 5.1.1 are functorial with respect to bounded homomorphisms $\psi : R \to S$ in which S is another perfect uniform Banach $\mathbb{F}_p$-algebra with norm β. If ψ is strict injective, then ψ is isometric by Remark 3.1.6, and it is evident that the functoriality maps induced by ψ are strict injective, for all of the topologies named in Definition 5.1.3. (The case when ψ is injective but not strict is more subtle; we do not treat it here.)

Similarly, if ψ is strict surjective, then the functoriality maps induced by ψ are again strict surjective, by the following argument. Choose $c > 0$ such that any $\overline{y} \in S$ admits a lift $\overline{x} \in R$ with $\alpha(\overline{x}) \leqslant c\beta(\overline{y})$. (In fact any $c > 1$ has this property by Remark 3.1.6, but we do not need this here.) By lifting each Teichmüller element separately, we can lift each $y \in \tilde{\mathcal{R}}_S^{\mathrm{int},r}$ to some $x \in \tilde{\mathcal{R}}_S^{\mathrm{int},r}$ for which $\lambda(\alpha^s)(x) \leqslant c^r\lambda(\beta^s)(y)$ for all $s \in (0, r]$. From this, the claim follows. (See Lemma 5.5.2 for a similar argument.)

Lemma 5.1.7. — *For some $0 < s \leqslant r$, let M be a finite projective module over $\tilde{\mathcal{R}}_R^{[s,r]}$. Choose $\beta \in \mathcal{M}(R)$, and choose $\mathbf{e}_1, \ldots, \mathbf{e}_n \in M$ to form a set of module generators of $M \otimes_{\tilde{\mathcal{R}}_R^{[s,r]}} \tilde{\mathcal{R}}_{\mathcal{H}(\beta)}^{[s,r]}$. Then there exists a rational localization $(R, R^+) \to (R', R'^+)$ encircling β such that $\mathbf{e}_1, \ldots, \mathbf{e}_n$ also form a set of module generators of $M \otimes_{\tilde{\mathcal{R}}_R^{[s,r]}} \tilde{\mathcal{R}}_{R'}^{[s,r]}$.*

Proof. — For each $\gamma \in \mathcal{M}(\tilde{\mathcal{R}}_{\mathcal{H}(\beta)}^{[s,r]})$, by Nakayama's lemma, $\mathbf{e}_1, \ldots, \mathbf{e}_n$ form a set of module generators of $M \otimes_{\tilde{\mathcal{R}}_R^{[s,r]}} S_\gamma$ for some rational localization $\tilde{\mathcal{R}}_R^{[s,r]} \to S_\gamma$ encircling γ. We may cover $\mathcal{M}(\tilde{\mathcal{R}}_{\mathcal{H}(\beta)}^{[s,r]})$ with finitely many of the $\mathcal{M}(S_\gamma)$; by Remark 2.3.15 (b), these also cover $\mathcal{M}(\tilde{\mathcal{R}}_{R'}^{[s,r]})$ for some rational localization $R \to R'$ encircling β. It follows that $\mathbf{e}_1, \ldots, \mathbf{e}_n$ also form a set of module generators of $M \otimes_{\tilde{\mathcal{R}}_R^{[s,r]}} \mathcal{H}(\gamma)$ for each $\gamma \in \mathcal{M}(\tilde{\mathcal{R}}_{R'}^{[s,r]})$; this implies the claim using Lemma 2.3.12. □

Remark 5.1.8. — By construction, the ring $\tilde{\mathcal{R}}_R^{[s,r]}$ is a Banach algebra over the analytic field $\mathbb{Q}_p$. By contrast, the ring $\tilde{\mathcal{R}}_R^{\mathrm{int},r}$ is complete with respect to the norm $\lambda(\alpha^r)$, but is not a Banach algebra over any analytic field. Nonetheless, it is a Banach ring according to our conventions: for any topologically nilpotent (resp. uniform) unit $\overline{z} \in R$, $z = [\overline{z}]$ is a topologically nilpotent unit (resp. uniform unit) in $\tilde{\mathcal{R}}_R^{\mathrm{int},r}$.

Remark 5.1.9. — The ring $\tilde{\mathcal{R}}_R^r$ is by construction the inverse limit of the rings $\tilde{\mathcal{R}}_R^{[s,r]}$ for all $s \in (0, r]$. As such, it behaves much like a *Fréchet-Stein algebra* in the sense of

Schneider and Teitelbaum [**111**]; in particular, it enjoys some cohomological properties more typical of Banach algebras than of general Fréchet algebras.

5.2. Reality checks

The operation of Fréchet completion in Definition 5.1.1 leaves the structure of the resulting rings a bit mysterious. To clarify these, we make some calculations akin to the *reality checks* of [**83**, §2.5].

Lemma 5.2.1. — *For each* $x \in \tilde{\mathcal{R}}_R^{[s,r]}$, *the function* $t \mapsto \log \lambda(\alpha^t)(x)$ *is continuous and convex. In particular,* $\max\{\lambda(\alpha^r), \lambda(\alpha^s)\} = \sup\{\lambda(\alpha^t) : t \in [s,r]\}$.

Proof. — As in Lemma 4.2.3. □

Lemma 5.2.2. — *For* $x \in \tilde{\mathcal{R}}_R$, *we have* $x \in \tilde{\mathcal{R}}_R^{\mathrm{bd}}$ *if and only if for some* $r > 0$, $\lambda(\alpha^s)(x)$ *is bounded for* $s \in (0,r]$.

Proof. — As in Lemma 4.2.4. □

Using this criterion, we obtain a generalization of Corollary 4.2.5.

Corollary 5.2.3. — *Any unit in* $\tilde{\mathcal{R}}_R$ *is also a unit in* $\tilde{\mathcal{R}}_R^{\mathrm{bd}}$.

Proof. — Suppose $x \in \tilde{\mathcal{R}}_R$ is a unit with inverse y. Choose $r > 0$ so that $x, y \in \tilde{\mathcal{R}}_R^r$. For each $\beta \in \mathcal{M}(R)$, the function $\log \lambda(\beta^s)(x)$ is affine in s, as in the proof of Corollary 4.2.5. Write this affine function as $a_\beta s + b_\beta$; we then have

$$b_\beta = 2\log\lambda(\beta^{r/2})(x) - \log\lambda(\beta^r)(x) = 2\log\lambda(\beta^{r/2})(x) + \log\lambda(\beta^r)(y)$$

and

$$a_\beta = \frac{1}{r}\left(\log\lambda(\beta^r)(x) - b_\beta\right) = \frac{2}{r}\log\lambda(\beta^{r/2})(y) + \frac{2}{r}\log\lambda(\beta^r)(x)$$

and so

$$\begin{aligned}\log\lambda(\beta^s)(x) &= a_\beta s + b_\beta \\ &\leqslant \left(\frac{2}{r}\log\lambda(\alpha^{r/2})(y) + \frac{2}{r}\log\lambda(\alpha^r)(x)\right)s + 2\log\lambda(\alpha^{r/2})(x) + \log\lambda(\alpha^r)(y).\end{aligned}$$

Taking suprema over $\mathcal{M}(R)$ yields a similar upper bound for $\log\lambda(\alpha^s)(x)$, so $x \in \tilde{\mathcal{R}}_R^{\mathrm{bd}}$ by Lemma 5.2.2. Similarly, $y \in \tilde{\mathcal{R}}_R^{\mathrm{bd}}$, so x is a unit in $\tilde{\mathcal{R}}_R^{\mathrm{bd}}$ as desired. □

Corollary 5.2.4. — *We have* $\tilde{\mathcal{E}}_R^{\varphi} = \tilde{\mathcal{R}}_R^{\varphi} = W(R^{\overline{\varphi}})[p^{-1}]$. *In particular, by Corollary 3.1.4,* $W(R^{\overline{\varphi}})[p^{-1}] = \mathbb{Q}_p$ *if and only if* R *is connected.*

Proof. — It is clear that $\tilde{\mathcal{E}}_R^{\varphi} = W(R^{\overline{\varphi}})[p^{-1}] \subseteq \tilde{\mathcal{R}}_R^{\varphi}$. We have $\tilde{\mathcal{R}}_R^{\varphi} = (\tilde{\mathcal{R}}_R^{\mathrm{bd}})^{\varphi} \subseteq \tilde{\mathcal{E}}_R^{\varphi}$ by Lemma 5.2.2, as in the proof of Lemma 4.2.10. □

Lemma 5.2.5. — *For* $0 < s_1 \leqslant s_2 \leqslant r_2 \leqslant r_1$, *the natural restriction map* $\tilde{\mathcal{R}}_R^{[s_1,r_1]} \to \tilde{\mathcal{R}}_R^{[s_2,r_2]}$ *is injective.*

Proof. — Since R is uniform, the spectral norm α is equal to the supremum norm on $\prod_{\beta \in \mathcal{M}(R)} \mathcal{H}(\beta)$ by Theorem 2.3.10. Thus it is straightforward to see that the natural map

$$\tilde{\mathcal{R}}_R^I \longrightarrow \prod_{\beta \in \mathcal{M}(R)} \tilde{\mathcal{R}}_{\mathcal{H}(\beta)}^I$$

is injective for any closed interval $I \subset (0, \infty)$. Thus it reduces to show the lemma in the case when $R = L$ is an analytic field. We deduce by Lemma 4.2.3 that for $x \in \tilde{\mathcal{R}}_L^{[s,r]}$, the function $t \mapsto \log \lambda(\alpha^t)(x)$ is continuous and convex on $[s, r]$. This implies that if $\lambda(\alpha^t)(x) = 0$ for some $t \in [s, r]$, then $\lambda(\alpha^t)(x) = 0$ for all $t \in [s, r]$; thus $x = 0$. The lemma then follows. □

Lemma 5.2.6. — *For $0 < r \leqslant r'$, inside $\tilde{\mathcal{R}}_R^{[r,r]}$ we have*

$$\tilde{\mathcal{R}}_R^{\mathrm{int},r} \cap \tilde{\mathcal{R}}_R^{[r,r']} = \tilde{\mathcal{R}}_R^{\mathrm{int},r'}.$$

Proof. — The case $r' = r$ is trivial, so assume that $r' > r$. Take $x \in \tilde{\mathcal{R}}_R^{\mathrm{int},r} \cap \tilde{\mathcal{R}}_R^{[r,r']}$, and write x as the limit in $\tilde{\mathcal{R}}_R^{[r,r']}$ of a sequence $x_0, x_1, \dots$ with $x_i \in \tilde{\mathcal{R}}_R^{\mathrm{bd},r'}$. For each positive integer j, we can find $N_j > 0$ such that

$$\lambda(\alpha^s)(x_i - x) \leqslant p^{-j} \qquad (i \geqslant N_j,\ s \in [r, r']).$$

Write $x_i = \sum_{l=m(i)}^{\infty} p^l [\overline{x}_{il}]$ and put $y_i = \sum_{l=0}^{\infty} p^l [\overline{x}_{il}] \in \tilde{\mathcal{R}}_R^{\mathrm{int},r'}$. For $i \geqslant N_j$, having $x \in \tilde{\mathcal{R}}_R^{\mathrm{int},r}$ and $\lambda(\alpha^r)(x_i - x) \leqslant p^{-j}$ implies that $\lambda(\alpha^r)(p^l[\overline{x}_{il}]) \leqslant p^{-j}$ for $l < 0$. That is,

$$\alpha(\overline{x}_{il}) \leqslant p^{(l-j)/r} \qquad (i \geqslant N_j,\ l < 0).$$

Since $p^{-l} p^{(l-j)r'/r} \leqslant p^{1+(1-j)r'/r}$ for $l \leqslant -1$, we deduce that $\lambda(\alpha^{r'})(x_i - y_i) \leqslant p^{1+(1-j)r'/r}$ for $i \geqslant N_j$. Consequently, the sequence $y_0, y_1, \dots$ converges to x under $\lambda(\alpha^{r'})$; it follows that $x \in \tilde{\mathcal{R}}_R^{\mathrm{int},r'}$. This proves the claim. □

Remark 5.2.7. — In both Lemma 4.2.4 and Lemma 5.2.6, the key step was to split an element of $\tilde{\mathcal{R}}_R^{\mathrm{bd}}$ into what one might call an *integral part* and a *fractional part.* One cannot directly imitate the construction for elements of $\tilde{\mathcal{R}}_R^{[s,r]}$ because they cannot be expressed as sums of Teichmüller elements. One can give presentations of a slightly less restrictive form with which one can make similar arguments (the *semiunit presentations* of [**83**, §2]); the *stable presentations* of [**87**, §5] are similar. We will instead rely on Lemma 5.2.8 (see below) to simulate splittings into integral and fractional parts.

As noted in Remark 5.2.7, the following lemma extends to $\tilde{\mathcal{R}}_R^{[s,r]}$ the splitting argument for elements of $\tilde{\mathcal{R}}_R^{\mathrm{bd}}$ used previously. Its formulation is modeled on [**83**, Lemma 2.5.11].

Lemma 5.2.8. — *For* $0 < s \leqslant r$ *and* $n \in \mathbb{Z}$, *any* $x \in \tilde{\mathcal{R}}_R^{[s,r]}$ *can be written as* $y + z$ *with* $y \in p^n \tilde{\mathcal{R}}_R^{\mathrm{int},r}$, $z \in \cap_{r' \geqslant r} \tilde{\mathcal{R}}_R^{[s,r']}$, *and*

$$\lambda(\alpha^t)(z) \leqslant p^{(1-n)(1-t/r)} \lambda(\alpha^r)(x)^{t/r} \qquad (t \geqslant r). \tag{5.2.8.1}$$

Proof. — In case $x \in \tilde{\mathcal{R}}_R^{\mathrm{bd}}$, write $x = \sum_{i=m(x)}^{\infty} p^i [\overline{x}_i]$, and put $y = \sum_{i=n}^{\infty} p^i [\overline{x}_i]$ and $z = y - x$. This works because for $i \leqslant n-1$ and $t \geqslant r$,

$$\begin{aligned} \lambda(\alpha^t)(p^i[\overline{x}_i]) &= p^{-i} \alpha(\overline{x}_i)^t \\ &= p^{-i(1-t/r)} \lambda(\alpha^r)(p^i[\overline{x}_i])^{t/r} \leqslant p^{(1-n)(1-t/r)} \lambda(\alpha^r)(p^i[\overline{x}_i])^{t/r}. \end{aligned} \tag{5.2.8.2}$$

To handle the general case, choose $x_0, x_1, \dots \in \tilde{\mathcal{R}}_R^{\mathrm{bd},r}$ so that

$$\lambda(\alpha^t)(x - x_0 - \cdots - x_i) \leqslant p^{-i-1} \lambda(\alpha^t)(x) \qquad (i = 0, 1, \dots; t \in [s,r]).$$

The series $\sum_{i=0}^{\infty} x_i$ converges to x under $\lambda(\alpha^t)$ for $t \in [s,r]$, and $\lambda(\alpha^t)(x_i) \leqslant p^{-i} \lambda(\alpha^t)(x)$ for $i = 0, 1, \dots$ and $t \in [s,r]$. Split each x_i as $y_i + z_i$ as above. Since the sum $\sum_{i=0}^{\infty} y_i$ converges under $\lambda(\alpha^r)$ and consists of elements of $\tilde{\mathcal{R}}_R^{\mathrm{int},r}$, it converges under $\lambda(\alpha^t)$ for all $t \in (0,r]$ and defines an element y of $\tilde{\mathcal{R}}_R^{\mathrm{int},r}$. Put $z = y - x$; then the series $\sum_{i=0}^{\infty} z_i$ converges to z under $\lambda(\alpha^t)$ for $t \in [s,r]$. On the other hand, for $t \geqslant r$, by (5.2.8.2) we have

$$\lambda(\alpha^t)(z_i) \leqslant p^{(1-n)(1-t/r)} \lambda(\alpha^r)(x_i)^{t/r} \leqslant p^{(1-n)(1-t/r)} p^{-i(t/r)} \lambda(\alpha^r)(x)^{t/r}.$$

Consequently, $\sum_{i=0}^{\infty} z_i$ also converges to z under $\lambda(\alpha^t)$, and (5.2.8.1) holds. □

We will also have use for the following variant, where we separate in terms of α rather than the p-adic norm.

Lemma 5.2.9. — *For* $c < 1$ *and* $0 < s \leqslant r$, *each* $x \in \tilde{\mathcal{R}}_R^{[s,r]}$ *can be written as* $y + z$ *with* $y \in \tilde{\mathcal{R}}_R^{\mathrm{bd},r}$, $z \in \tilde{\mathcal{R}}_{R^+}^{[s,r]}$ *and*

$$\begin{aligned} \lambda(\overline{\alpha}^t)(y), \lambda(\overline{\alpha}^t)(z) &\leqslant \lambda(\overline{\alpha}^t)(x) && (t \in [s,r]) \\ y &\in p^n \tilde{\mathcal{R}}_R^{\mathrm{int},r} && (n = \lceil -\log_p(\lambda(\overline{\alpha}^s)(x) c^{-s}) \rceil) \\ \lambda(\overline{\alpha}^t)(z) &\leqslant c^{t-r} \lambda(\overline{\alpha}^r)(z) && (t > r). \end{aligned}$$

In particular, for any positive integer a, *if we put* $q = p^a$, *then for all* $t \in [s,r]$,

$$\begin{aligned} \lambda(\alpha^t)(y), \lambda(\alpha^t)(z) &\leqslant \lambda(\alpha^t)(x), \\ \lambda(\alpha^t)(\varphi^{-a}(y)) &\leqslant c^{-(q-1)t/q} \lambda(\alpha^t)(x), \\ \lambda(\alpha^t)(\varphi^a(z)) &\leqslant c^{(q-1)t} \lambda(\alpha^t)(x). \end{aligned}$$

Proof. — By approximating x as in the proof of Lemma 5.2.8, it is enough to consider the case $x \in \tilde{\mathcal{R}}_R^{\mathrm{bd},r}$. In this case, write $x = \sum_{i=m}^{\infty} p^i [\overline{x}_i]$, let y be the sum of $p^i[\overline{x}_i]$ over all indices i for which $\alpha(\overline{x}_i) > c$, and put $z = x - y$. □

As an immediate application of Lemma 5.2.8, we extend Lemma 5.2.6 as follows.

Lemma 5.2.10. — *For $0 < s \leqslant s' \leqslant r \leqslant r'$, inside $\tilde{\mathcal{R}}_R^{[s',r]}$ we have*
$$\tilde{\mathcal{R}}_R^{[s,r]} \cap \tilde{\mathcal{R}}_R^{[s',r']} = \tilde{\mathcal{R}}_R^{[s,r']}.$$

Proof. — Given x in the intersection, apply Lemma 5.2.8 to write $x = y + z$ with $y \in \tilde{\mathcal{R}}_R^{\mathrm{int},r}$, $z \in \tilde{\mathcal{R}}_R^{[s,r']}$. By Lemma 5.2.6,
$$y = z - x \in \tilde{\mathcal{R}}_R^{\mathrm{int},r} \cap \tilde{\mathcal{R}}_R^{[s',r']} = \tilde{\mathcal{R}}_R^{\mathrm{int},r'} \subseteq \tilde{\mathcal{R}}_R^{[s,r']},$$
so $x \in \tilde{\mathcal{R}}_R^{[s,r']}$ as desired. □

Lemma 5.2.11. — *Suppose that $R^+ = \mathfrak{o}_R$.*

(a) *An element $x \in \tilde{\mathcal{R}}_R^{\infty}$ belongs to $\tilde{\mathcal{R}}_R^+$ if and only if*
$$\limsup_{r \to +\infty} \lambda(\alpha^r)(x)^{1/r} \leqslant 1.$$

(b) *For $r > 0$ and $x \in \tilde{\mathcal{R}}_R^r$, we have $x = y + z$ for some $y \in \tilde{\mathcal{R}}_R^{\mathrm{bd},r}$ and $z \in \tilde{\mathcal{R}}_R^+$.*

(c) *We have $\tilde{\mathcal{R}}_R^{\mathrm{int},r} \cap \tilde{\mathcal{R}}_R^+ = \tilde{\mathcal{R}}_R^{\mathrm{int},+}$.*

Proof. — We first prove (a) for $x = \sum_{i=m}^{\infty} p^i [\overline{x}_i] \in \tilde{\mathcal{R}}_R^{\mathrm{bd}} \cap \tilde{\mathcal{R}}_R^{\infty}$, by observing that
$$\limsup_{r \to +\infty} \lambda(\alpha^r)(x)^{1/r} = \limsup_{r \to +\infty} \sup_i \{p^{-i/r} \alpha(\overline{x}_i)\}.$$
If $x \in \tilde{\mathcal{R}}_R^{\mathrm{bd},+}$, then for some m, we can bound the quantity $p^{-i/r}\alpha(\overline{x}_i)$ from above by $p^{m/r}$, and so the limit superior in question is at most 1. Conversely, if $x \notin \tilde{\mathcal{R}}_R^{\mathrm{bd},+}$, then there exists an index i for which $\alpha(\overline{x}_i) > 1$; we can then find $\epsilon > 0$ so that $p^{-i/r}\alpha(\overline{x}_i) > 1 + \epsilon$ for r large, so the limit superior is at least $1 + \epsilon$. This proves (a) for such x.

We next prove (b). For $r > 0$ and $x \in \tilde{\mathcal{R}}_R^r$, choose $n \in \mathbb{Z}$ so that $n < 1$ and $p^{(n-1)/(2r)} \lambda(\alpha^r)(x)^{1/r} < 1$. Set notation as in the proof of Lemma 5.2.8; for $t \geqslant 2r$, we have $\lambda(\alpha^t)(z_i)^{1/t} \leqslant p^{(1-n)(1/t-1/r)} \lambda(\alpha^r)(x)^{1/r} \leqslant 1$, so $z_i \in \tilde{\mathcal{R}}_R^{\mathrm{bd},+}$ by the previous paragraph. Hence $z \in \tilde{\mathcal{R}}_R^+$ as needed.

We next return to (a). If $x \in \tilde{\mathcal{R}}_R^+$, then $\limsup_{r \to +\infty} \lambda(\alpha^r)(x)^{1/r} \leqslant 1$ by the first paragraph. Conversely, if $x \in \tilde{\mathcal{R}}_R^{\infty}$ and $\limsup_{r \to +\infty} \lambda(\alpha^r)(x)^{1/r} \leqslant 1$, apply (b) to write $x = y + z$ with $y \in \tilde{\mathcal{R}}_R^{\mathrm{bd}}$ and $z \in \tilde{\mathcal{R}}_R^+$. We may then apply the first paragraph to y to deduce that $x \in \tilde{\mathcal{R}}_R^+$.

To deduce (c), use Lemma 5.2.6 to deduce that $\tilde{\mathcal{R}}_R^{\mathrm{int},r} \cap \tilde{\mathcal{R}}_R^+ \subseteq \cap_{s>0} \tilde{\mathcal{R}}_R^{\mathrm{int},s}$, then argue as in the proof of (a). □

Corollary 5.2.12. — *For n a nonnegative integer, d a positive integer, $q = p^d$, and $r > 0$, the inclusions*
$$\{x \in \tilde{\mathcal{R}}_R^+ : \varphi^d(x) = p^n x\} \subseteq \{x \in \tilde{\mathcal{R}}_R : \varphi^d(x) = p^n x\},$$
$$\{x \in \tilde{\mathcal{R}}_R^+ : \varphi^d(x) = p^n x\} \subseteq \{x \in \tilde{\mathcal{R}}_R^{[r/q,r]} : \varphi^d(x) = p^n x\}$$
are bijective.

Proof. — Suppose first that $x \in \tilde{\mathcal{R}}_R^\infty$ and $\varphi^d(x) = p^n x$. For each $r > 0$,

$$\lambda(\alpha^{rq})(x) = \lambda(\alpha^r)(\varphi^d(x)) = p^{-n}\lambda(\alpha^r)(x).$$

It follows that for any fixed $s > 0$,

$$\limsup_{r\to+\infty} \lambda(\alpha^r)(x)^{1/r} \leqslant \sup_{r\in[s,qs]} \{\limsup_{n\to+\infty} p^{-n/(rq^n)}\lambda(\alpha^r)(x)^{1/(rq^n)}\} \leqslant 1,$$

so if $R^+ = \mathfrak{o}_R$ then $x \in \tilde{\mathcal{R}}_R^+$ by Lemma 5.2.11 (a). To treat the general case, note that x is now known to belong to the Fréchet completion of $W(\mathfrak{o}_R)[p^{-1}]$. Under the projection from this ring to $W(\kappa_R)[p^{-1}]$, x maps to zero if $n > 0$ and to $W(\kappa_R^{\overline{\varphi}^d})[p^{-1}]$ if $n = 0$; in either case it follows that $x \in \tilde{\mathcal{R}}_R^+$.

Given $x \in \tilde{\mathcal{R}}_R$ for which $\varphi^d(x) = p^n x$, there exists $r > 0$ for which $x \in \tilde{\mathcal{R}}_R^r$, but then $x = p^{-n}\varphi^{-d}(x) \in \tilde{\mathcal{R}}_R^{rq}$. Consequently, $x \in \tilde{\mathcal{R}}_R^\infty$, so by the previous paragraph, $x \in \tilde{\mathcal{R}}_R^+$.

Given $x \in \tilde{\mathcal{R}}_R^{[r/q,r]}$ for which $\varphi^d(x) = p^n x$, for each positive integer m we also have $x \in \tilde{\mathcal{R}}_R^{[r/q^m,r]}$. Namely, this holds for $m = 1$, and given the statement for some m, we also have $x = p^{-n}\varphi^d(x) \in \tilde{\mathcal{R}}_R^{[r/q^{m+1},r/q]}$, so $x \in \tilde{\mathcal{R}}_R^{[r/q^{m+1},r]}$ by Lemma 5.2.10. It follows that $x \in \tilde{\mathcal{R}}_R^r$, so by the previous paragraph, $x \in \tilde{\mathcal{R}}_R^+$. □

Remark 5.2.13. — Let $R \subseteq S' \subseteq S$ and $R \subseteq S'' \subseteq S$ be strict (and hence isometric, by Remark 3.1.6) inclusions of perfect uniform Banach $\mathbb{F}_p$-algebras, and suppose that within S we have $S' \cap S'' = R$. It is easy to see that

$$*_{S'} \cap *_{S''} = *_R \qquad * = \tilde{\mathcal{E}}, \tilde{\mathcal{R}}^{\mathrm{int},r}, \tilde{\mathcal{R}}^{\mathrm{int},+}, \tilde{\mathcal{R}}^{\mathrm{int}}, \tilde{\mathcal{R}}^{\mathrm{bd},r}, \tilde{\mathcal{R}}^{\mathrm{bd},+}, \tilde{\mathcal{R}}^{\mathrm{bd}},$$

with the intersection taking place within $*_S$; namely, $W(S') \cap W(S'') = W(R)$ within $W(S)$.

Now suppose additionally that the morphism $S' \oplus S'' \to S$ taking (s', s'') to $s' - s''$ is strict. Then

$$*_{S'} \cap *_{S''} = *_R \qquad * = \tilde{\mathcal{R}}^{[s,r]}, \tilde{\mathcal{R}}^r, \tilde{\mathcal{R}}^+, \tilde{\mathcal{R}}.$$

If the inclusions are not strict, it is unclear whether such maps as $*_R \to *_{S'}$ are even injective.

Remark 5.2.14. — It would be useful to have the following refinement of Lemma 5.2.2: an element $x \in \tilde{\mathcal{R}}_R$ belongs to $\tilde{\mathcal{R}}_R^{\mathrm{bd}}$ if and only if for each $\beta \in \mathcal{M}(R)$, the image of x in $\tilde{\mathcal{R}}_{\mathcal{H}(\beta)}$ belongs to $\tilde{\mathcal{R}}_{\mathcal{H}(\beta)}^{\mathrm{bd}}$. However, it is unclear to us whether to expect this to hold.

5.3. Sheaf properties

We next investigate the sheaf-theoretic properties of the preceding constructions.

Definition 5.3.1. — For (R, R^+) a perfect uniform adic Banach algebra and

$$* = \tilde{\mathcal{E}}^{\mathrm{int}}, \tilde{\mathcal{E}}, \tilde{\mathcal{R}}^{\mathrm{int},r}, \tilde{\mathcal{R}}^{\mathrm{int},+}, \tilde{\mathcal{R}}^{\mathrm{int}}, \tilde{\mathcal{R}}^{\mathrm{bd},r}, \tilde{\mathcal{R}}^{\mathrm{bd},+}, \tilde{\mathcal{R}}^{\mathrm{bd}}, \tilde{\mathcal{R}}^{[s,r]}, \tilde{\mathcal{R}}^r, \tilde{\mathcal{R}}^+, \tilde{\mathcal{R}},$$

construct the presheaf $*$ on $\mathrm{Spa}(R, R^+)$ assigning to each open subset U the inverse limit of $*_S$ over each rational localization $(R, R^+) \to (S, S^+)$ for which $\mathrm{Spa}(S, S^+) \subseteq U$.

Lemma 5.3.2. — *With notation as in Lemma 3.1.12, the sequence*

$$0 \longrightarrow R \longrightarrow R_1 \oplus R_2 \longrightarrow R_{12} \longrightarrow 0 \tag{5.3.2.1}$$

remains exact, and the morphisms remain almost optimal, when R_* is replaced by any of $\tilde{\mathcal{E}}^{\mathrm{int}}_{R_*}$ or $\tilde{\mathcal{E}}_{R_*}$ (for the p-adic norm), $\tilde{\mathcal{R}}^{\mathrm{int},r}_{R_*}$ (for the norm $\lambda(\alpha^r)$), $\tilde{\mathcal{R}}^{\mathrm{bd},r}_{R_*}$ (for the maximum of $\lambda(\alpha^r)$ and the p-adic norm), $\tilde{\mathcal{R}}^{[s,r]}_{R_*}$ (for the norm $\max\{\lambda(\alpha^r), \lambda(\alpha^s)\}$), or $\tilde{\mathcal{R}}^r_{R_*}$ (omitting the statement about norms).

Proof. — This is a straightforward consequence of Lemma 3.1.12 except for the case of $\tilde{\mathcal{R}}^r_{R_*}$, for which we must separately check exactness on the right. This calculation may be viewed as giving a vanishing of a $\varprojlim^1$ term, as suggested by Remark 5.1.9.

Given $x \in \tilde{\mathcal{R}}^r_{R_{12}}$, we may construct elements $x_{n,j} \in \tilde{\mathcal{R}}^{\mathrm{bd},r}_{R_j}$ for $n = 0, 1, \dots$ and $j = 1, 2$ such that for $y_n = x - \sum_{m=0}^{n-1}(x_{m,1} + x_{m,2})$, we have

$$\lambda(\alpha^s)(y_n) \leqslant p^{-n}\lambda(\alpha^s)(x) \qquad (s \in [p^{-n}r, r])$$
$$\lambda(\alpha^s)(x_{n,j}) \leqslant (1 + p^{-n})\lambda(\alpha^s)(y_n) \qquad (s \in [p^{-n}r, r]; j \in \{1, 2\}).$$

Namely, given y_n, we split y_n in $\tilde{\mathcal{R}}^{[p^{-n-1}r,r]}_{R_*}$ using Lemma 3.1.12, then approximate the results suitably well with elements of $\tilde{\mathcal{R}}^{\mathrm{bd},r}_{R_*}$. The sums $\sum_{n=0}^{\infty} x_{n,j}$ for $j = 1, 2$ then converge and give the desired splitting of x. □

Theorem 5.3.3. — *All of the presheaves defined in Definition 5.3.1 are sheaves. Moreover, for $* = \tilde{\mathcal{E}}^{\mathrm{int}}, \tilde{\mathcal{E}}, \tilde{\mathcal{R}}^{\mathrm{int},r}, \tilde{\mathcal{R}}^{\mathrm{int}}, \tilde{\mathcal{R}}^{\mathrm{bd},r}, \tilde{\mathcal{R}}^{\mathrm{bd}}, \tilde{\mathcal{R}}^{[s,r]}, \tilde{\mathcal{R}}^r, \tilde{\mathcal{R}}$, the resulting sheaf is acyclic (*i.e.*, satisfies the Tate sheaf property).*

Proof. — By Proposition 2.4.21, we may deduce the claim from Lemma 5.3.2. □

The Kiehl property is somewhat more elusive; we only obtain it for a few of the sheaves in question.

Lemma 5.3.4. — *Choose $0 < s \leqslant r$.*

(a) *A multiplicative seminorm β on $\tilde{\mathcal{R}}^{[s,r]}_R$ is dominated by $\max\{\lambda(\alpha^s), \lambda(\alpha^r)\}$ if and only if it is dominated by $\lambda(\alpha^t)$ for some unique $t \in [s, r]$.*

(b) *For β, t as in (a), put $\gamma = \mu(\beta)^{1/t} \in \mathcal{M}(R)$. Then β extends uniquely to a multiplicative seminorm on $\tilde{\mathcal{R}}^{[s,r]}_{\mathcal{H}(\gamma)}$.*

Proof. — We first address (a). If β is dominated by $\lambda(\alpha^t)$ for some $t \in [s, r]$, then β is dominated by $\max\{\lambda(\alpha^s), \lambda(\alpha^r)\}$ by Lemma 5.2.1. Conversely, suppose that β is dominated by $\max\{\lambda(\alpha^s), \lambda(\alpha^r)\}$. Write R as a Banach algebra over some analytic field K, and pick $\overline{\pi} \in K^\times$ with $\alpha(\overline{\pi}) < 1$. Then $\beta([\overline{\pi}]) \in [\alpha^r(\overline{\pi}), \alpha^s(\overline{\pi})]$, so there exists $t \in [s, r]$ such that $\beta([\overline{\pi}]) = \alpha^t(\overline{\pi})$. For $\overline{x} \in R$ such that $\alpha(\overline{x}) \leqslant 1$, we also have $\beta([\overline{x}]) \leqslant 1$. If we take $\overline{x} = \overline{y}^m\overline{\pi}^{-n}$ for m a positive integer and n an arbitrary integer, we deduce that if $\alpha(\overline{y})^m\alpha(\overline{\pi})^{-n} \leqslant 1$, then $\beta([\overline{y}])^m\beta([\overline{\pi}])^{-n} \leqslant 1$. That is, if $\alpha(\overline{y}) \leqslant \alpha(\overline{\pi})^{n/m}$, then $\beta([\overline{y}]) \leqslant \alpha(\overline{\pi})^{nt/m}$. Since n/m can be chosen to be any rational number, it follows that $\beta([\overline{y}]) \leqslant \alpha(\overline{y})^t$.

To deduce (b), note first that β extends uniquely to the localization of $\tilde{\mathcal{R}}_R^{[s,r]}$ at the multiplicative set consisting of $[\overline{x}]$ for each $x \in R \setminus \mathfrak{p}_\gamma$, and that this extension is dominated by $\max\{\lambda(\gamma^s), \lambda(\gamma^r)\}$. Then observe that the separated completion under $\max\{\lambda(\gamma^s), \lambda(\gamma^r)\}$ of this localization is precisely $\tilde{\mathcal{R}}_{\mathcal{H}(\gamma)}^{[s,r]}$. □

Lemma 5.3.5. — *With notation as in Lemma 3.1.12, for any $r > 0$, the diagrams*

$$\begin{array}{ccc} R & \longrightarrow & R_1 \\ \downarrow & & \downarrow \\ R_2 & \longrightarrow & R_{12} \end{array} \qquad \begin{array}{ccc} \tilde{\mathcal{R}}_R^{\mathrm{int},r} & \longrightarrow & \tilde{\mathcal{R}}_{R_1}^{\mathrm{int},r} \\ \downarrow & & \downarrow \\ \tilde{\mathcal{R}}_{R_2}^{\mathrm{int},r} & \longrightarrow & \tilde{\mathcal{R}}_{R_{12}}^{\mathrm{int},r} \end{array}$$

are glueing squares in the sense of Definition 2.7.3. (Note that the rings in the second diagram are Banach rings by virtue of Remark 5.1.8.)

Proof. — Note that $R_2 \to R_{12}$ has dense image because f is already invertible in R_2, and that by construction, $\mathcal{M}(R_1 \oplus R_2) \to \mathcal{M}(R)$ is surjective. Hence $R \to R_1$, $R \to R_2$, $R \to R_{12}$ form a glueing square by Lemma 5.3.2. The other assertion follows similarly, keeping in mind Lemma 5.3.4 in order to get surjectivity of $\mathcal{M}(\tilde{\mathcal{R}}_{R_1}^{\mathrm{int},r} \oplus \tilde{\mathcal{R}}_{R_2}^{\mathrm{int},r}) \to \mathcal{M}(\tilde{\mathcal{R}}_R^{\mathrm{int},r})$. □

Theorem 5.3.6. — *The sheaves $\tilde{\mathcal{E}}^{\mathrm{int}}$ and $\tilde{\mathcal{R}}^{\mathrm{int},r}$ satisfy the Kiehl glueing property.*

Proof. — We check both claims using Proposition 2.4.20. The case of $\tilde{\mathcal{R}}^{\mathrm{int},r}$ follows from Lemma 5.3.5 and Proposition 2.7.5. The case of $\tilde{\mathcal{E}}^{\mathrm{int}}$ does not follow in this manner (see Remark 5.3.7); we instead directly check the hypotheses of Lemma 1.3.9 (b) as follows. Let M_1, M_2, M_{12} be a finite projective glueing datum. Reducing modulo p, we obtain a finite projective glueing datum for the rings R_*, which by Theorem 3.1.13 arises from a finite projective module $\overline{M}$ over R. For any $\overline{\mathbf{v}} \in \overline{M}$, we may lift $\overline{\mathbf{v}}$ to an element $\mathbf{v}$ in the kernel of $M_1 \oplus M_2 \to M_{12}$ by successive approximations: if $\mathbf{v} \in M_1 \oplus M_2$ maps to $p^n\mathbf{w} \in M_{12}$, we lift $\mathbf{w}$ to $\mathbf{v}' \in M_1 \oplus M_2$ and replace $\mathbf{v}$ with $\mathbf{v} - p^n\mathbf{v}'$. In particular, we may lift a finite generating set of $\overline{M}$ to a generating set of M_1. This verifies that the hypothesis of Lemma 1.3.8 holds for finite projective glueing data; the fact that $\mathrm{Spec}(W(R_1) \oplus W(R_2))$ covers $\mathrm{Maxspec}(W(R))$ follows from Lemma 5.3.5 and the identification $\mathrm{Maxspec}(W(*)) \cong \mathrm{Maxspec}(*)$. □

Remark 5.3.7. — The reason the proof of Theorem 5.3.6 for the sheaf $\tilde{\mathcal{E}}^{\mathrm{int}}$ fails to follow the model of $\tilde{\mathcal{R}}^{\mathrm{int},r}$ is that $\tilde{\mathcal{E}}_{R_2}^{\mathrm{int}}$ is not dense in $\tilde{\mathcal{E}}_{R_{12}}^{\mathrm{int}}$ for the p-adic topology, only for the weak topology. It should be possible to adapt the concept of a glueing square in Definition 2.7.3 to apply to a topology not specified in terms of a norm, but we did not verify this.

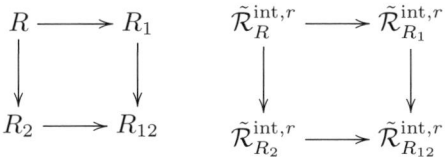

However, no such adaptation can exist for $\tilde{\mathcal{E}}$ or $\tilde{\mathcal{R}}^{\mathrm{bd},r}$, because the Kiehl property fails in these cases; see Example 8.5.17. Somewhat confusingly, the rings $\tilde{\mathcal{E}}_R$ and $\tilde{\mathcal{R}}_R^{\mathrm{bd},r}$ are themselves sheafy (Theorem 5.3.9), but there is no inconsistency because localizations of these rings do not correspond directly to localizations of R (again because of the density issue).

While the proof of Theorem 5.3.6 can be carried through for $\tilde{\mathcal{R}}^{[s,r]}$, it is more useful to deduce the corresponding statement from a much stronger glueing property. Theorem 5.3.9 and its proof are taken from the PhD thesis of Ryan Rodriguez [**109**].

Lemma 5.3.8. — *Let A be a Banach algebra over $\mathbb{Q}_p$ and let K be the completion of $\mathbb{Q}_p(p^{p^{-\infty}})$. Then each element x of $A \widehat{\otimes}_{\mathbb{Q}_p} K$ admits a unique presentation as a convergent sum $\sum_i x_i \otimes p^i$, where i runs over $\mathbb{Z}[p^{-1}] \cap [0,1)$; moreover, the tensor product norm of x can be computed as $\max_i\{|x_i|\, p^{-i}\}$.*

Proof. — This is a straightforward consequence of the fact that the p^i form an orthogonal topological basis of K over $\mathbb{Q}_p$. $\square$

***Theorem 5.3.9* (Rodriguez)**. — *For $0 < s \leqslant r$, the Banach rings $\tilde{\mathcal{E}}_R$, $\tilde{\mathcal{R}}_R^{\mathrm{bd},r}$, and $\tilde{\mathcal{R}}_R^{[s,r]}$ are relatively perfectoid. In particular, by Theorem 3.7.4, any extension of one of these rings to an adic Banach ring is stably uniform and sheafy and satisfies the Tate sheaf and Kiehl glueing properties.*

Proof. — We treat only the case of $\tilde{\mathcal{R}}_R^{[s,r]}$ in detail, the other cases being easier (and not needed in what follows). Let K be the completion of $\mathbb{Q}_p(p^{p^{-\infty}})$, which is a perfectoid field (see Example 3.3.8). We first check that $S = \tilde{\mathcal{R}}_R^{[s,r]} \widehat{\otimes}_{\mathbb{Q}_p} K$ is uniform. For $t \in [s,r]$, let $|\bullet|_t$ be the tensor product norm on S induced by $\lambda(\alpha^t)$; we first check that $|\bullet|_t$ is power-multiplicative. It suffices to check the inequality $|x^2|_t \geqslant |x|_t^2$ for x running over a dense subset of S. We may thus assume that x has the form $\sum_{i=1}^n [\overline{x}_i] \otimes p^{j_i}$ for some $\overline{x}_i \in R$, $j_i \in \mathbb{Z}[p^{-1}]$ such that the $\overline{x}_i$ are nonzero and the j_i are pairwise distinct. Note that in this case, $|\bullet|_u$ is well-defined for all $u > 0$, not just $u \in [s,r]$. Let T_x be the set of $u > 0$ for which $\lambda(\alpha^u)(\overline{x}_i)p^{-j_i} = \lambda(\alpha^u)(\overline{x}_{i'})p^{-j_{i'}}$ for some $i \neq i'$; this set is finite. For each $u > 0$ with $u \notin T_x$, there is a unique index i maximizing $\lambda(\alpha)^u(\overline{x}_i)p^{-j_i}$; if we put $y = [\overline{x}_i] \otimes p^{j_i}$, then $|y^2|_u = |y|_u^2$ and $|y|_u > |x-y|_u$. Using Lemma 5.3.8, it follows easily that

$$|x^2|_u = |y^2 + (x+y)(x-y)|_u = |y|_u^2 = |x|_u^2 .$$

This proves the claim for the given x and all $u \notin T_x$; however, Lemma 5.2.1 implies that $\log |x|_u$ is a convex and hence continuous function of u, so we may interpolate the claim for $u = t$ even if $t \in T_x$. We thus conclude that $|\bullet|_t$ is power-multiplicative for each $t \in [s,r]$, and so $\max\{|\bullet|_s, |\bullet|_r\}$ (which by log-convexity again is the norm induced by $\max\{\lambda(\alpha^s), \lambda(\alpha^r)\}$) is also power-multiplicative. In particular, S is uniform.

We next check that for any perfectoid algebra A for which $T = \tilde{\mathcal{R}}_R^{[s,r]} \widehat{\otimes}_{\mathbb{Q}_p} A$ is uniform, T is also perfectoid. By Proposition 3.6.25, it suffices to show that $\varprojlim_{x \mapsto x^p} T$ generates a dense subring of T. Since A is perfectoid, $\varprojlim_{x \mapsto x^p} A$ generates a dense subring of A; we obtain a dense subring of T by adding the additional generator $\theta([\overline{x}])$ for each $\overline{x} \in R$.

To conclude, note that S is uniform (by the first paragraph) and hence perfectoid (by the second paragraph). For any perfectoid algebra A, it follows that T is uniform (by Proposition 3.7.3) and hence perfectoid (by the second paragraph again). □

We can use Theorem 5.3.9 to deduce the Kiehl property for the sheaf $\tilde{\mathcal{R}}^{[s,r]}$ by relating rational localizations of R and $\tilde{\mathcal{R}}_R^{[s,r]}$.

Definition 5.3.10. — Recall that Lemma 5.3.4 defines a map $\mathcal{M}(\tilde{\mathcal{R}}_R^{[s,r]}) \to \mathcal{M}(R)$. We lift this to a map of adic spectra as follows.

For v a semivaluation on $\tilde{\mathcal{R}}_R^{[s,r]}$, let β be the associated seminorm and define t and γ as in Lemma 5.3.4. Since β extends uniquely to a multiplicative seminorm on $\tilde{\mathcal{R}}_{\mathcal{H}(\gamma)}^{[s,r]}$, v extends to a semivaluation on this ring. The set of $\overline{x} \in \mathcal{H}(\gamma)$ for which $v([\overline{x}]) \leqslant 1$ is a valuation ring; we thus obtain a valuation w on R.

Let $\tilde{\mathcal{R}}_R^{[s,r],+}$ be the completion with respect to $\max\{\lambda(\alpha^s), \lambda(\alpha^r)\}$ of the subring of $\tilde{\mathcal{R}}_R^{[s,r]}$ generated by those x for which $\max\{\lambda(\alpha^s), \lambda(\alpha^r)\}(x) < 1$ and $[\overline{y}]$ for $\overline{y} \in R^+$. The previous paragraph then defines a map $\mu : \mathrm{Spa}(\tilde{\mathcal{R}}_R^{[s,r]}, \tilde{\mathcal{R}}_R^{[s,r],+}) \to \mathrm{Spa}(R, R^+)$.

Lemma 5.3.11. — *For $0 < s \leqslant r$ and $(R, R^+) \to (S, S^+)$ a rational localization, the morphism $(\tilde{\mathcal{R}}_R^{[s,r]}, \tilde{\mathcal{R}}_R^{[s,r],+}) \to (\tilde{\mathcal{R}}_S^{[s,r]}, \tilde{\mathcal{R}}_S^{[s,r],+})$ is a rational localization representing the inverse image of* $\mathrm{Spa}(S, S^+)$ *under the morphism of Definition 5.3.10.*

Proof. — Represent the rational subdomain represented by $(R, R^+) \to (S, S^+)$ as

$$\{v \in \mathrm{Spa}(R, R^+) : v(\overline{f}_i) \leqslant v(\overline{g}) \quad (i = 1, \ldots, n)\}.$$

Form the rational subspace

$$U = \{v \in \mathrm{Spa}(\tilde{\mathcal{R}}_R^{[s,r]}, \tilde{\mathcal{R}}_R^{[s,r],+}) : v([\overline{f}_i]) \leqslant v([\overline{g}_i]) \quad (i = 1, \ldots, n)\}.$$

Let $\mathrm{Spa}(\tilde{\mathcal{R}}_R^{[s,r]}, \tilde{\mathcal{R}}_R^{[s,r],+}) \to (T, T^+)$ be the corresponding rational localization. The image of $\mathrm{Spa}(\tilde{\mathcal{R}}_S^{[s,r]}, \tilde{\mathcal{R}}_S^{[s,r],+}) \to \mathrm{Spa}(\tilde{\mathcal{R}}_R^{[s,r]}, \tilde{\mathcal{R}}_R^{[s,r],+})$ is equal to U, so by the universal property of rational localizations, we obtain a morphism $(T, T^+) \to (\tilde{\mathcal{R}}_S^{[s,r]}, \tilde{\mathcal{R}}_S^{[s,r],+})$ which induces a bijection of adic spectra. By Theorem 5.3.9, (T, T^+) is uniform. By Theorem 2.3.10, the morphism $T \to \tilde{\mathcal{R}}_S^{[s,r]}$ is isometric for the spectral norms; however, the image of this morphism contains the dense subring generated over $\mathbb{Q}_p$ by $[\overline{x}]$ for $\overline{x} \in S$. This yields the claim. □

Corollary 5.3.12. — *For $0 < s \leqslant r$, if $\{(R, R^+) \to (R_i, R_i^+)\}_i$ is a rational covering, then $\{(\tilde{\mathcal{R}}_R^{[s,r]}, \tilde{\mathcal{R}}_R^{[s,r],+}) \to (\tilde{\mathcal{R}}_{R_i}^{[s,r]}, \tilde{\mathcal{R}}_{R_i}^{[s,r],+})\}_i$ is also a rational covering.*

We now recover an analogue of Theorem 5.3.6.

Theorem 5.3.13. — *The sheaf $\tilde{\mathcal{R}}^{[s,r]}$ on* $\mathrm{Spa}(R, R^+)$ *satisfies the Kiehl glueing property.*

Proof. — This is immediate from Theorem 5.3.9 and Corollary 5.3.12. □

From Theorem 5.3.9, we also obtain a glueing result with respect to the interval $[s, r]$.

Lemma 5.3.14. — *Choose* $0 < s \leqslant r$. *Let* K *be a perfectoid analytic field containing* $\mathbb{Q}_p(p^{p^{-\infty}})$ *with* $|K^\times| = \mathbb{R}^+$.

(a) *The tensor product norm on* $\tilde{\mathcal{R}}_R^{[s,r]} \widehat{\otimes}_{\mathbb{Q}_p} K$ *is power-multiplicative, and* $\tilde{\mathcal{R}}_R^{[s,r]} \widehat{\otimes}_{\mathbb{Q}_p} K$ *is perfectoid.*

(b) *Choose* $\overline{z} \in R$ *with* $\alpha(\overline{z}) < 1$ *and* $\alpha(\overline{z})\alpha(\overline{z}^{-1}) = 1$ *(possible because* R *is a Banach algebra over an analytic field). Then for* $0 < s \leqslant s' \leqslant r' \leqslant r$, $(\tilde{\mathcal{R}}_R^{[s,r]}, \tilde{\mathcal{R}}_R^{[s,r],+}) \widehat{\otimes}_{\mathbb{Q}_p} K \to (\tilde{\mathcal{R}}_R^{[s',r']}, \tilde{\mathcal{R}}_R^{[s',r'],+}) \widehat{\otimes}_{\mathbb{Q}_p} K$ *is the rational localization corresponding to*

$$\{v \in \operatorname{Spa}((\tilde{\mathcal{R}}_R^{[s,r]}, \tilde{\mathcal{R}}_R^{[s,r],+}) \widehat{\otimes}_{\mathbb{Q}_p} K) : v([\overline{z}]) \in [\alpha(\overline{z})^{r'}, \alpha(\overline{z})^{s'}]\}.$$

Proof. — Part (a) follows from Theorem 5.3.9 and Proposition 3.6.11. Part (b) follows by a similar argument as in the proof of Lemma 5.3.11. □

Corollary 5.3.15. — *For* K *as in Lemma 5.3.14, for* $I, I_1, \dots, I_n$ *closed subintervals of* $(0, +\infty)$ *satisfying* $I = I_1 \cup \dots \cup I_n$,

$$\{(\tilde{\mathcal{R}}_R^{I}, \tilde{\mathcal{R}}_R^{I,+}) \widehat{\otimes}_{\mathbb{Q}_p} K \longrightarrow (\tilde{\mathcal{R}}_R^{I_i}, \tilde{\mathcal{R}}_R^{I_i,+}) \widehat{\otimes}_{\mathbb{Q}_p} K\}_{i=1}^n$$

is a rational covering.

Theorem 5.3.16. — *For* $I, I_1, \dots, I_n$ *closed subintervals of* $(0, +\infty)$ *satisfying* $I = I_1 \cup \dots \cup I_n$, *the morphism* $\tilde{\mathcal{R}}_R^I \to \oplus_i \tilde{\mathcal{R}}_R^{I_i}$ *is an effective descent morphism for the category of finite projective modules over uniform Banach rings.*

Proof. — Using Lemma 2.2.9, we can tensor over $\mathbb{Q}_p$ with a perfectoid field K containing $\mathbb{Q}_p(p^{p^{-\infty}})$ with $|K^\times| = \mathbb{R}^+$; we may then deduce the claim from Theorem 5.3.9 and Corollary 5.3.15. Alternatively, one can reduce to the case $n = 2$ and check directly that one gets a glueing square in the sense of Definition 2.7.3: condition (a) holds by Theorem 2.3.10 (for exactness at the left), Lemma 5.2.10 (for exactness at the middle), and Lemma 5.2.9 (for exactness at the right); condition (b) is straightforward; condition (c) holds by Lemma 5.3.4. □

5.4. Some geometric observations

We mention some observations concerning the geometry of the spaces $\mathcal{M}(\tilde{\mathcal{R}}_R^{[s,r]})$, in the spirit of [**87**]. These results will be used later to build relative Fargues-Fontaine curves; see §8.7.

Theorem 5.4.1. — *Define* $\lambda : \mathcal{M}(R) \to \mathcal{M}(\tilde{\mathcal{R}}_R^{\mathrm{int},1})$, $\mu : \mathcal{M}(\tilde{\mathcal{R}}_R^{\mathrm{int},1}) \to \mathcal{M}(R)$ *as in Proposition 5.1.2.*

(a) *The maps* λ *and* μ *are continuous. Moreover, the inverse image under either map of a finite union of Weierstrass (resp. Laurent, rational) subdomains has the same form.*

(b) *For all* $\beta \in \mathcal{M}(R)$, $(\mu \circ \lambda)(\beta) = \beta$.

(c) *For all* $\gamma \in \mathcal{M}(W(R))$, $(\lambda \circ \mu)(\gamma) \geqslant \gamma$.

Proof. — The proof of [**87**, Theorem 4.5] carries over without change. □

Lemma 5.4.2. — *Let* $R \to S$ *be a bounded homomorphism of perfect uniform Banach* $\mathbb{F}_p$*-algebras such that* $\mathcal{M}(S) \to \mathcal{M}(R)$ *is surjective. Then for any* $r > 0$, *the map* $\mathcal{M}(\tilde{\mathcal{R}}_S^{\mathrm{int},r}) \to \mathcal{M}(\tilde{\mathcal{R}}_R^{\mathrm{int},r})$ *is also surjective.*

Proof. — Equip $R[T]$ and $S[T]$ with the p^{-1}-Gauss norm, and let R' and S' be the completions of $R[T]^{\mathrm{perf}}$ and $S[T]^{\mathrm{perf}}$. We may then identify S' with $S\widehat{\otimes}_R R'$; since $\mathcal{M}(S) \to \mathcal{M}(R)$ is surjective, so is $\mathcal{M}(S') \to \mathcal{M}(R')$ by [**87**, Lemma 1.20].

Given $\gamma \in \mathcal{M}(\tilde{\mathcal{R}}_R^{\mathrm{int},r})$, put $\beta = \mu(\gamma)$ and $\mathfrak{o} = \mathfrak{o}_{\mathcal{H}(\beta)}$, extend γ to $\tilde{\mathcal{R}}_{\mathcal{H}(\beta)}^{\mathrm{int},r}$ by continuity, then restrict to $W(\mathfrak{o})$. Let $\mathfrak{o}'$ be the completed perfect closure of $\mathfrak{o}[T]$ for the p^{-1}-Gauss norm; as in [**87**, Definition 7.5], we may extend γ from $W(\mathfrak{o})$ to a seminorm γ' on $W(\mathfrak{o}')$ in such a way that $\gamma'(p-T) = 0$. By [**87**, Remark 5.14], this extension computes the quotient seminorm on $W(\mathfrak{o}')/(p-T)$ induced by $\lambda(\beta')$ for $\beta' = \mu(\gamma')$.

Choose $\tilde{\beta} \in \mathcal{M}(S)$ lifting β, put $\tilde{\mathfrak{o}} = \mathfrak{o}_{\mathcal{H}(\tilde{\beta})}$, and let $\tilde{\mathfrak{o}}'$ be the completed perfect closure of $\tilde{\mathfrak{o}}[T]$ for the p^{-1}-Gauss norm. We may then identify $\tilde{\mathfrak{o}}'$ with $\tilde{\mathfrak{o}}\widehat{\otimes}_{\mathfrak{o}}\mathfrak{o}'$; by [**87**, Lemma 1.20], the map $\mathcal{M}(\tilde{\mathfrak{o}}') \to \mathcal{M}(\mathfrak{o}')$ is surjective.

We can thus lift β' to a seminorm $\tilde{\beta}'$ on $\tilde{\mathfrak{o}}'$. Let $\tilde{\beta}'$ be the quotient norm on $W(\tilde{\mathfrak{o}}')/(p-T)$ induced by $\lambda(\tilde{\beta}')$, viewed as a seminorm on $W(\tilde{\mathfrak{o}}')$. We may then restrict $\tilde{\beta}'$ to a seminorm $\tilde{\beta}$ on $W(\tilde{\mathfrak{o}})$, extend multiplicatively to $\tilde{\mathcal{R}}_{\mathcal{H}(\tilde{\beta})}^{\mathrm{int},r}$, then restrict to $\tilde{\mathcal{R}}_S^{\mathrm{int},r}$. This proves the claim. □

Definition 5.4.3. — Choose $r > 0$ and $\gamma \in \tilde{\mathcal{R}}_R^{\mathrm{int},r}$. For $\beta = \mu(\gamma)^{1/r}$, we may extend γ to $\tilde{\mathcal{R}}_{\mathcal{H}(\beta)}^{\mathrm{int},r}$ and then restrict to $W(\mathfrak{o}_{\mathcal{H}(\beta)})$. We may then define the multiplicative seminorm $H(\gamma, t)$ on $W(\mathfrak{o}_{\mathcal{H}(\beta)})$ as in [**87**, Definition 7.5], extend multiplicatively to $\tilde{\mathcal{R}}_{\mathcal{H}(\beta)}^{\mathrm{int},r}$, then restrict back to $\tilde{\mathcal{R}}_R^{\mathrm{int},r}$. From [**87**, Theorem 7.8], the construction has the following properties.

(a) We have $H(\gamma, 0) = \gamma$.

(b) We have $H(\gamma, 1) = (\lambda \circ \mu)(\gamma)$.

(c) For $t \in [0, 1]$, $\mu(H(\gamma, t)) = \mu(\gamma)$.

(d) For $t, u \in [0, 1]$, $H(H(\gamma, t), u) = H(\gamma, \max\{t, u\})$.

Theorem 5.4.4. — *For any* $r > 0$, *the map* $H : \mathcal{M}(\tilde{\mathcal{R}}_R^{\mathrm{int},r}) \times [0, 1] \to \mathcal{M}(\tilde{\mathcal{R}}_R^{\mathrm{int},r})$ *is continuous.*

Proof. — Equip R with the spectral norm. For the trivial norm on $\mathfrak{o}_R$, the map $H : \mathcal{M}(W(\mathfrak{o}_R)) \times [0, 1] \to \mathcal{M}(W(\mathfrak{o}_R))$ is continuous by [**87**, Theorem 7.8]. By identifying $\mathcal{M}(\tilde{\mathcal{R}}_R^{\mathrm{int},r})$ with a closed subspace of $\mathcal{M}(W(\mathfrak{o}_R))$, we deduce the claim. □

Remark 5.4.5. — One can go further with analysis of this sort; for instance, one can show that the fibres of μ bear a strong resemblance to the spectra of one-dimensional affinoid algebras over an analytic field. See [**87**, §8].

Proposition 5.4.6. — *Define the topological space*

$$T_R = \bigcup_{0<s<r} \mathcal{M}(\tilde{\mathcal{R}}_R^{[s,r]}).$$

(a) *For each $\beta \in T_R$, there is a unique value $t \in (0,+\infty)$ for which α^t dominates $\mu(\beta)$ (or equivalently, $\lambda(\alpha^t)$ dominates β).*

(b) *Let $t : T_R \to (0,+\infty)$ be the map described in (a). Then the formula $\beta \mapsto (\mu(\beta)^{1/t(\beta)}, t(\beta))$ defines a continuous map $T_R \to \mathcal{M}(R) \times (0,+\infty)$. In particular, t is continuous.*

(c) *The group $(\varphi^*)^{\mathbb{Z}}$ acts properly discontinuously on T_R with compact quotient X_R. (Note that the map in (b) induces a continuous map $X_R \to \mathcal{M}(R) \times S^1$.)*

(d) *The map $T_R \to \mathcal{M}(R) \times (0,+\infty)$ is a strong deformation retract, and induces a strong deformation retract $X_R \to \mathcal{M}(R) \times S^1$.*

Proof. — To check (a), we appeal to Lemma 5.3.4. To check (b), choose $\beta_0 \in T_R$ and put $t_0 = t(\beta_0)$. Let U be any open neighborhood of $\mu(\beta_0)$ of the form $\{\gamma \in \mathcal{M}(R) : \gamma(\overline{f}_1) \in I_1, \dots, \gamma(\overline{f}_n) \in I_n\}$ for some $\overline{f}_1, \dots, \overline{f}_n \in R$ and some open intervals $I_1, \dots, I_n$. Let (a,b) be any open subinterval of $(0,+\infty)$ containing t_0. Choose $\overline{z} \in R$ for which $0 < \mu(\beta_0)(\overline{z}) < 1$, and put $z = [\overline{z}]$. Choose $\delta > 1$ such that $a < t_0/\delta, t_0\delta < b$. For $i = 1, \dots, n$, choose an open neighborhood J_i of $\beta_0([\overline{f}_i])$ such that for all $x \in J_i$ and all $u \in [1/\delta, \delta]$, $x^{1/(ut_0)} \in I_i$. Put

$$V = \{\gamma \in T_R : \gamma([\overline{f}_1]) \in J_1, \dots, \gamma([\overline{f}_n]) \in J_n, \gamma(z) \in (\beta_0(z)^{t_0/\delta}, \beta_0(z)^{t_0\delta})\};$$

this is an open subset of T_R with the property that for any $\gamma \in V$, $\mu(\gamma)^{1/t(\gamma)} \in U$ and $t(\gamma) \in (a,b)$. This gives the desired continuity.

To check (c), first apply (b) after observing that for all $\beta \in T_R$, $t(\varphi^*(\beta)) = pt(\beta)$. We see from this that the action is properly discontinuous, so X_R is Hausdorff. Then note that for any $r > 0$, the projection $\mathcal{M}(\tilde{\mathcal{R}}_R^{[r/p,r]}) \to X_R$ is surjective. Since X_R receives a surjective continuous map from a compact space, it is quasicompact (Remark 2.3.15 (a)) and hence compact.

To check (d), argue as in the proof of Theorem 5.4.4 to produce a continuous map $H : \mathcal{M}(\tilde{\mathcal{R}}_R^{\mathrm{int},r}) \times [0,1] \to \mathcal{M}(\tilde{\mathcal{R}}_R^{\mathrm{int},r})$. Then observe by Lemma 5.3.4 that the image of $\mathcal{M}(\tilde{\mathcal{R}}_R^{\mathrm{int},r}) \times \{1\}$ may be identified with $\mathcal{M}(R) \times (0,r]$ by mapping (γ, s) to $\lambda(\gamma^s)$, and that the resulting map $\mathcal{M}(\tilde{\mathcal{R}}_R^{\mathrm{int},r}) \to \mathcal{M}(R) \times (0,r]$ is precisely T_R. □

Remark 5.4.7. — The space X_R will later appear as the maximal Hausdorff quotient of the relative Fargues-Fontaine curve over R; see §8.7. For now, we note that when $\mathcal{M}(R)$ is contractible, Proposition 5.4.6 asserts that X_R has the homotopy type of a circle, T_R is the universal covering space of X_R, and φ^* acts on

T_R as a deck transformation generating the fundamental group. For instance, this is the case when $R = L$ is an analytic field; in this case, the profinite étale fundamental group of X_R is the product of the absolute Galois groups of $\mathbb{Q}_p$ and L (see [**126**]), but the étale fundamental group (once it is suitably defined, which we will not do here) should differ from this. In particular, the $\mathbb{Z}$-covering coming from $T_R \to X_R$ corresponds to the subgroup of the unramified Galois group of $\mathbb{Q}_p$ generated by Frobenius.

When L is a finite extension of $\mathbb{F}_p((\overline{\pi}))$, this suggests a relationship with the Weil group of the field $\tilde{\mathcal{R}}_L^{\mathrm{int},1}/(z)$ for $z = \sum_{i=0}^{p-1}[1+\overline{\pi}]^{i/p}$ (this field being a finite extension of the completion of $\mathbb{Q}_p(\mu_{p^\infty})$). However, this relationship remains to be clarified; see Remark 8.7.16 for further discussion.

5.5. Compatibility with finite étale extensions

We next establish a compatibility between the construction of extended Robba rings and formation of finite étale ring extensions. As promised earlier, this yields a variant of Faltings's almost purity theorem, thus refining the perfectoid correspondence introduced in §3.6.

Convention 5.5.1. — For $S \in \mathbf{F\acute{E}t}(R)$, we will always view S as a finite Banach R-algebra as per Proposition 2.8.16. By Lemma 3.1.9, S is then also a perfect uniform Banach $\mathbb{F}_p$-algebra.

Lemma 5.5.2. — *Let $\psi : R \to S$ be a bounded homomorphism from R to a perfect uniform $\mathbb{F}_p$-algebra S with spectral norm β. Use ψ to view S as an R-algebra. Let $x_1, \dots, x_n$ be elements of $\tilde{\mathcal{R}}_S^{\mathrm{int}}$ whose reductions $\overline{x}_1, \dots, \overline{x}_n$ modulo p generate S as an R-module. Then for all sufficiently small $r > 0$, $x_1, \dots, x_n$ generate $\tilde{\mathcal{R}}_S^{\mathrm{int},r}$ as a module over $\tilde{\mathcal{R}}_R^{\mathrm{int},r}$.*

Proof. — It is harmless to assume that $\overline{x}_1, \dots, \overline{x}_n$ are all nonzero. Since the surjection $R^n \to S$ is strict by Theorem 2.2.8, we can find $c \geqslant 1$ such that for each $\overline{z} \in S$, there exist $\overline{a}_1, \dots, \overline{a}_n \in R$ for which $\overline{z} = \sum_{i=1}^n \overline{a}_i \overline{x}_i$ and $\alpha(\overline{a}_i)\beta(\overline{x}_i) \leqslant c\beta(\overline{z})$ for $i = 1, \dots, n$.

Given $z \in \tilde{\mathcal{R}}_S^{\mathrm{int}}$, for $l = 0, 1, \dots$ we choose $z_l \in \tilde{\mathcal{R}}_S^{\mathrm{int}}$ and $a_{l,1}, \dots, a_{l,n} \in \tilde{\mathcal{R}}_R^{\mathrm{int}}$ as follows. Put $z_0 = 0$. Given z_l, let $\overline{z}_l$ be its reduction modulo p, and invoke the previous paragraph to construct $\overline{a}_{l,1}, \dots, \overline{a}_{l,n} \in R$ with $\alpha(\overline{a}_{l,i})\beta(\overline{x}_i) \leqslant c\beta(\overline{z}_l)$ for $i = 1, \dots, n$ such that $\overline{z}_l = \sum_{i=1}^n \overline{a}_{l,i}\overline{x}_i$. Then put $a_{l,i} = [\overline{a}_{l,i}]$ and $z_{l+1} = p^{-1}(z_l - \sum_{i=1}^n a_{l,i}x_i)$.

Choose $r > 0$ such that $x_1, \dots, x_n \in \tilde{\mathcal{R}}_S^{\mathrm{int},r}$ and $\lambda(\beta^r)(x_i - [\overline{x}_i]) < \beta(\overline{x}_i)^r$. For $z \in \tilde{\mathcal{R}}_S^{\mathrm{int},r}$, we then have $\lambda(\beta^r)(z_{l+1}) \leqslant c^r p \lambda(\beta^r)(z_l)$, and so $\lambda(\beta^r)(z_l) \leqslant (c^r p)^l \lambda(\beta^r)(z_0)$. In particular,

$$\alpha(\overline{a}_{l,i}) \leqslant c\beta(\overline{x}_i)^{-1}\lambda(\beta^r)(z_0)^{1/r}(cp^{1/r})^l.$$

For s sufficiently small (depending on r), we have $c^s p^{s/r-1} < 1$, and so the series $\sum_{l=0}^{\infty} p^l a_{l,i}$ converges under $\lambda(\alpha^s)$.

We now specialize the previous construction to the case $z = [\overline{z}]$. In this case, we can write $[\overline{z}] = \sum_{i=1}^{n} a_i x_i$ with $a_i \in \tilde{\mathcal{R}}_R^{\mathrm{int},s}$ and $\lambda(\alpha^s)(a_i) \leqslant c_1 \alpha(\overline{z})^s$ for some $c_1 > 0$ not depending on $\overline{z}$. By writing a general element of $\tilde{\mathcal{R}}_R^{\mathrm{int},s}$ as $z = \sum_{n=0}^{\infty} p^n [\overline{z}_n]$, we see that $x_1, \dots, x_n$ generate $\tilde{\mathcal{R}}_S^{\mathrm{int},s}$ as a module over $\tilde{\mathcal{R}}_R^{\mathrm{int},s}$. This proves the claim. □

Proposition 5.5.3. — *Choose any $r > 0$.*

(a) *The base extension functors*

$$\varphi^{-1}\text{-}\mathbf{F\acute{E}t}(\tilde{\mathcal{R}}_R^{\mathrm{int},r}) \longrightarrow \mathbf{F\acute{E}t}(\tilde{\mathcal{R}}_R^{\mathrm{int}}) \longrightarrow \mathbf{F\acute{E}t}(W(R)) \longrightarrow \mathbf{F\acute{E}t}(R)$$

are tensor equivalences, where φ^{-1}-$\mathbf{F\acute{E}t}(*)$ denote the category of finite étale $*$-algebras equipped with isomorphisms with their φ^{-1}-pullbacks.

(b) *The composition φ^{-1}-$\mathbf{F\acute{E}t}(\tilde{\mathcal{R}}_R^{\mathrm{int},r}) \to \mathbf{F\acute{E}t}(R)$ admits a quasi-inverse taking S to $\tilde{\mathcal{R}}_S^{\mathrm{int},r}$.*

Proof. — We first prove a weak version of (a): the functors $\mathbf{F\acute{E}t}(\tilde{\mathcal{R}}_R^{\mathrm{int}}) \to \mathbf{F\acute{E}t}(W(R)) \to \mathbf{F\acute{E}t}(R)$ are tensor equivalences. For each $r > 0$, $\tilde{\mathcal{R}}_R^{\mathrm{int},r}$ is complete with respect to the maximum of $\lambda(\alpha^r)$ and the p-adic norm. For these norms, the maps $\tilde{\mathcal{R}}_R^{\mathrm{int},r} \to \tilde{\mathcal{R}}_R^{\mathrm{int},s}$ for $0 < s \leqslant r$ are submetric by Lemma 5.2.1. Consequently, Lemma 2.2.3 (b) implies that the pair $(\tilde{\mathcal{R}}_R^{\mathrm{int}}, (p))$ is henselian. We may thus conclude using Theorem 1.2.8.

We next prove a weak version of (b): for $S \in \mathbf{F\acute{E}t}(R)$, the corresponding element U_S of $\mathbf{F\acute{E}t}(\tilde{\mathcal{R}}_R^{\mathrm{int}})$ may be identified with $\tilde{\mathcal{R}}_S^{\mathrm{int}}$. We may identify $U_S/(p)$ and $\tilde{\mathcal{R}}_S^{\mathrm{int}}/(p)$ with S; the p-adic completions of U_S and $\tilde{\mathcal{R}}_S^{\mathrm{int}}$ may then be identified with $W(S)$ by the uniqueness property of the latter.

Let $\pi_1, \pi_2 : S \to S \otimes_R S$ denote the structure morphisms. Put $V = U_S \otimes_{\tilde{\mathcal{R}}_R^{\mathrm{int}}} \tilde{\mathcal{R}}_S^{\mathrm{int}}$, and let $\tilde{\pi}_1 : U_S \to V$ and $\tilde{\pi}_2 : \tilde{\mathcal{R}}_S^{\mathrm{int}} \to V$ denote the structure morphisms. Note that $\tilde{\pi}_2$ is the distinguished lift of π_2 from $\mathbf{F\acute{E}t}(S)$ to $\mathbf{F\acute{E}t}(\tilde{\mathcal{R}}_S^{\mathrm{int}})$ constructed above. Consequently, if we view the multiplication map $\mu : S \otimes_R S \to S$ as a map in $\mathbf{F\acute{E}t}(S)$ by equipping $S \otimes_R S$ with the structure morphism π_2, then (a) provides a lift $\tilde{\mu}$ of μ to $\mathbf{F\acute{E}t}(\tilde{\mathcal{R}}_S^{\mathrm{int}})$. The composition $\psi = \tilde{\mu} \circ \tilde{\pi}_1 : U_S \to \tilde{\mathcal{R}}_S^{\mathrm{int}}$ lifts the identity map modulo p. As noted above, the injection $U_S \to W(S)$ factors through ψ, so ψ is injective; since ψ is $\tilde{\mathcal{R}}_R^{\mathrm{int}}$-linear, it is also surjective by Lemma 5.5.2 and Convention 5.5.1. This proves the claim.

We next verify that for $S \in \mathbf{F\acute{E}t}(R)$, we have $\tilde{\mathcal{R}}_S^{\mathrm{int},r} \in \mathbf{F\acute{E}t}(\tilde{\mathcal{R}}_R^{\mathrm{int},r})$. We first observe that $\tilde{\mathcal{R}}_S^{\mathrm{int},r}$ is finitely generated as a module over $\mathbf{F\acute{E}t}(\tilde{\mathcal{R}}_R^{\mathrm{int},r})$: this holds for small $r > 0$ by Lemma 5.5.2, and hence for all $r > 0$ by repeated application of φ^{-1}. We next note that given any $\tilde{\mathcal{R}}_R^{\mathrm{int},r}$-linear surjection of a finite free module onto $\tilde{\mathcal{R}}_S^{\mathrm{int},r}$, using the fact that $\tilde{\mathcal{R}}_S^{\mathrm{int}} \in \mathbf{F\acute{E}t}(\tilde{\mathcal{R}}_R^{\mathrm{int}})$ we can find a $\tilde{\mathcal{R}}_R^{\mathrm{int},s}$-linear splitting for some

$0 < s \leqslant r$. It follows that $\tilde{\mathcal{R}}_S^{\mathrm{int},r}$ is finite projective as a module over $\tilde{\mathcal{R}}_R^{\mathrm{int},r}$ for $r > 0$ small, and hence again for all $r > 0$ using φ^{-1}. To conclude, it remains to check that the natural map

$$\tilde{\mathcal{R}}_S^{\mathrm{int},r} \otimes_{\tilde{\mathcal{R}}_R^{\mathrm{int},r}} \tilde{\mathcal{R}}_S^{\mathrm{int},r} \longrightarrow \tilde{\mathcal{R}}_{S\otimes_R S}^{\mathrm{int},r}$$

is an isomorphism (as then we may conclude that $\tilde{\mathcal{R}}_S^{\mathrm{int},r}$ is also finite projective over $\tilde{\mathcal{R}}_S^{\mathrm{int},r} \otimes_{\tilde{\mathcal{R}}_R^{\mathrm{int},r}} \tilde{\mathcal{R}}_S^{\mathrm{int},r}$); this follows from it being a map with dense image between finite projective $\tilde{\mathcal{R}}_R^{\mathrm{int},r}$-modules of the same rank.

We now note that φ^{-1}- $\mathbf{F\acute{E}t}(\tilde{\mathcal{R}}_R^{\mathrm{int},r}) \to \mathbf{F\acute{E}t}(\tilde{\mathcal{R}}_R^{\mathrm{int}})$ is fully faithful (because any morphism in $\mathbf{F\acute{E}t}(\tilde{\mathcal{R}}_R^{\mathrm{int}})$ is automatically φ-equivariant by virtue of the equivalence with $\mathbf{F\acute{E}t}(R)$) and has a right quasi-inverse (by the previous paragraph). This completes the proof of both (a) and (b). □

Proposition 5.5.4. — *Let S be a (faithfully) finite étale R-algebra. Then for*

$$* \in \{\tilde{\mathcal{E}}, \tilde{\mathcal{R}}^{\mathrm{int},r}, \tilde{\mathcal{R}}^{\mathrm{int}}, \tilde{\mathcal{R}}^{\mathrm{bd},r}, \tilde{\mathcal{R}}^{\mathrm{bd}}, \tilde{\mathcal{R}}^{[s,r]}, \tilde{\mathcal{R}}^{r}, \tilde{\mathcal{R}}\},$$

the natural homomorphism $*_R \to *_S$ is (faithfully) finite étale.

Proof. — The cases $* = \tilde{\mathcal{E}}, \tilde{\mathcal{R}}^{\mathrm{int},r}, \tilde{\mathcal{R}}^{\mathrm{int}}, \tilde{\mathcal{R}}^{\mathrm{bd}}, \tilde{\mathcal{R}}^{\mathrm{bd}}$ follow at once from Proposition 5.5.3. To handle the cases $* = \tilde{\mathcal{R}}^{[s,r]}, \tilde{\mathcal{R}}^{r}, \tilde{\mathcal{R}}$, it suffices to check that the natural map $\tilde{\mathcal{R}}_R^r \otimes_{\tilde{\mathcal{R}}_R^{\mathrm{bd},r}} \tilde{\mathcal{R}}_S^{\mathrm{bd},r} \to \tilde{\mathcal{R}}_S^r$ is an isometric isomorphism with respect to $\lambda(\alpha^r)$. We proceed by first noticing that $\tilde{\mathcal{R}}_R^r \otimes_{\tilde{\mathcal{R}}_R^{\mathrm{bd},r}} \tilde{\mathcal{R}}_S^{\mathrm{bd},r} = \tilde{\mathcal{R}}_R^r \widehat{\otimes}_{\tilde{\mathcal{R}}_R^{\mathrm{bd},r}} \tilde{\mathcal{R}}_S^{\mathrm{bd},r}$ because $\tilde{\mathcal{R}}_S^{\mathrm{bd},r}$ is a finite projective $\tilde{\mathcal{R}}_R^{\mathrm{bd},r}$-module (see Definition 1.2.1). We may then argue as in Lemma 5.5.2 that $\tilde{\mathcal{R}}_R^r \otimes_{\tilde{\mathcal{R}}_R^{\mathrm{bd},r}} \tilde{\mathcal{R}}_S^{\mathrm{bd},r} \to \tilde{\mathcal{R}}_S^r$ is a strict surjection for sufficiently small $r > 0$, and hence for all $r > 0$ by applying φ^{-1} as needed. In particular, there exists $c > 0$ (depending on r) for which any element $z \in \tilde{\mathcal{R}}_S^r$ can be lifted to $\sum_i x_i \otimes y_i \in \tilde{\mathcal{R}}_R^r \otimes_{\tilde{\mathcal{R}}_R^{\mathrm{bd},r}} \tilde{\mathcal{R}}_S^{\mathrm{bd},r}$ with $\max_i\{\lambda(\alpha^r)(x_i y_i)\} \leqslant c\lambda(\alpha^r)(z)$. Finally, given a nonzero element $z \in \tilde{\mathcal{R}}_S^r$, choose $z_0 \in \tilde{\mathcal{R}}_S^{\mathrm{bd},r}$ with $\lambda(\alpha^r)(z - z_0) < c^{-1}\lambda(\alpha^r)(z)$, and lift $z - z_0$ to $\sum_i x_i \otimes y_i$ with $\max_i\{\lambda(\alpha^r)(x_i y_i)\} \leqslant c\lambda(\alpha^r)(z - z_0)$. The representation $1 \otimes z_0 + \sum_i x_i \otimes y_i$ of z then shows that the map $\tilde{\mathcal{R}}_R^r \otimes_{\tilde{\mathcal{R}}_R^{\mathrm{bd},r}} \tilde{\mathcal{R}}_S^{\mathrm{bd},r} \to \tilde{\mathcal{R}}_S^r$, which is evidently submetric, is in fact isometric. In particular, it is injective, completing the argument. □

To link these results to almost purity, we use the following extension of Lemma 3.3.6.

Lemma 5.5.5. — *Suppose that $z \in W(R^+)$ is primitive of degree 1. Choose any $r \geqslant 1$.*

(a) *Any $x \in \tilde{\mathcal{R}}_R^{\mathrm{int},r}/(z)$ lifts to $y = \sum_{i=0}^{\infty} p^i[\overline{y}_i] \in W(R)$ with $\alpha(\overline{y}_0) \geqslant \alpha(\overline{y}_i)$ for all i.*

(b) *For any closed interval I of $(0, +\infty)$ containing 1 and any positive integer m, each of the arrows in the diagram*

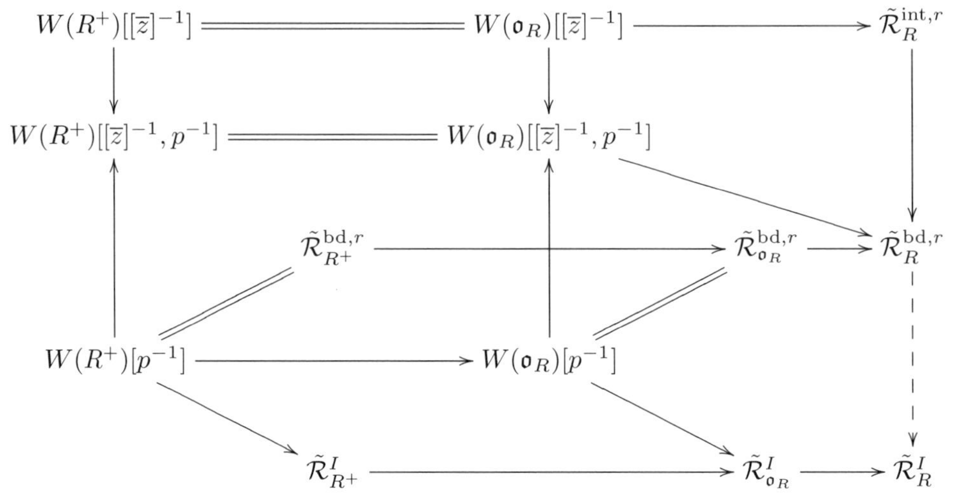

(including $\tilde{\mathcal{R}}_R^{\mathrm{bd},r} \to \tilde{\mathcal{R}}_R^I$ only if $I \subseteq (0, r]$) becomes an isomorphism upon quotienting by the ideal (z^m).

Proof. — By hypothesis, we have $p = y(z - [\overline{z}])$ for some unit $y \in W(R^+)$. Given $x = \sum_{i=0}^{\infty} p^i [\overline{x}_i] \in \tilde{\mathcal{R}}_R^{\mathrm{int},r}$, the series $\sum_{i=0}^{\infty} (-y)^i [\overline{x}_i \overline{z}^i]$ converges to an element of $W(\mathfrak{o}_R)[[\overline{z}]^{-1}]$ congruent to x modulo z. This proves surjectivity of the map

$$W(\mathfrak{o}_R)[[\overline{z}]^{-1}]/(z) \longrightarrow \tilde{\mathcal{R}}_R^{\mathrm{int},r}/(z);$$

the existence of lifts as in (a) follows by applying Lemma 3.3.6 to $x[\overline{z}]^n$ for a large positive integer n.

To prove (b), note that since z is not a zero-divisor in any of the rings appearing in the diagram (by consideration of Newton polygons as in Definition 4.2.8), the claim reduces at once to the case $m = 1$. Since $y \in W(R^+)$, when working modulo z, inverting $[\overline{z}]$ is equivalent to inverting p. Consequently, all of the vertical arrows in the first and third column induce isomorphisms modulo z, as does the arrow $\tilde{\mathcal{R}}_R^{\mathrm{int},r} \to \tilde{\mathcal{R}}_R^{\mathrm{bd},r}$.

We next check that $W(R^+)[p^{-1}] \to \tilde{\mathcal{R}}_{R^+}^I$ induces a surjection modulo z; this will imply the same for $W(\mathfrak{o}_R)[p^{-1}] \to \tilde{\mathcal{R}}_{\mathfrak{o}_R}^I$. Any element of $\tilde{\mathcal{R}}_{R^+}^I$ can be written as a convergent sum $\sum_{n=0}^{\infty} [\overline{x}_n] p^{-i_n}$ for some $\overline{x}_n \in R^+$ and $i_n \in \mathbb{Z}$. Convergence with respect to $\lambda(\alpha)$ means that there exists some $n_0 \geqslant 0$ such that for all $n \geqslant n_0$, $\alpha(\overline{x}_n) p^{i_n} < 1$. Let S be the set of integers $n \geqslant n_0$ for which $i_n > 0$, and let T be the complement of S in $\{0, 1, \dots\}$. We may thus represent the same class in $\tilde{\mathcal{R}}_{R^+}^I/(z)$ by the sum

$$\sum_{n \in S} (-y)^{-i_n} [\overline{x}_n \overline{z}^{-i_n}] + \sum_{n \in T} x_n p^{-i_n}$$

which converges in $W(R^+)[p^{-1}]$ with respect to $\lambda(\alpha)$.

We next check that $\tilde{\mathcal{R}}^I_{R^+} \to \tilde{\mathcal{R}}^I_R$ induces a surjection modulo z; this will imply the same for $\tilde{\mathcal{R}}^I_{\mathfrak{o}_R} \to \tilde{\mathcal{R}}^I_R$. By Lemma 5.2.9, for t the right endpoint of I, any class in $\tilde{\mathcal{R}}^I_{R^+}/(z)$ can be represented as the sum of a class arising from $\tilde{\mathcal{R}}^I_{R^+}/(z)$ and a class arising from $\tilde{\mathcal{R}}^{\mathrm{bd},t}_R/(z)$. By what we have already shown, the latter class can also be found in $W(R^+)[p^{-1}]/(z)$ and hence in $\tilde{\mathcal{R}}^I_{R^+}/(z)$. This proves the claim.

To conclude, it suffices to check that $W(R^+)[p^{-1}] \to \tilde{\mathcal{R}}^I_R$ induces an injection modulo z (as this will formally imply the same for $W(R^+)[p^{-1}] \to \tilde{\mathcal{R}}^I_{R^+}$, and similarly with R^+ replaced by $\mathfrak{o}_R$). Choose $x \in W(R^+)[p^{-1}]$ which maps to zero in $\tilde{\mathcal{R}}^I_R/(z)$. We wish to check that x is divisible by z in $W(R^+)[p^{-1}]$. For this purpose, there is no harm to assume that $x \in W(R^+)[p^{-1}]$ or to modify x by a multiple of z; by (a), we may thus assume that $x = \sum_{n=0}^{\infty}[\overline{x}_n]p^n$ with $\overline{\alpha}(\overline{y}_0) \geqslant \overline{\alpha}(\overline{x}_n)$ for all $n \geqslant 0$. If x is nonzero, we can choose $\beta \in \mathcal{M}(R)$ such that $\beta(\overline{y}_0) > p^{-1}\alpha(\overline{y}_0)$; then the divisibility of x by z in $\tilde{\mathcal{R}}^I_{\mathcal{H}(\beta)}$ yields a contradiction by consideration of Newton polygons. □

Corollary 5.5.6. — *For any $z \in W(\mathfrak{o}_R)$ primitive of degree 1 and any $r \geqslant 1$, for $A = W(\mathfrak{o}_R)[[\overline{z}]^{-1}]/(z) = \tilde{\mathcal{R}}^{\mathrm{int},r}_R/(z)$ (by Lemma 5.5.5), we have a 2-commuting diagram*

(5.5.6.1)
$$\begin{array}{ccc} \varphi^{-1}\text{-}\mathbf{F\acute{E}t}(\tilde{\mathcal{R}}^{\mathrm{int},r}_R) & \longrightarrow & \mathbf{F\acute{E}t}(A) \\ \downarrow & \swarrow & \\ \mathbf{F\acute{E}t}(R) & & \end{array}$$

in which the solid arrows are base extensions and the dashed arrow is the one provided by Theorem 3.6.21. In particular, each of these is a tensor equivalence.

Proof. — The commutativity comes from Lemma 5.5.5. The dashed arrow is a tensor equivalence by Theorem 3.6.21, while the vertical arrow is an equivalence by Proposition 5.5.3. □

Remark 5.5.7. — Corollary 5.5.6 makes it possible to study the effect of Frobenius on the perfectoid correspondence, leading to the almost purity theorem (Theorem 5.5.9). Arthur Ogus has asked about an alternate approach to Theorem 3.6.21 obtained by directly establishing essential surjectivity of $\mathbf{F\acute{E}t}(\tilde{\mathcal{R}}^{\mathrm{int},r}_R) \to \mathbf{F\acute{E}t}(A)$ using lifting arguments for smooth algebras, as in the work of Elkik [**40**] and Arabia [**3**]; however, it is not immediately clear how to obtain φ^{-1}-equivariance in such an approach.

To assert an almost purity theorem, we need a few definitions from almost ring theory; for these we follow [**52**].

Definition 5.5.8. — Let A be a uniform Banach algebra equipped with its spectral norm. Suppose that for each $\epsilon > 1$, there exists $\lambda \in A$ with $1 < |\lambda| < \epsilon$ and $|\lambda|\,|\lambda^{-1}| = 1$. For instance, this holds if A is a Banach algebra over an analytic field with nondiscrete norm.

An $\mathfrak{o}_A$-module is *almost zero* if it is killed by $\mathfrak{m}_A$. The category of *almost modules* over $\mathfrak{o}_A$ is the localization of the category of $\mathfrak{o}_A$-modules at the class of morphisms with almost zero kernel and cokernel.

An A-module B is *almost finite projective* if for each $t \in \mathfrak{m}_A$, there exist a finite free A-module F and some morphisms $B \to F \to B$ of A-modules whose composition is multiplication by t. We say that B is *uniformly almost finite projective* if there is a positive integer m so that we can always choose F to be free of rank m. (See [**52**, Lemma 2.4.15] for some equivalent formulations.)

For B an A-algebra whose underlying A-module is almost finite projective, there is a well-defined trace map $\mathrm{Trace} : B \to A$ in the category of almost modules over $\mathfrak{o}_A$ [**52**, §4.1.7]; we say that B is *almost finite étale* if the trace pairing induces an almost isomorphism $B \to \mathrm{Hom}_A(B, A)$. This is not the definition used in [**52**], but is equivalent to it *via* [**52**, Theorem 4.1.14].

We are now ready to fulfill the promise made in Remark 3.6.24. The key new ingredient provided by relative Robba rings is an action of Frobenius (or more precisely its inverse) in characteristic 0, which can be used in much the same way that Frobenius can be used to give a cheap proof of almost purity in positive characteristic (see Remark 3.1.11 and then [**52**, Chapter 3]).

***Theorem 5.5.9* (Almost purity).** — *Let A be a perfectoid algebra. Then for any $B \in$ $\mathbf{F\acute{E}t}(A)$, $\mathfrak{o}_B$ is uniformly almost finite projective and almost étale over $\mathfrak{o}_A$.*

Proof. — By Theorem 3.6.21, B is also perfectoid. Equip A and B with their spectral norms. Apply Lemma 3.6.3 to construct $z \in W(\mathfrak{o}_A^{\mathrm{frep}})$ primitive of degree 1 generating the kernel of $\theta : W(\mathfrak{o}_A^{\mathrm{frep}}) \to \mathfrak{o}_A$. For n a nonnegative integer, put $y_n = [\overline{z}^{-p^{-n}}]z$. Let R, S correspond to A, B *via* Theorem 3.6.5, so that $S \in \mathbf{F\acute{E}t}(A)$ by Theorem 3.6.21 again. Write $\mathfrak{o}_*^r$ as shorthand for $\mathfrak{o}_{\tilde{\mathcal{R}}_*^{\mathrm{int},r}}$; by Lemma 5.5.5, for each nonnegative integer n, we have isomorphisms $\mathfrak{o}_R^{p^n}/y_n\mathfrak{o}_R^{p^n} \cong \mathfrak{o}_A$, $\mathfrak{o}_S^{p^n}/y_n\mathfrak{o}_S^{p^n} \cong \mathfrak{o}_B$.

By Proposition 5.5.3, $\tilde{\mathcal{R}}_S^{\mathrm{int},1}$ is the object of φ^{-1}-$\mathbf{F\acute{E}t}(\tilde{\mathcal{R}}_R^{\mathrm{int},1})$ corresponding to S. In particular, $\tilde{\mathcal{R}}_S^{\mathrm{int},1}$ is a finite projective $\tilde{\mathcal{R}}_R^{\mathrm{int},1}$-module, so for some positive integer m there exist morphisms $\tilde{\mathcal{R}}_S^{\mathrm{int},1} \to (\tilde{\mathcal{R}}_R^{\mathrm{int},1})^m \to \tilde{\mathcal{R}}_S^{\mathrm{int},1}$ of $\tilde{\mathcal{R}}_R^{\mathrm{int},1}$-modules whose composition is the identity. For a suitable $\lambda \in \mathfrak{o}_R$ as in Definition 5.5.8, we may multiply through to obtain morphisms $\mathfrak{o}_S^1 \to (\mathfrak{o}_R^1)^m \to \mathfrak{o}_S^1$ of $\mathfrak{o}_R^1$-modules whose composition is multiplication by $[\lambda]$. By applying φ^{-n} and then quotienting by y_n, we obtain morphisms $\mathfrak{o}_B \to \mathfrak{o}_A^m \to \mathfrak{o}_B$ of $\mathfrak{o}_A$-modules whose composition is multiplication by $\theta([\lambda^{p^{-n}}])$. Since the norm of $\theta([\lambda^{p^{-n}}])$ tends to 1 as $n \to \infty$, it follows that $\mathfrak{o}_B$ is uniformly almost finite projective over $\mathfrak{o}_A$. Similarly, starting from the perfectness of the trace pairing on $\tilde{\mathcal{R}}_S^{\mathrm{int},1}$ over $\tilde{\mathcal{R}}_R^{\mathrm{int},1}$, then applying φ^{-n}, and finally quotienting by y_n, we deduce that the trace pairing defines an almost isomorphism $\mathfrak{o}_B \to \mathrm{Hom}_{\mathfrak{o}_A}(\mathfrak{o}_B, \mathfrak{o}_A)$. $\square$

Remark 5.5.10. — The original almost purity theorem of Faltings [**42, 43**] differs a bit in form from Theorem 5.5.9, in that it refers to a specific construction to pass from a suitable affinoid algebra over a complete discretely valued field of mixed characteristics to a perfectoid algebra. We will encounter this construction, which uses toric local coordinates, in a subsequent paper.

After Faltings introduced the concept of almost purity, and the broader context of almost ring theory, an abstract framework for such results has been introduced by Gabber and Ramero [**52**], and used by them to establish certain generalizations of Faltings's almost purity theorem [**53**].

Theorem 5.5.9 appears to be much stronger than the main result of [**53**], and the proof is simpler. In place of some complicated analysis in the style of Grothendieck's proof of Zariski-Nagata purity (as in the original work of Faltings), the proof of Theorem 5.5.9 ultimately rests on the local nature of the perfectoid correspondence.

An independent derivation of Theorem 5.5.9, based on the same set of ideas, has been given by Scholze [**114**]. Scholze goes further, extending the perfectoid correspondence and the almost purity theorem to a certain class of adic analytic spaces; he then uses these to establish relative versions of the de Rham-étale comparison isomorphism in p-adic Hodge theory [**115**]. We will make contact with the latter results later in this series.

CHAPTER 6

φ-MODULES

We now introduce φ-modules over the rings introduced in §5. In order to avoid some headaches later when working in the relative setting, we expend some energy here to relate φ-modules to more geometrically defined concepts, including a relative analogue of the vector bundles considered by Fargues and Fontaine [**45**].

Hypothesis 6.0.1. — Throughout §6, continue to retain Hypothesis 5.0.1. In addition, let a denote a positive integer, and put $q = p^a$.

6.1. φ-modules and φ-bundles

Definition 6.1.1. — A *φ^a-module* over $W(R)$ (resp. $\tilde{\mathcal{E}}_R$, $\tilde{\mathcal{R}}_R^{\mathrm{int}}$, $\tilde{\mathcal{R}}_R^{\mathrm{bd}}$, $\tilde{\mathcal{R}}_R$, $\tilde{\mathcal{R}}_R^{+}$, $\tilde{\mathcal{R}}_R^{\infty}$) is a finite locally free module M equipped with a semilinear φ^a-action. For example, one may take the direct sum of one or more copies of the base ring and use the action of φ^a on the ring; any such φ^a-module is said to be *trivial.*

For $* \in \{\tilde{\mathcal{R}}^{\mathrm{int}}, \tilde{\mathcal{R}}^{\mathrm{bd}}, \tilde{\mathcal{R}}\}$ and $r > 0$, note that any φ^a-module over $*_R$ descends uniquely to a finite locally free module M_r over $*_R^r$ equipped with an isomorphism $(\varphi^a)^* M_r \cong M_r \otimes_{*_R^r} *_R^{r/q}$ of modules over $*_R^{r/q}$. (The argument is by applying φ^{-1} as in the proof of Proposition 5.5.3.) We call M_r the *model* of M over $*_R^r$.

Unfortunately, it is not straightforward to deal with φ^a-modules over $\tilde{\mathcal{R}}_R$ because of the complicated nature of the base ring. We are thus forced to introduce an auxiliary definition with a more geometric flavor.

Definition 6.1.2. — For $0 < s \leqslant r/q$, a *φ^a-module* over $\tilde{\mathcal{R}}_R^{[s,r]}$ is a finite locally free module M equipped with an isomorphism $(\varphi^a)^* M \otimes_{\tilde{\mathcal{R}}_R^{[s/q,r/q]}} \tilde{\mathcal{R}}_R^{[s,r/q]} \cong M \otimes_{\tilde{\mathcal{R}}_R^{[s,r]}} \tilde{\mathcal{R}}_R^{[s,r/q]}$ of modules over $\tilde{\mathcal{R}}_R^{[s,r/q]}$. A *$\varphi^a$-bundle* over $\tilde{\mathcal{R}}_R$ consists of a φ^a-module M_I over $\tilde{\mathcal{R}}_R^I$ for every interval $I = [s,r]$ with $0 < s \leqslant r/q$, together with isomorphisms $\psi_{I,I'} : M_I \otimes_{\tilde{\mathcal{R}}_R^I} \tilde{\mathcal{R}}_R^{I'} \cong M_{I'}$ for every pair of intervals I, I' with $I' \subseteq I$, satisfying the cocycle condition $\psi_{I',I''} \circ \psi_{I,I'} = \psi_{I,I''}$. We refer to M_I as the *model* of the φ^a-bundle over $\tilde{\mathcal{R}}_R^I$; we may freely pass between φ^a-modules and models using

Lemma 6.1.5 below. We define base extensions and exact sequences of φ^a-modules in terms of models.

For $M = \{M_I\}$ a φ^a-bundle over $\tilde{\mathcal{R}}_R$ and any interval $I' = [s,r]$ with $0 < s < r < +\infty$ (not necessarily satisfying $s \leqslant r/q$), define $M_{I'} = M_I \otimes_{\tilde{\mathcal{R}}^I_R} \tilde{\mathcal{R}}^{I'}_R$ where M_I is a model of M with $I' \subseteq I$; this is independent of the choice of M_I. Moreover, it is clear that one gets the isomorphisms $\psi_{I,I'} : M_I \otimes_{\tilde{\mathcal{R}}^I_R} \tilde{\mathcal{R}}^{I'}_R \cong M_{I'}$ for every pair of intervals I, I' with $I' \subseteq I$, satisfying the cocycle condition, and the isomorphisms between $(\varphi^a)^* M_I$ and $M_{I/q}$ which commute with the $\psi_{I,I'}$. A *global section* of M consists of an element $\mathbf{v}_I \in M_I$ for each I such that $\psi_{I,I'}(\mathbf{v}_I) = \mathbf{v}_{I'}$; note that φ^a acts on the module of global sections.

Remark 6.1.3. — The kernel of a surjective morphism of finite projective modules over a ring is itself a finite projective module. Consequently, the kernel of a surjective morphism of φ^a-modules or φ^a-bundles (where surjectivity in the latter case means that each map of models is surjective) is again a φ^a-module or φ^a-bundle, respectively.

The following lemma makes it unambiguous to say that a φ^a-bundle is generated by a given finite set of global sections. The proof is loosely modeled on that of [**91**, Satz 2.4].

Lemma 6.1.4. — *Let $M = \{M_I\}$ be a φ^a-bundle over $\tilde{\mathcal{R}}_R$. Suppose that $\mathbf{v}_1, \dots, \mathbf{v}_n$ are global sections of M which generate M_I as a module over $\tilde{\mathcal{R}}^I_R$ for every closed interval $I \subset (0,+\infty)$. Then $\mathbf{v}_1, \dots, \mathbf{v}_n$ also generate the set of global sections of M as a module over $\tilde{\mathcal{R}}^\infty_R$.*

Proof. — For each nonnegative integer l, choose a morphism $\psi_l : M_{[p^{-l},p^l]} \to (\tilde{\mathcal{R}}^{[p^{-l},p^l]}_R)^n$ of $\tilde{\mathcal{R}}^{[p^{-l},p^l]}_R$-modules whose composition with the map $(\tilde{\mathcal{R}}^{[p^{-l},p^l]}_R)^n \to M_{[p^{-l},p^l]}$ defined by $\mathbf{v}_1, \dots, \mathbf{v}_n$ is the identity. By Lemma 2.2.12, we can choose $c_l > 0$ such that the subspace norm on $M_{[p^{-l},p^l]}$ defined by ψ_{l+1} (or rather its base extension from $\tilde{\mathcal{R}}^{[p^{-l+1},p^{l+1}]}_R$ to $\tilde{\mathcal{R}}^{[p^{-l},p^l]}_R$) and the quotient norm on $M_{[p^{-l},p^l]}$ differ by a multiplicative factor of at most c_l.

Given a global section $\mathbf{w}$ of M, we choose elements $a_{il} \in \tilde{\mathcal{R}}^{[p^{-l},p^l]}_R$, $b_{il} \in \tilde{\mathcal{R}}^\infty_R$ for $i = 1, \dots, n$, $l = 0, 1, \dots$ as follows.

- Given the b_{ij} for $j < l$, use ψ_l to construct a_{il} so that $\mathbf{w} - \sum_i \sum_{j<l} b_{ij}\mathbf{v}_i = \sum_i a_{il}\mathbf{v}_i$.
- Given the a_{il}, choose the b_{il} so that $\lambda(\alpha^t)(b_{il} - a_{il}) \leqslant p^{-1}c_l^{-1}\lambda(\alpha^t)(a_{il})$ for $i = 1, \dots, n$ and $t \in [p^{-l}, p^l]$.

Note that for $t \in [p^{-l}, p^l]$, we have $\max_i\{\lambda(\alpha^t)(a_{i(l+1)})\} \leqslant p^{-1}\max_i\{\lambda(\alpha^t)(a_{il})\}$. Consequently, the series $\sum_l a_{il}$ converges to a limit $a_i \in \tilde{\mathcal{R}}^\infty_R$ satisfying $\mathbf{w} = \sum_i a_i\mathbf{v}_i$; this proves the claim. □

Lemma 6.1.5. — *For $0 < s \leqslant r/q$, the projection functor from φ^a-bundles over $\tilde{\mathcal{R}}_R$ to φ^a-modules over $\tilde{\mathcal{R}}^{[s,r]}_R$ is a tensor equivalence.*

Proof. — For each nonnegative integer n, we may uniquely lift a φ^a-module over $\tilde{\mathcal{R}}_R^{[s,r]}$ to $\tilde{\mathcal{R}}_R^{[sq^{-n},rq^n]}$ by pulling back along positive and negative powers of φ^a, then glueing using Theorem 5.3.16. The claim follows at once. □

Remark 6.1.6. — There is a natural functor from φ^a-modules over $\tilde{\mathcal{R}}_R$ to φ^a-bundles over $\tilde{\mathcal{R}}_R$: given a φ^a-module over $\tilde{\mathcal{R}}_R$, form its model M_r over $\tilde{\mathcal{R}}_R^r$ and then base extend to $\tilde{\mathcal{R}}_R^{[s,r]}$ to obtain a φ^a-module over $\tilde{\mathcal{R}}_R^{[s,r]}$. We may recover M_r as the set of $(0,r]$-sections of the resulting φ^a-bundle, so this functor is fully faithful. It also turns out to be essentially surjective; see Theorem 6.3.12 below.

6.2. Construction of φ-invariants

We next introduce some calculations that allow us to construct φ^a-invariants. These are relative analogues of results from [**83**, §4] which were used as part of the construction of slope filtrations.

Definition 6.2.1. — For M a φ^a-module or φ^a-bundle and $n \in \mathbb{Z}$, define the twist $M(n)$ of M to be the same underlying module or bundle with the φ^a-action multiplied by p^{-n}.

Proposition 6.2.2. — *Let $M = \{M_I\}$ be a φ^a-bundle over $\tilde{\mathcal{R}}_R$. Then there exists an integer N such that for $n \geqslant N$ and $0 < s \leqslant r$, the map $\varphi^a - 1 : M_{[s,rq]}(n) \to M_{[s,r]}(n)$ is surjective. Moreover, if M arises from a φ^a-module over $\tilde{\mathcal{R}}_R^{\mathrm{int}}$, we may take $N = 1$.*

Proof. — We first assume that $r/s \leqslant q^{1/2}$; by applying a suitable power of φ^a as needed, we may also reduce to the case where $r \in [1,q]$. Choose module generators $\mathbf{v}_1, \dots, \mathbf{v}_m$ of $M_{[s/q,rq]}$ and representations $\varphi^{-a}(\mathbf{v}_j) = \sum_i A_{ij}\mathbf{v}_i$, $\varphi^a(\mathbf{v}_j) = \sum_i B_{ij}\mathbf{v}_i$ with $A_{ij} \in \tilde{\mathcal{R}}_R^{[s,rq]}$, $B_{ij} \in \tilde{\mathcal{R}}_R^{[s/q,r]}$. Put

$$c_1 = \sup\{\lambda(\alpha^t)(A) : t \in [s,rq]\}, \qquad c_2 = \sup\{\lambda(\alpha^t)(B) : t \in [s/q,r]\}.$$

We take N large enough so that

$$p^{-N}c_1 < 1, \qquad p^{N(1-q^{1/2})}c_1^{q^{1/2}}c_2 < 1; \tag{6.2.2.1}$$

note that we may take $N = 1$ if M arises from a φ^a-module over $\tilde{\mathcal{R}}_R^{\mathrm{int}}$, as then we can ensure that $c_1 = c_2 = 1$. The choice of N ensures that for $n \geqslant N$, we can choose $c \in (0,1)$ so that

$$\epsilon = \max\{p^{-n}c_1c^{-(q-1)r/q}, p^n c_2 c^{(q-1)s}\} < 1. \tag{6.2.2.2}$$

Given $(x_1, \dots, x_m) \in (\tilde{\mathcal{R}}_R^{[s,r]})^m$, apply Lemma 5.2.9 to write $x_i = y_i + z_i$ with $y_i \in \tilde{\mathcal{R}}_R^{[s/q,r]}$, $z_i \in \tilde{\mathcal{R}}_R^{[s,rq]}$ such that for $t \in [s,r]$,

$$\begin{aligned}\lambda(\alpha^t)(y_i), \lambda(\alpha^t)(z_i) &\leqslant \lambda(\alpha^t)(x_i),\\ \lambda(\alpha^t)(\varphi^{-a}(y_i)) &\leqslant c^{-(q-1)t/q}\lambda(\alpha^t)(x_i),\\ \lambda(\alpha^t)(\varphi^a(z_i)) &\leqslant c^{(q-1)t}\lambda(\alpha^t)(x_i).\end{aligned}$$

Put

$$x'_i = p^n \sum_j A_{ij} \varphi^{-a}(y_j) + p^{-n} \sum_j B_{ij} \varphi^a(z_j),$$

so that

$$\sum_i x'_i \mathbf{v}_i = p^n \varphi^{-a} \left(\sum_i y_i \mathbf{v}_i \right) + p^{-n} \varphi^a \left(\sum_i z_i \mathbf{v}_i \right)$$

and

$$\max_i \{\lambda(\alpha^t)(x'_i)\} \leqslant \epsilon \max_i \{\lambda(\alpha^t)(x_i)\} \qquad (t \in [s,r]).$$

Let us view $y = (y_1, \dots, y_m)$, $z = (z_1, \dots, z_m)$, and $x' = (x'_1, \dots, x'_m)$ as functions of $x = (x_1, \dots, x_m)$. Given $x_{(0)} \in (\tilde{\mathcal{R}}_R^{[s,r]})^m$, define $x_{(l+1)} = x'(x_{(l)})$, and put

$$\mathbf{v} = \sum_{l=0}^{\infty} \left(-p^n \varphi^{-a} \left(\sum_i y(x_{(l)})_i \mathbf{v}_i \right) + \sum_i z(x_{(l)})_i \mathbf{v}_i \right). \tag{6.2.2.3}$$

This series converges to an element of $M_{[s,rq]}$ satisfying $\mathbf{v} - p^{-n}\varphi^a(\mathbf{v}) = \mathbf{w}$ for $\mathbf{w} = \sum_{i=1}^m x_{(0),i} \mathbf{v}_i \in M_{[s,r]}$. More precisely, from the choice of the y_i and z_i, $\sum_l \varphi^{-a}(y(x_{(l)})_i)$ converges under $\lambda(\alpha^t)$ in the ranges $t \in [s,r]$ and $t \in [sq, rq]$, hence for $t \in [s, rq]$ by Lemma 5.2.1; a similar argument applies to $\sum_l z(x_{(l)})_i$.

For general r, s, given $\mathbf{w} \in M_{[s,r]}$, the previous paragraph produces $\mathbf{v} \in M_{[t,rq]}$ with $t = \max\{rq^{-1/2}, s\}$ such that $\mathbf{v} - p^{-n}\varphi^a(\mathbf{v}) = \mathbf{w}$. By rewriting this equation as $\mathbf{w} + p^{-n}\varphi^a(\mathbf{v}) = \mathbf{v}$ and invoking Lemma 5.2.10, we find that $\mathbf{v} \in M_{[t',r]} \cap M_{[t,rq]} = M_{[t',rq]}$ for $t' = \max\{t/q, s\}$. Repeating this argument, we eventually obtain $\mathbf{v} \in M_{[s,rq]}$ as desired. □

Corollary 6.2.3. — *Let $0 \to M_1 \to M \to M_2 \to 0$ be an exact sequence of φ^a-bundles over $\tilde{\mathcal{R}}_R$. Then there exists an integer N such that for $n \geqslant N$, the sequence*

$$0 \longrightarrow M_1(n)^{\varphi^a} \longrightarrow M(n)^{\varphi^a} \longrightarrow M_2(n)^{\varphi^a} \longrightarrow 0$$

is again exact.

Proof. — Apply Proposition 6.2.2 and the snake lemma. □

Proposition 6.2.4. — *Let $M = \{M_I\}$ be a φ^a-bundle over $\tilde{\mathcal{R}}_R$. For N as in Proposition 6.2.2 and $n \geqslant N$, there exist finitely many φ^a-invariant global sections of $M(n)$ which generate M. (If M is obtained from a φ^a-module over $\tilde{\mathcal{R}}_R^{\mathrm{int}}$ generated by m elements, then we may take $N = 1$ and use only $2m$ global sections.)*

Proof. — Pick any $r > 0$, and set notation as in the proof of Proposition 6.2.2 with $s = rq^{-1/2}$. By hypothesis, R is a Banach algebra over some analytic field L; choose $\overline{\pi} \in L$ with $0 < \alpha(\overline{\pi}) < 1$. For any $n \geqslant N$, we can find a positive rational number

$u \in \mathbb{Z}[p^{-1}]$ so that $c = \alpha(\overline{\pi}^u)$ satisfies (6.2.2.2). For $i = 1, \dots, m$, define $\mathbf{w}_i$ to be the sum of a series as in (6.2.2.3) in which

$$x_{(0)} = 0, \qquad (y(x_{(0)})_j, z(x_{(0)})_j) = \begin{cases} (-[\overline{\pi}^u], [\overline{\pi}^u]) & (j = i) \\ (0, 0) & (j \neq i); \end{cases}$$

this gives an element of $M_{[rq^{-1/2}, rq]}$ killed by $\varphi^a - 1$, and hence a φ^a-invariant global section of M. We can write $\mathbf{w}_j = [\overline{\pi}^s]\mathbf{v}_j + \sum_i X_{ij}\mathbf{v}_i$ with $\lambda(\alpha^t)(X_{ij}) \leqslant \epsilon\alpha(\overline{\pi})^{st}$ for all i, j and all $t \in [rq^{-1/2}, r]$. It follows that the matrix $1 + X$ is invertible over $\tilde{\mathcal{R}}_R^{[rq^{-1/2}, r]}$, so $\mathbf{w}_1, \dots, \mathbf{w}_m$ generate $M_{[rq^{-1/2}, r]}$.

By repeating the argument with r replaced by $rq^{-1/2}$, we obtain φ^a-invariant global sections $\mathbf{w}'_1, \dots, \mathbf{w}'_m$ which generate $M_{[rq^{-1}, rq^{-1/2}]}$. By applying powers of φ^a and invoking Lemma 2.3.12 and Lemma 5.3.4, we see that $\mathbf{w}_1, \dots, \mathbf{w}_m, \mathbf{w}'_1, \dots, \mathbf{w}'_m$ generate M_I for any I. □

Remark 6.2.5. — One cannot hope to refine the calculation in Proposition 6.2.2 to cover the entire interval $[r/q, r]$ and thus prove Proposition 6.2.4 in one step. This approach is obstructed by the following observation: take $R = L$ and obtain M from a trivial φ^a-module with φ^a-invariant basis $\mathbf{v}_1, \dots, \mathbf{v}_m$. If it were possible to refine the construction in Proposition 6.2.2 so that the elements $\mathbf{w}_1, \dots, \mathbf{w}_m$ produced in Proposition 6.2.4 were generators of M, we would have produced two isomorphic φ-modules with different degrees, a contradiction.

Remark 6.2.6. — The proof of Proposition 6.2.4 uses in a crucial way the running hypothesis that R is a Banach algebra over an analytic field. However, the proof can be modified to also treat the case where R is free of trivial spectrum (see Remark 2.3.9): one produces elements which generate the fiber of M over a point of $\mathcal{M}(R)$, notes that these also generate the fibers in some neighborhood, and argues by compactness. Similarly, the results of §6.3 can be extended to the case where R is free of trivial spectrum.

6.3. Vector bundles à la Fargues-Fontaine

We now make contact with the new perspective on p-adic Hodge theory provided by the work of Fargues and Fontaine [**45**] (see Remark 6.3.20). After we introduce adic spaces, we will be able to restate these results: see §8.7.

Definition 6.3.1. — Define the reduced graded ring $P = \oplus_{n=0}^{\infty} P_n$ by

$$P_n = \{x \in \tilde{\mathcal{R}}_R^+ : \varphi^a(x) = p^n x\} = \{x \in \tilde{\mathcal{R}}_R : \varphi^a(x) = p^n x\} \qquad (n = 0, 1, \dots).$$

The last equality holds by Corollary 5.2.12. (We will write P_R instead of P in case it becomes necessary to specify R.) For $d > 0$ and $f \in P_d$, let $P[f^{-1}]_0$ denote the degree zero subring of $P[f^{-1}]$; the affine schemes $D_+(f) = \operatorname{Spec}(P[f^{-1}]_0)$ glue to define a reduced separated scheme $\operatorname{Proj}(P)$ as in [**57**, Proposition 2.4.2]. The points of Proj(P) may be identified with the homogeneous prime ideals of P not containing $P_+ = \oplus_{n>0} P_n$, with $D_+(f)$ consisting of those ideals not containing f.

Definition 6.3.2. — For $f \in P_d$ for some $d > 0$, for $M = \{M_I\}$ a φ^a-bundle over $\tilde{\mathcal{R}}_R$, define

$$M_f = \bigcup_{n\in\mathbb{Z}} f^{-n} M(dn)^{\varphi^a} \tag{6.3.2.1}$$

as a module over $P[f^{-1}]_0$. (In other words, $M_f = M[f^{-1}]^{\varphi^a}$.) For any closed interval $I \subset (0, +\infty)$, we have a natural map

$$M_f \otimes_{P[f^{-1}]_0} \tilde{\mathcal{R}}_R^I[f^{-1}] \longrightarrow M_I \otimes_{\tilde{\mathcal{R}}_R^I} \tilde{\mathcal{R}}_R^I[f^{-1}]. \tag{6.3.2.2}$$

Lemma 6.3.3. — *Choose $f \in P_d$ for some $d > 0$. Let $0 \to M_1 \to M \to M_2 \to 0$ be a short exact sequence of φ^a-bundles over $\tilde{\mathcal{R}}_R$. Then the sequence*

$$0 \longrightarrow M_{1,f} \longrightarrow M_f \longrightarrow M_{2,f} \longrightarrow 0 \tag{6.3.3.1}$$

is exact.

Proof. — This follows from (6.3.2.1) and Corollary 6.2.3. □

Corollary 6.3.4. — *For $f \in P_d$ for some $d > 0$, M a φ^a-bundle over $\tilde{\mathcal{R}}_R$, and M_1, M_2 two φ^a-subbundles of M for which $M_1 + M_2$ is again a φ^a-subbundle, the natural inclusion $M_{1,f} + M_{2,f} \to (M_1 + M_2)_f$ within M_f is an equality.*

Proof. — Apply Lemma 6.3.3 to the surjection $M_1 \oplus M_2 \to M_1 + M_2$ (after extending this to an exact sequence using Remark 6.1.3). □

Definition 6.3.5. — For A an integral domain, the following conditions are equivalent [**20**, IV.2, Exercise 12].

(a) Every finitely generated ideal of A is projective.

(b) Every finitely generated torsion-free module over A is projective.

(c) For all ideals I_1, I_2, I_3 of A, the inclusion $I_1 \cap I_2 + I_1 \cap I_3 \to I_1 \cap (I_2 + I_3)$ is an equality. Note that it is sufficient to test finitely generated ideals.

An integral domain satisfying any of these conditions is called a *Prüfer domain.* For example, any Bézout domain is a Prüfer domain (but not conversely). Note that a noetherian Prüfer domain is a Dedekind domain, analogously to the fact that a noetherian Bézout domain is a principal ideal domain.

Lemma 6.3.6. — *Suppose that $R = L$ is an analytic field. Then for $f \in P_d$ for some $d > 0$, the ring $P[f^{-1}]_0 = (\tilde{\mathcal{R}}_L[f^{-1}])^{\varphi^a}$ is a Prüfer domain.*

Proof. — We check criterion (c) of Definition 6.3.5. Let I_1, I_2, I_3 be three finitely generated ideals of $P[f^{-1}]_0$. For $j = 1, 2, 3$, $I_j \otimes_{P[f^{-1}]_0} \tilde{\mathcal{R}}_L[f^{-1}]$ can be generated by a finite set of elements $x_{j,1}, \dots, x_{j,m} \in \tilde{\mathcal{R}}_L$ such that for each i, $\varphi^a(x_{j,i}) = p^h x_{j,i}$ for some $h \in \mathbb{Z}$. Let M_j be the ideal of $\tilde{\mathcal{R}}_L$ generated by $x_{j,1}, \dots, x_{j,m}$; it is principal (because $\tilde{\mathcal{R}}_L$ is a Bézout domain by Lemma 4.2.6) and φ^a-stable (because each $x_{j,i}$

generates a φ^a-stable ideal), and hence a φ^a-module over $\tilde{\mathcal{R}}_L$. Since $\tilde{\mathcal{R}}_L$ is a Bézout domain and hence a Prüfer domain, by Definition 6.3.5, the inclusion

$$M_1 \cap M_2 + M_1 \cap M_3 \longrightarrow M_1 \cap (M_2 + M_3)$$

is surjective. By Lemma 6.3.3, the map

$$(M_1 \cap M_2 + M_1 \cap M_3)_f \longrightarrow (M_1 \cap (M_2 + M_3))_f$$

is also surjective. The operation $M \mapsto M_f$ clearly distributes across intersections; it also distributes across sums thanks to Corollary 6.3.4 and the fact that the sum of two φ^a-submodules of $\tilde{\mathcal{R}}_L$ is again a φ^a-submodule (by the Bézout property again). By identifying I_j with $M_{j,f}$, we verify criterion (c) of Definition 6.3.5, so $P[f^{-1}]_0$ is a Prüfer domain as desired. □

Lemma 6.3.7. — *The following statements are true.*

(a) *For any $d > 0$, P_d generates the unit ideal in $\tilde{\mathcal{R}}_R^\infty$. In particular, we may choose $f_1, \dots, f_m \in P_d$ which generate the unit ideal in $\tilde{\mathcal{R}}_R^\infty$.*

(b) *For any such elements, the ideal in P generated by $f_1, \dots, f_m$ is saturated (*i.e.*, its radical equals P_+). Consequently, the schemes $D_+(f_1), \dots, D_+(f_m)$ cover* Proj(P)*, so* Proj(P) *is quasicompact.*

Proof. — To prove (a), apply Proposition 6.2.4 and Lemma 6.1.4 to $\tilde{\mathcal{R}}_R^\infty$ viewed as a trivial φ^a-bundle. To prove (b), choose any homogeneous $f \in P_+$, then apply Lemma 6.3.3 to show that the map $P[f^{-1}]_0^m \to P[f^{-1}]_0$ defined by $f_1, \dots, f_m$ contains 1 in its image. We then obtain an expression for some power of f as an element of the ideal of P generated by $f_1, \dots, f_m$, proving the claim. □

Lemma 6.3.8. — *Choose $r > 0$ and $f \in P_d$ for some $d > 0$. Let $\mathfrak{p}$ be any maximal ideal of $P[f^{-1}]_0$, and let $\mathfrak{q}$ be the corresponding homogeneous prime ideal of P not containing f.*

(a) *The ideal in $\tilde{\mathcal{R}}_R^{[r/q,r]}$ generated by $\mathfrak{q}$ and f is trivial.*

(b) *The ideal in $\tilde{\mathcal{R}}_R^{[r/q,r]}$ generated by $\mathfrak{q}$ is not trivial.*

Proof. — The homogeneous ideal in P generated by $\mathfrak{q}$ and f contains P_{dn} for some $n > 0$. By Lemma 6.3.7, $\mathfrak{q}$ and f generate the trivial ideal in $\tilde{\mathcal{R}}_R^\infty$ and hence also in $\tilde{\mathcal{R}}_R^{[r/q,r]}$. This yields (a).

Suppose now that $\mathfrak{q}$ contains elements $g_1, \dots, g_m$ which generate the unit ideal in $\tilde{\mathcal{R}}_R^{[r/q,r]}$. We may as well assume $g_1, \dots, g_m \in P_{dn}$ for some $n > 0$; then these elements define a map $(\tilde{\mathcal{R}}_R^\infty(-dn))^{\oplus m} \to \tilde{\mathcal{R}}_R^\infty$ of φ-modules. This map is surjective by Lemma 6.1.5; by Remark 6.1.3 and Lemma 6.3.3, we again get a surjective map upon inverting f and taking φ-invariants. But this implies that $f \in \mathfrak{q}$, a contradiction. This yields (b). □

Theorem 6.3.9. — *Choose $f \in P_d$ for some $d > 0$. Let $M = \{M_I\}$ be a φ^a-bundle over $\tilde{\mathcal{R}}_R$. Then M_f is a finite projective module over $P[f^{-1}]_0$ and (6.3.2.2) is bijective for every interval I.*

Proof. — Apply Proposition 6.2.4 to construct an integer n and a finite set $\mathbf{w}_1, \dots, \mathbf{w}_m$ of φ^a-invariant global sections of $M(dn)$ which generate M. We may then view $f^{-n}\mathbf{w}_1, \dots, f^{-n}\mathbf{w}_m$ as elements of M_f; this implies that (6.3.2.2) is surjective.

Let M' be the φ^a-bundle associated to the φ^a-module $(\tilde{\mathcal{R}}_R(-dn))^{\oplus m}$, so that $\mathbf{w}_1, \dots, \mathbf{w}_m$ define a surjection $M'(dn) \to M(dn)$ and hence a surjection $M' \to M$. By Remark 6.1.3, we obtain an exact sequence $0 \to M'' \to M' \to M \to 0$ of φ^a-bundles. By Lemma 6.3.3, the sequence

$$0 \longrightarrow M''_f \longrightarrow M'_f \longrightarrow M_f \longrightarrow 0 \qquad (6.3.9.1)$$

is exact. Consequently, M_f is a finitely generated module over $P[f^{-1}]_0$. For any interval I, we obtain a commuting diagram

$$\begin{array}{ccccccccc} & & M''_f \otimes_{P[f^{-1}]_0} \tilde{\mathcal{R}}^I_R[f^{-1}] & \longrightarrow & M'_f \otimes_{P[f^{-1}]_0} \tilde{\mathcal{R}}^I_R[f^{-1}] & \longrightarrow & M_f \otimes_{P[f^{-1}]_0} \tilde{\mathcal{R}}^I_R[f^{-1}] & \longrightarrow & 0 \\ & & \downarrow & & \downarrow & & \downarrow & & \\ 0 & \longrightarrow & M''_I \otimes_{\tilde{\mathcal{R}}^I_R} \tilde{\mathcal{R}}^I_R[f^{-1}] & \longrightarrow & M'_I \otimes_{\tilde{\mathcal{R}}^I_R} \tilde{\mathcal{R}}^I_R[f^{-1}] & \longrightarrow & M_I \otimes_{\tilde{\mathcal{R}}^I_R} \tilde{\mathcal{R}}^I_R[f^{-1}] & \longrightarrow & 0 \end{array}$$

with exact rows. (The left exactness in the last row follows from the exactness of localization.) Since the left vertical arrow is surjective (by the first part of the proof) and the middle vertical arrow is bijective, by the five lemma the right vertical arrow is injective. Hence (6.3.2.2) is a bijection. We may also repeat the arguments with M replaced by M'' to deduce that M_f is finitely presented.

The exact sequence $0 \to M'' \to M' \to M \to 0$ corresponds to an element of $H^1_{\varphi^a}(M^\vee \otimes M'')$. By Proposition 6.2.2, for m sufficiently large, we have $H^1_{\varphi^a}(M^\vee \otimes M''(dm)) = 0$. That is, if we form the commutative diagram

$$\begin{array}{ccccccccc} 0 & \longrightarrow & M'' & \longrightarrow & M' & \longrightarrow & M & \longrightarrow & 0 \\ & & \downarrow & & \downarrow & & \downarrow & & \\ 0 & \longrightarrow & f^{-m}M'' & \longrightarrow & N & \longrightarrow & M & \longrightarrow & 0 \end{array}$$

by pushing out, then the exact sequence in the bottom row splits in the category of φ^a-bundles. By Lemma 6.3.3, we obtain a split exact sequence

$$0 \longrightarrow (f^{-m}M'')_f \longrightarrow N_f \longrightarrow M_f \longrightarrow 0$$

of modules over $P[f^{-1}]_0$; however, $M'[f^{-1}] \cong N[f^{-1}]$ and so $M'_f = M'[f^{-1}]^\varphi \cong N[f^{-1}]^\varphi = N_f$. Since the construction of M' guarantees that M'_f is a free module over $P[f^{-1}]_0$, the same is true of N_f; it follows that M_f is a projective module over $P[f^{-1}]_0$, as desired. □

Theorem 6.3.9 can be reformulated in terms of an equivalence of categories between φ^a-bundles and vector bundles on $\mathrm{Proj}(P)$.

Definition 6.3.10. — Let V be a quasicoherent finite locally free sheaf on $\mathrm{Proj}(P)$. For each homogeneous $f \in P_+$, form the module $\Gamma(D_+(f), V) \otimes_{P[f^{-1}]_0} \tilde{\mathcal{R}}_R^\infty[f^{-1}]$. Since P_+ generates the unit ideal in $\tilde{\mathcal{R}}_R^\infty$ by Lemma 6.3.7, we may glue to obtain a quasicoherent finite locally free sheaf on $\mathrm{Spec}(\tilde{\mathcal{R}}_R^\infty)$. Let $M(V)$ be the module of global sections of this sheaf; then $V \rightsquigarrow M(V)$ defines an exact functor from quasicoherent finite locally free sheaves on $\mathrm{Proj}(P)$ to φ^a-modules over $\tilde{\mathcal{R}}_R^\infty$.

Definition 6.3.11. — Let M be a φ^a-bundle over $\tilde{\mathcal{R}}_R$. By Theorem 6.3.9, for each $f \in P_+$, M_f is a finite locally free module over $P[f^{-1}]_0$ and the natural map (6.3.2.2) is an isomorphism. In particular, the M_f glue to define a quasicoherent finite locally free sheaf $V(M)$ on $\mathrm{Proj}(P)$ (which one might call a *vector bundle* on $\mathrm{Proj}(P)$).

Theorem 6.3.12. — *The following tensor categories are equivalent.*

(a) *The category of quasicoherent finite locally free sheaves on* $\mathrm{Proj}(P)$.

(b) *The category of φ^a-modules over $\tilde{\mathcal{R}}_R^\infty$.*

(c) *The category of φ^a-modules over $\tilde{\mathcal{R}}_R$.*

(d) *The category of φ^a-bundles over $\tilde{\mathcal{R}}_R$.*

More precisely, the functor from (a) to (b) is the functor $V \rightsquigarrow M(V)$ given in Definition 6.3.10, the functor from (b) to (c) is base extension, the functor from (c) to (d) is the one indicated in Remark 6.1.6, and the functor from (d) to (a) is the functor $M \rightsquigarrow V(M)$ given in Definition 6.3.11.

Proof. — This is immediate from Theorem 6.3.9. □

Corollary 6.3.13. — *For any rational covering $\{(R, R^+) \to (R_i, R_i^+)\}_i$, the morphism $R \to \oplus_i R_i$ is an effective descent morphism for φ^a-modules over $\tilde{\mathcal{R}}_*$.*

Proof. — This is immediate from Theorem 5.3.13 and Theorem 6.3.12. □

It is worth mentioning the following refinement of Theorem 6.3.12 in the case of an analytic field.

Theorem 6.3.14. — *Suppose that $R = L$ is an analytic field. Then the construction of Definition 6.3.10 defines an equivalence of categories between the category of coherent sheaves on* $\mathrm{Proj}(P)$ *and the category of finitely presented $\tilde{\mathcal{R}}_L$-modules equipped with semilinear φ^a-actions.*

Proof. — Since $\tilde{\mathcal{R}}_L$ is a Bézout domain by Lemma 4.2.6 and hence a Prüfer domain, any finitely presented module over $\tilde{\mathcal{R}}_L$ is automatically coherent. The claim thus follows from Theorem 6.3.12. □

Remark 6.3.15. — The obstruction to generalizing Theorem 6.3.14 is that it is unclear whether the rings $P[f^{-1}]_0$ have the property that every finitely generated ideal is finitely presented (*i.e.*, whether these rings are *coherent*). This is most likely not true in general; however, we do not know what to expect if R is restricted to being the completed perfection of an affinoid algebra over an analytic field.

Remark 6.3.16. — One can improve the formal analogy between $\mathrm{Proj}(P)$ and the projective line over a field by defining $\mathcal{O}(n)$ for $n \in \mathbb{Z}$ as the invertible sheaf on $\mathrm{Proj}(P)$ corresponding *via* Theorem 6.3.12 to the φ^a-module over $\tilde{\mathcal{R}}_R$ free on one generator $\mathbf{v}$ satisfying $\varphi^a(\mathbf{v}) = p^{-n}\mathbf{v}$. For V a quasicoherent sheaf on $\mathrm{Proj}(P)$, write $V(n)$ for $V \otimes_{\mathcal{O}} \mathcal{O}(n)$; we may then naturally identify $M(V(n))$ with $M(V)(n)$. For V a quasicoherent finite locally free sheaf on $\mathrm{Proj}(P)$, $V(n)$ is generated by finitely many global sections for n large (by Theorem 6.3.12 and Proposition 6.2.4). For a vanishing theorem for H^1 in the same vein, see Corollary 8.7.14.

It will be useful for subsequent developments to explain how to add topologies to both types of objects appearing in Theorem 6.3.12.

Lemma 6.3.17. — *Let $M = \{M_I\}$ be a φ^a-bundle over $\tilde{\mathcal{R}}_R$. Choose $r > 0$, and induce from $\lambda(\alpha^r)$ a norm on $M_{[r/q,r]}$ as in Lemma 2.2.12. Then for each $n \in \mathbb{Z}$, the equivalence class of the restriction of this norm to $\{\mathbf{v} \in M_{[r/q,r]} : \varphi^a(\mathbf{v}) = p^{-n}\mathbf{v}\} \cong M(n)^{\varphi^a}$ is independent of r and of the choice of the norm on M. (However, the construction is not uniform in n.)*

Proof. — It is clear that the choice of the norm on M makes no difference up to equivalence, so we need only check the dependence on r. For any $s \in (0, r/q]$, by fixing a set of generators for $M_{[s,r]}$, we obtain norms $|\cdot|_t$ induced by $\lambda(\alpha^t)$ for all $t \in [s, r]$. We can choose $c_1, c_2 > 0$ so that these norms satisfy $c_1|\mathbf{v}|_t \leqslant |\varphi^a(\mathbf{v})|_{t/q} \leqslant c_2|\mathbf{v}|_t$ for all $\mathbf{v} \in M_{[s,r]}$ and all $t \in [sq, r]$.

For $\mathbf{v} \in M_{[r/q,r]}$ with $\varphi^a(\mathbf{v}) = p^{-n}\mathbf{v}$, we have $|\varphi^a(\mathbf{v})|_{r/q} = p^n|\mathbf{v}|_{r/q}$. Consequently, $|\cdot|_r$ and $|\cdot|_{r/q}$ have equivalent restrictions. By induction, we see that r and rq^m give equivalent norms for any $m \in \mathbb{Z}$. To complete the proof, it is enough to observe that for t in the interval between r and rq^m (inclusive), we have $|\cdot|_t \leqslant \max\{|\cdot|_r, |\cdot|_{rq^m}\}$ by Lemma 5.2.1; this then implies that the norm induced by r dominates the norm induced by s, and *vice versa* by symmetry. □

Definition 6.3.18. — Let G be a profinite group acting continuously on $\tilde{\mathcal{R}}^r_R$ for each $r > 0$ and commuting with φ^a; then G also acts on P (but not continuously). Let M be a φ^a-module over $\tilde{\mathcal{R}}_R$, and apply Theorem 6.3.12 to construct a corresponding quasicoherent finite locally free sheaf V on $\mathrm{Proj}(P)$. Then the following conditions on an action of G on V (or equivalently on M) are equivalent.

(a) The action of G on M is continuous for the LF topology.

(b) For each $n \in \mathbb{Z}$, the action of G on $M(n)^{\varphi^a} = \Gamma(\mathrm{Proj}(P), V(n))$ is continuous for any norm as in Lemma 6.3.17. (This implies (a) by Proposition 6.2.4.)

If these equivalent conditions are satisfied, we say the action is *continuous*.

We record a consequence of Theorem 6.3.12 for the cohomology of φ-modules.

Proposition 6.3.19. — *Let M be a φ^a-module over $\tilde{\mathcal{R}}_R^\infty$.*

(a) *For $r > 0$, put $M_r = M \otimes_{\tilde{\mathcal{R}}_R^\infty} \tilde{\mathcal{R}}_R^r$. Then the vertical arrows in the diagram*

$$\begin{array}{ccccccc} 0 & \longrightarrow & M & \xrightarrow{\varphi^a - 1} & M & \longrightarrow & 0 \\ & & \downarrow & & \downarrow & & \\ 0 & \longrightarrow & M_r & \xrightarrow{\varphi^a - 1} & M_{r/q} & \longrightarrow & 0 \end{array}$$

induce an isomorphism on the cohomology of the horizontal complexes. In particular, the lower complex computes $H^i_{\varphi^a}(M)$.

(b) *The map $M \to M \otimes_{\tilde{\mathcal{R}}_R^\infty} \tilde{\mathcal{R}}_R$ induces an isomorphism on cohomology.*

(c) *For r, s with $0 < s \leqslant r/q$, put $M_{[s,r]} = M \otimes_{\tilde{\mathcal{R}}_R^\infty} \tilde{\mathcal{R}}_R^{[s,r]}$. Then the vertical arrows in the diagram*

$$\begin{array}{ccccccc} 0 & \longrightarrow & M & \xrightarrow{\varphi^a - 1} & M & \longrightarrow & 0 \\ & & \downarrow & & \downarrow & & \\ 0 & \longrightarrow & M_{[s,r]} & \xrightarrow{\varphi^a - 1} & M_{[s,r/q]} & \longrightarrow & 0 \end{array}$$

induce an isomorphism on the cohomology of the horizontal complexes. In particular, the lower complex computes $H^i_{\varphi^a}(M)$.

(d) *In (c), the map $\varphi^a - 1 : M_{[s,r]} \to M_{[s,r/q]}$ is strict. Consequently, for $i = 0, 1$, $H^i_{\varphi^a}(M)$ admits the structure of a Banach space over $\mathbb{Q}_p$.*

Proof. — Note that (b) follows from (a) by taking direct limits, so we need only treat (a) and (c). Write $H^i_{\varphi^a}(M_r)$ and $H^i_{\varphi^a}(M_{[s,r]})$ as shorthand for the kernel and cokernel of the second row in (a), (c), and (d), respectively. Since M is finite projective over $\tilde{\mathcal{R}}_R^\infty$, the maps $M \to M_r$, $M \to M_{[s,r]}$ are injective; consequently, the maps $H^0_{\varphi^a}(M) \to H^0_{\varphi^a}(M_r)$, $H^0_{\varphi^a}(M) \to H^0_{\varphi^a}(M_{[s,r]})$ are injective. Conversely, for $\mathbf{v} \in H^0_{\varphi^a}(M_r)$, we also have $\mathbf{v} = \varphi^{-a}(\mathbf{v}) \in M_{rq}$. By induction, we have $\mathbf{v} \in M_{rq^n}$ for all n and so $\mathbf{v} \in M$; that is, $H^0_{\varphi^a}(M) \to H^0_{\varphi^a}(M_r)$ is surjective. Similarly, for $\mathbf{v} \in H^0_{\varphi^a}(M_{[s,r]})$, we may apply powers of φ^a and invoke Lemma 5.2.10 to deduce that $\mathbf{v} \in M$; that is, $H^0_{\varphi^a}(M) \to H^0_{\varphi^a}(M_{[s,r]})$ is surjective.

By similar reasoning, if $\mathbf{v} \in M_r$ (resp. $\mathbf{v} \in M_{[s,r]}$) is such that $(\varphi^a - 1)(\mathbf{v}) \in M_{r/q}$, then $\mathbf{v} \in M$. Consequently, the maps $H^1_{\varphi^a}(M) \to H^1_{\varphi^a}(M_r)$, $H^1_{\varphi^a}(M) \to H^1_{\varphi^a}(M_{[s,r]})$ are injective. To see that $H^1_{\varphi^a}(M) \to H^1_{\varphi^a}(M_r)$ is surjective, note that any class in the target defines an extension of φ^a-modules over $\tilde{\mathcal{R}}_R$, which lifts to an extension of φ^a-modules over $\tilde{\mathcal{R}}_R^\infty$ by Theorem 6.3.12. The argument for $H^1_{\varphi^a}(M) \to H^1_{\varphi^a}(M_{[s,r]})$ is similar.

To prove (d), apply Proposition 6.2.2 to find a nonnegative integer n such that $\varphi^a - 1 : M(n) \to M(n)$ is surjective. Using Proposition 6.2.4 we may construct a strict injective morphism $M \to M(n)$ of φ^a-modules; we then obtain a commutative diagram

$$\begin{array}{ccccccccc} 0 & \longrightarrow & M_{[s,r]} & \longrightarrow & M_{[s,r]}(n) & \longrightarrow & M_{[s,r]}(n)/M_{[s,r]} & \longrightarrow & 0 \\ & & \downarrow & & \downarrow & & \downarrow & & \\ 0 & \longrightarrow & M_{[s,r/q]} & \longrightarrow & M_{[s,r/q]}(n) & \longrightarrow & M_{[s,r/q]}(n)/M_{[s,r/q]} & \longrightarrow & 0 \end{array}$$

of Banach spaces in which the vertical arrows are induced by $\varphi^a - 1$. The second vertical arrow is surjective by the choice of n plus (c), and hence strict by the open mapping theorem (Theorem 2.2.8). Consequently, the connecting homomorphism

$$\ker\left(\varphi^a - 1 : M_{[s,r]}(n)/M_{[s,r]} \longrightarrow M_{[s,r/q]}(n)/M_{[s,r/q]}\right) \longrightarrow$$
$$\longrightarrow \operatorname{coker}(\varphi^a - 1 : M_{[s,r]} \longrightarrow M_{[s,r/q]})$$

is also strict surjective. By (c) again, it follows that $H^i_{\varphi^a}(M)$ is a Banach space over $\mathbb{Q}_p$ for $i = 1$; this is also clear for $i = 0$. By the open mapping theorem again, we deduce that $\varphi^a - 1 : M_{[s,r]} \to M_{[s,r/q]}$ is strict. □

Remark 6.3.20. — Theorem 6.3.14 is essentially due to Fargues and Fontaine [**45**], who have further studied the structure of the scheme $\operatorname{Proj}(P)$ when $R = L$ is an analytic field (see also [**44, 46**]). They show that it is a *complete absolute curve* in the sense of being noetherian, connected, separated, and regular of dimension 1, with each closed point having a well-defined degree and the total degree of any principal divisor being 0. (If L is algebraically closed, then the degrees are all equal to 1.) One corollary is that the rings $P[f^{-1}]_0$ for $f \in P_+$ homogeneous are not just Prüfer domains but Dedekind domains.

Many of the basic notions in p-adic Hodge theory can be interpreted in terms of the theory of vector bundles on $\operatorname{Proj}(P_L)$; this is the viewpoint developed in [**45**], which we find appealing and suggestive. For instance, the slope polygon of a φ^a-module over $\tilde{\mathcal{R}}_L$ can be interpreted as the Harder-Narasimhan polygon of the corresponding vector bundle, with étale φ^a-modules corresponding to semistable vector bundles of degree 0 *via* the slope filtration theorem over $\tilde{\mathcal{R}}_L$ (Theorem 4.2.12). The correspondence between étale φ^a-modules and étale local systems then bears a remarkable formal similarity to the correspondence between stable vector bundles on compact Riemann surfaces and irreducible unitary representations of the fundamental group, due to Narasimhan and Seshadri [**102**]. (A materially equivalent construction was given by Berger [**11**] in the somewhat less geometric language of *B-pairs*.)

For general R, we expect the scheme $\operatorname{Proj}(P)$ to exhibit much less favorable behavior (see for instance Remark 6.3.16). However, it may still be profitable to view the relationship between étale local systems and φ-modules through the optic of vector bundles over $\operatorname{Proj}(P)$. One possibly surprising aspect we will encounter later (see §8.7

for further discussion) is that étale $\mathbb{Q}_p$-local systems on the analytic space associated to R, which need not descend to $\mathrm{Spec}(R)$ (as in Example 8.5.17), will nonetheless give rise to algebraic vector bundles on $\mathrm{Proj}(P)$ *via* Theorem 8.7.8 and Corollary 8.7.10.

CHAPTER 7

SLOPES IN FAMILIES

When one considers φ-modules over relative Robba rings, one has not one slope polygon but a whole family of polygons indexed by the base analytic space. We now study the variation of the slope polygon in such families. Throughout §7, continue to retain Hypothesis 6.0.1.

7.1. An approximation argument

Much of our analysis of slopes in families depends on the following argument for spreading out certain bases of φ-modules, modeled on [**83**, Lemma 6.1.1 and Proposition 6.2.2].

Lemma 7.1.1. — *Let M be a φ^a-module over $\tilde{\mathcal{R}}_R$. For $r > 0$, let M_r be the model of M over $\tilde{\mathcal{R}}_R^r$. Let $\{M_I\}$ be the φ^a-bundle associated to M. Suppose that there exists a basis $\mathbf{v}_1, \dots, \mathbf{v}_n$ of $M_{[r/q,r]}$ on which φ acts* via *an invertible matrix F over $\tilde{\mathcal{R}}_R^{r/q}$. Then $\mathbf{v}_1, \dots, \mathbf{v}_n$ is a basis of M_r.*

Proof. — As in Lemma 6.1.4, it suffices to prove that $\mathbf{v}_1, \dots, \mathbf{v}_n$ is a basis of $M_{[r/q^{l+1},r]}$ for each nonnegative integer l. As the case $l = 0$ is given, we may proceed by induction on l.

Suppose that $l > 0$ and that the claim is known for $l - 1$, so that $\mathbf{v}_1, \dots, \mathbf{v}_n$ form a basis of $M_{[r/q^l,r]}$. Then $\varphi^a(\mathbf{v}_1), \dots, \varphi^a(\mathbf{v}_n)$ is a basis of $M_{[r/q^{l+1},r/q]}$. By hypothesis, $\mathbf{v}_j$ can be written as a $\tilde{\mathcal{R}}_R^{r/q}$-linear combination of the $\varphi^a(\mathbf{v}_i)$ and *vice versa*, so the $\mathbf{v}_j$ also form a basis of $M_{[r/q^{l+1},r/q]}$. By Lemma 2.3.12 and Lemma 5.3.4, $\mathbf{v}_1, \dots, \mathbf{v}_n$ form a basis of $M_{[r/q^{l+1},r]}$ as desired. □

Lemma 7.1.2. — *Let M be a φ^a-module over $\tilde{\mathcal{R}}_R$. For $r > 0$, let M_r be the model of M over $\tilde{\mathcal{R}}_R^r$. Let $\{M_I\}$ be the associated φ^a-bundle. Suppose that there exist a nonnegative integer h, a diagonal matrix D with diagonal entries $p^{d_1}, \dots, p^{d_n}$ for some $d_1, \dots, d_n \in \mathbb{Z}$ no two of which differ by more than h, and a basis $\mathbf{e}_1, \dots, \mathbf{e}_n$ of $M_{[r/q,r]}$ on which φ^a acts* via *a matrix F over $\tilde{\mathcal{R}}_R^{[r/q,r/q]}$ for which*

$\lambda(\alpha^{r/q})(FD-1) < p^{-h}$. *Then there exists a basis* $\mathbf{v}_1, \dots, \mathbf{v}_n$ *of* M_r *on which* φ^a *acts* via *a matrix* F' *over* $\tilde{\mathcal{R}}_R^{[r/q,r/q]}$ *such that* $F'D-1$ *has entries in* $p\tilde{\mathcal{R}}_R^{\mathrm{int},r/q}$, *and for which the invertible matrix* U *over* $\tilde{\mathcal{R}}_R^{[r/q,r]}$ *defined by* $\mathbf{v}_j = \sum_i U_{ij}\mathbf{e}_i$ *satisfies* $\lambda(\alpha^{r/q})(U-1), \lambda(\alpha^r)(D^{-1}UD-1) < p^{-h}$.

Proof. — Put $c_0 = p^h\lambda(\alpha^{r/q})(FD-1) < 1$. We construct a sequence of invertible $n \times n$ matrices $U_0, U_1, \dots$ over $\tilde{\mathcal{R}}_R^{[r/q,r]}$ such that the following conditions hold for $l = 0, 1, \dots$.

(a) We have $\lambda(\alpha^{r/q})(U_l - 1), \lambda(\alpha^r)(D^{-1}U_lD-1) \leqslant c_0p^{-h}$.

(b) For $F_l = U_l^{-1}F\varphi^a(U_l)$ (which has entries in $\tilde{\mathcal{R}}_R^{[r/q,r/q]}$ and satisfies $\lambda(\alpha^{r/q})(F_lD-1) \leqslant c_0p^{-h}$), there exists a matrix X_l over $\tilde{\mathcal{R}}_R^{[r/q,r/q]}$ such that $F_lD - X_l - 1$ has entries in $p\tilde{\mathcal{R}}_R^{\mathrm{int},r/q}$ and $\lambda(\alpha^{r/q})(X_l) \leqslant c_0^{l+1}p^{-h}$.

For $l = 0$, we may take $U_0 = 1$ and $X_0 = F_0D-1$. Given U_l for some $l \geqslant 0$, by applying Lemma 5.2.8 to the entries of $p^{-1}X_l$, we construct a matrix Z_l over $\tilde{\mathcal{R}}_R^{[r/q,r]}$ such that $F_lD - Z_l - 1$ has entries in $p\tilde{\mathcal{R}}_R^{\mathrm{int},r/q}$, $\lambda(\alpha^{r/q})(Z_l) \leqslant c_0^{l+1}p^{-h}$, and $\lambda(\alpha^r)(Z_l) \leqslant c_0^{q(l+1)}p^{-qh}$. Since

$$\lambda(\alpha^{r/q})(D^{-1}\varphi^a(Z_l)D) \leqslant p^h\lambda(\alpha^r)(Z_l) \leqslant p^hc_0^{q(l+1)}p^{-qh} \leqslant c_0^{l+2}p^{-h}, \tag{7.1.2.1}$$

both $1+Z_l$ and $1+D^{-1}\varphi^a(Z_l)D$ are invertible over $\tilde{\mathcal{R}}_R^{[r/q,r/q]}$. We may thus put

$$\begin{aligned} U_{l+1} &= U_l(1+Z_l) \\ F_{l+1} &= U_{l+1}^{-1}F\varphi^a(U_l) = (1+Z_l)^{-1}F_l(1+\varphi^a(Z_l)) \\ X_{l+1} &= F_{l+1}D - 1 - (F_lD - Z_l - 1), \end{aligned}$$

so that $F_{l+1}D - X_{l+1} - 1 = F_lD - Z_l - 1$ has entries in $p\tilde{\mathcal{R}}_R^{\mathrm{int},r/q}$. By writing

$$X_{l+1} = (1+Z_l)^{-1}(F_lD)(D^{-1}\varphi^a(Z_l)D) + Z_l^2(1+Z_l)^{-1}F_lD + Z_l(1-F_lD),$$

we see that $\lambda(\alpha^{r/q})(X_{l+1}) \leqslant c_0^{l+2}p^{-h}$. Hence U_{l+1} has the desired properties.

The matrices U_l converge to an invertible matrix U over $\tilde{\mathcal{R}}_R^{[r/q,r]}$ for which the matrix $G = U^{-1}F\varphi^a(U)D$ has entries in $\tilde{\mathcal{R}}_R^{\mathrm{int},r/q}$ and $G-1$ has entries in $p\tilde{\mathcal{R}}_R^{\mathrm{int},r/q}$ (because G is the limit of the Cauchy sequence F_lD with respect to $\lambda(\alpha^{r/q})$). If we put $\mathbf{v}_j = \sum_i U_{ij}\mathbf{e}_i$, we obtain a basis of $M_{[r/q,r]}$ on which φ^a acts *via* GD^{-1}. By Lemma 7.1.1, $\mathbf{v}_1, \dots, \mathbf{v}_n$ form a basis of M_r, proving the desired result. □

7.2. Rank, degree, and slope

Definition 7.2.1. — Define the *rank* and *degree* of a φ^a-module M over $W(R)$ (resp. $\tilde{\mathcal{E}}_R$, $\tilde{\mathcal{R}}_R^{\mathrm{int}}$, $\tilde{\mathcal{R}}_R^{\mathrm{bd}}$, $\tilde{\mathcal{R}}_R$) as the functions $\mathrm{rank}(M,\cdot) : \mathcal{M}(R) \to \mathbb{Z}$ and $\deg(M,\cdot) : \mathcal{M}(R) \to \frac{1}{a}\mathbb{Z}$ whose values at $\beta \in \mathcal{M}(R)$ are the rank and degree, respectively, of the φ^a-module obtained by base extension from M by passing from R to $\mathcal{H}(\beta)$ (recalling Convention 4.1.13 in case $a > 1$). Note that by our definitions, the degree of a φ^a-module over $W(R)$ or $\tilde{\mathcal{R}}_R^{\mathrm{int}}$ is identically zero.

Lemma 7.2.2. — *The rank and degree of a φ^a-module over any of the rings allowed in Definition 6.1.1 are continuous on $\mathcal{M}(R)$. In other words, the set of all points at which the rank or degree takes any given value is closed and open in $\mathcal{M}(R)$.*

Proof. — The rank function is continuous for the Zariski topology on $\operatorname{Spec}(R)$, and hence also for the topology on $\mathcal{M}(R)$. Continuity of the degree follows from Corollary 5.2.3 and Remark 5.0.2. □

Definition 7.2.3. — Let M be a φ^a-module over one of the rings allowed in Definition 6.1.1, of nowhere zero rank (that is, $\operatorname{rank}(M, \cdot)$ never takes the value 0). The *slope* of M is then defined as the function $\mu(M, \cdot) : \mathcal{M}(R) \to \mathbb{Q}$ given by $\mu(M, \beta) = \deg(M, \beta)/\operatorname{rank}(M, \beta)$. By Lemma 7.2.2, $\mu(M, \cdot)$ is continuous for the discrete topology on $\mathbb{Q}$.

The *pure locus* (resp. *étale locus*) of M is the set of $\beta \in \mathcal{M}(R)$ for which M becomes pure (resp. étale) upon passing from R to $\mathcal{H}(\beta)$. If this locus is all of $\mathcal{M}(R)$, we say that M is *pointwise pure* (resp. *pointwise étale*). These conditions have a more global interpretation; see Corollary 7.3.9.

Convention 7.2.4. — At a point where a φ^a-module has rank 0, it is considered to be pointwise pure of *every* slope. This is the correct convention for defining the categories of pure and étale φ-modules, so that they admit kernels for surjective morphisms (by Remark 6.1.3).

7.3. Pure models

Definition 7.3.1. — Fix integers c, d with d a positive multiple of a. Let M be a φ^a-module over $\tilde{\mathcal{E}}_R$ (resp. $\tilde{\mathcal{R}}_R^{\mathrm{bd}}$, $\tilde{\mathcal{R}}_R$). A (c, d)-*pure model* of M is a $W(R)$-submodule (resp. $\tilde{\mathcal{R}}_R^{\mathrm{int}}$-submodule, $\tilde{\mathcal{R}}_R^{\mathrm{int}}$-submodule) M_0 of M which is bounded (*i.e.*, there exists a finitely generated submodule N_0 of M over the same subring such that $p^n M_0 \subseteq N_0$, $p^n N_0 \subseteq M_0$ for some $n \geqslant 0$) with $M_0 \otimes_{W(R)} \tilde{\mathcal{E}}_R \cong M$ (resp. with $M_0 \otimes_{\tilde{\mathcal{R}}_R^{\mathrm{int}}} \tilde{\mathcal{R}}_R^{\mathrm{bd}} \cong M$, with $M_0 \otimes_{\tilde{\mathcal{R}}_R^{\mathrm{int}}} \tilde{\mathcal{R}}_R \cong M$), such that the φ^a-action on M induces an isomorphism $(p^c \varphi^d)^* M_0 \cong M_0$. The existence of such a model implies that M is pointwise pure of constant slope c/d. A pure model is *locally free* or *free* if its underlying module is finite locally free or finite free; by Proposition 3.2.13, any finitely presented pure model is locally free.

For $\beta \in \mathcal{M}(R)$, a *(locally free, free) local* (c, d)-*pure model* of M at β consists of a rational localization $R \to R'$ encircling β and a (locally free, free) (c, d)-pure model of $M \otimes_{\tilde{\mathcal{E}}_R} \tilde{\mathcal{E}}_{R'}$ (resp. $M \otimes_{\tilde{\mathcal{R}}_R^{\mathrm{bd}}} \tilde{\mathcal{R}}_{R'}^{\mathrm{bd}}$, $M \otimes_{\tilde{\mathcal{R}}_R} \tilde{\mathcal{R}}_{R'}$).

A $(0, d)$-pure model will also be called an *étale model*, and likewise with the modifiers *local*, *locally free*, or *free* in place.

Remark 7.3.2. — Note that if M is a φ^a-module over $\tilde{\mathcal{R}}_R$ and M_0 is a (c, d)-pure model, it is only assumed that $M_0[p^{-1}]$ is stable under the action of φ^d, rather than φ^a. For a locally free (c, d)-pure model, stability under φ^a will follow later from

Theorem 8.5.6, which will imply that the isomorphism $(\varphi^a)^* M \cong M$ descends to $M_0[p^{-1}]$. Until then, we will not assume this stability.

Lemma 7.3.3. — *Keep notation as in Definition 7.3.1.*

(a) *The φ^a-module M admits a locally free local (c,d)-pure model at β if and only if it admits a free local (c,d)-pure model at β.*

(b) *If M is a φ^a-module over $\tilde{\mathcal{R}}_R^{\mathrm{bd}}$, then M admits a free local (c,d)-pure model at β if and only if $M \otimes_{\tilde{\mathcal{R}}_R^{\mathrm{bd}}} \tilde{\mathcal{E}}_R$ does.*

Proof. — Part (a) follows at once from the fact that the direct limit of $W(S)$ (resp. $\tilde{\mathcal{R}}_S^{\mathrm{int}}$) over all rational localizations $R \to S$ encircling β is a local ring: namely, it contains p in its Jacobson radical, and its quotient by the ideal (p) is the local ring R_β (see Lemma 2.4.17).

To deduce (b), we may assume that $M' = M \otimes_{\tilde{\mathcal{R}}_R^{\mathrm{bd}}} \tilde{\mathcal{E}}_R$ admits a free (c,d)-pure model M'_0. Let $\mathbf{e}_1, \dots, \mathbf{e}_n$ be a basis of M'_0, let $\mathbf{v}_1, \dots, \mathbf{v}_m$ be generators of M, and write $\mathbf{e}_j = \sum_i B_{ij} \mathbf{v}_i$, $\mathbf{v}_j = \sum_i C_{ij} \mathbf{e}_i$ with $B_{ij}, C_{ij} \in \tilde{\mathcal{E}}_R$. Note that $CB = 1$ and that $\sum_i (1 - BC)_{ij} \mathbf{v}_i = 0$.

Since $\tilde{\mathcal{R}}_R^{\mathrm{bd}}$ is dense in $\tilde{\mathcal{E}}_R$ for the p-adic topology, we can find elements $B'_{ij} \in \tilde{\mathcal{R}}_R^{\mathrm{bd}}$ so that $C(B' - B)$ has entries in $pW(R)$, and so $X = 1 + C(B' - B)$ is invertible over $W(R)$. Define elements $\mathbf{e}'_1, \dots, \mathbf{e}'_n$ of M by the formula $\mathbf{e}'_j = \sum_i B'_{ij} \mathbf{v}_i$. Then $\mathbf{e}'_j = \sum_i (CB')_{ij} \mathbf{e}_i = \sum_i X_{ij} \mathbf{e}_i$, so $\mathbf{e}'_1, \dots, \mathbf{e}'_n$ form another basis of M'_0 and hence of M'.

Note that for each maximal ideal $\mathfrak{m}$ of $\tilde{\mathcal{R}}_R^{\mathrm{bd}}$, we can find a maximal ideal $\mathfrak{m}'$ of $\tilde{\mathcal{E}}_R$ containing $\mathfrak{m}$. (Otherwise, $\mathfrak{m}$ would generate the unit ideal in $\tilde{\mathcal{E}}_R$, and so would contain an element of $\tilde{\mathcal{R}}_R^{\mathrm{int}}$ congruent to 1 modulo p. But the latter would be a unit, contradiction.) Since $\mathbf{v}_1, \dots, \mathbf{v}_m$ generate M, there exists an n-element subset J of $\{1, \dots, m\}$ such that the $\mathbf{v}_j$ for $j \in J$ form a basis of $M/\mathfrak{m}M$. Since $\mathbf{e}'_1, \dots, \mathbf{e}'_n$ form a basis of M', the maximal minor of B' corresponding to J is nonzero in $\tilde{\mathcal{E}}_R/\mathfrak{m}'$ and hence also in $\tilde{\mathcal{R}}_R^{\mathrm{bd}}/\mathfrak{m}$. By Nakayama's lemma, $\mathbf{e}'_1, \dots, \mathbf{e}'_n$ generate the localization of M at $\mathfrak{m}$; since this holds for all $\mathfrak{m}$, $\mathbf{e}'_1, \dots, \mathbf{e}'_n$ generate M. This proves the desired result. □

Definition 7.3.4. — Let M be a φ^a-module over one of $\tilde{\mathcal{E}}_R, \tilde{\mathcal{R}}_R^{\mathrm{bd}}, \tilde{\mathcal{R}}_R$. For $\beta \in \mathcal{M}(R)$ and $s \in \mathbb{Q}$, we say M is *pure (of slope s)* at β if M admits a locally free local (c,d)-pure model at β for some pair (c,d) of integers with d a positive multiple of a and $c/d = s$. This forces $s = \mu(M)$ if $\operatorname{rank}(M, \beta) > 0$; if $\operatorname{rank}(M, \beta) = 0$, then M is pure of *every* slope at β (see Convention 7.2.4). We say M is *pure* if it is pure at each $\beta \in \mathcal{M}(R)$; in this case, we can cover all $\beta \in \mathcal{M}(R)$ using finitely many local pure models thanks to the compactness of $\mathcal{M}(R)$. In these definitions, we regard *étale* as a synonym for *pure of slope* 0.

In case we need to be precise about the choice of c and d, we will say that a module is *(c,d)-pure* rather than *pure of slope s*. This will ultimately be rendered unnecessary by the observation that purity for one pair (c, d) with $d > 0$ and $c/d = s$ implies the same for any other pair (Corollary 8.5.13).

We also say that M is *globally pure/étale* if it admits a locally free pure/étale model. For more on this condition, see Remark 7.3.5.

Remark 7.3.5. — Consider the following conditions on a φ^a-module M over one of $\tilde{\mathcal{E}}_R, \tilde{\mathcal{R}}_R^{\mathrm{bd}}, \tilde{\mathcal{R}}_R$.

(a) The φ^a-module M is globally pure (*i.e.*, admits a locally free pure model).

(b) The φ^a-module M admits a pure model.

(c) The φ^a-module M is pure (*i.e.*, admits locally free local pure models).

(d) The φ^a-module M admits local pure models.

(e) The φ^a-module M is pointwise pure.

There are some trivial implications among these conditions: (a) implies (b) and (c), which in turn each imply (d), which in turn implies (e). We will see eventually that there are more implications as follows.

- Over $\tilde{\mathcal{E}}_R$ or $\tilde{\mathcal{R}}_R^{\mathrm{bd}}$, Corollary 8.5.14 will show that (e) implies (c), while Example 8.5.18 will show that (c) does not imply (a). Also, (d) implies (b): given some local pure models on a finite covering of $\mathcal{M}(R)$, the elements of M which restrict into each local pure model form a pure model. Consequently, (a) strictly implies (b) and (b)–(e) are equivalent.
- Over $\tilde{\mathcal{R}}_R$, Corollary 7.3.9 will show that (e) implies (c), while Example 8.5.18 will show that (c) does not imply (a) and Example 8.5.17 will show that (c) does not imply (b). Consequently, (a) strictly implies (b), (b) strictly implies (c), and (c)–(e) are equivalent.

Proposition 7.3.6. — *Let M be a φ^a-module over $\tilde{\mathcal{E}}_R$ (resp. $\tilde{\mathcal{R}}_R^{\mathrm{bd}}$, $\tilde{\mathcal{R}}_R$) admitting a free (c, d)-pure model M_0 for some $c, d \in \mathbb{Z}$ with d a positive multiple of a. Then there exists an R-algebra S which is the completed direct limit of some faithfully finite étale R-subalgebras, such that $M_0 \otimes_{W(R)} W(S)$ (resp. $M_0 \otimes_{\tilde{\mathcal{R}}_R^{\mathrm{int}}} \tilde{\mathcal{R}}_S^{\mathrm{int}}$, $M_0 \otimes_{\tilde{\mathcal{R}}_R^{\mathrm{int}}} \tilde{\mathcal{R}}_S^{\mathrm{int}}$) admits a basis fixed by $p^c \varphi^d$.*

Proof. — The assertion over $\tilde{\mathcal{E}}_R$ follows from Lemma 3.2.6. To handle the other cases, invoke the case $n = 1$ of Lemma 3.2.6 to reduce to the case where M_0 admits a basis $\mathbf{e}_1, \dots, \mathbf{e}_n$ on which $p^c \varphi^d$ acts *via* a matrix A over $\tilde{\mathcal{R}}_R^{\mathrm{int}}$ congruent to 1 modulo p. Choose S so that $M_0 \otimes_{\tilde{\mathcal{R}}_R^{\mathrm{int}}} \tilde{\mathcal{E}}_S$ contains a basis $\mathbf{v}_1, \dots, \mathbf{v}_n$ fixed by $p^c \varphi^d$ with $\mathbf{v}_i \equiv \mathbf{e}_i \pmod{p}$. Since $A - 1$ has p-adic absolute value less than 1, we have $\lambda(\alpha^r)(A - 1) < 1$ for all sufficiently small $r > 0$. For any such r, the matrix U over $W(S)$ defined by $\mathbf{v}_j = \sum_i U_{ij} \mathbf{e}_i$ is congruent to 1 modulo p and satisfies $A \varphi^d(U) = U$.

We now argue as in [**84**, Proposition 2.5.8]. We may assume $a = d$, so $q = p^d$. Put $C = \max\{p^{-1}, \lambda(\alpha^r)(A - 1)\} < 1$. We prove by induction that for each positive integer m, U is congruent modulo p^m to an invertible matrix V_m over $\tilde{\mathcal{R}}_S^{\mathrm{int}, rq}$ with

$$\lambda(\alpha^r)(V_m - 1), \lambda(\alpha^{rq})(V_m - 1) \leqslant C.$$

This is obvious for $m = 1$ by taking $V_m = 1$. Given the claim for some m, U is congruent modulo p^{m+1} to a matrix $V_m + p^m X$ in which each entry X_{ij} is the Teichmüller lift of some $\overline{X}_{ij} \in S$. We have

$$\varphi^d(X) - X \equiv p^{-m}(V_m - \varphi^d(V_m) - (A-1)\varphi^d(V_m)) \pmod{p},$$

from which it follows that

$$\alpha(\overline{X}_{ij})^r \leqslant \max\{1, (p^m C)^{q^{-1}}\} = (p^m C)^{q^{-1}}.$$

If we put $V_{m+1} = V_m + p^m X$, then

$$\lambda(\alpha^r)(V_{m+1} - 1) \leqslant \max\{C, p^{-m}(p^m C)^{q^{-1}}\} \leqslant C$$
$$\lambda(\alpha^{rq})(V_{m+1} - 1) \leqslant \max\{C, p^{-m}(p^m C)\} = C$$

as desired.

From the previous induction, we conclude that U has entries in $\tilde{\mathcal{R}}_S^{\mathrm{int},rq}$ and satisfies $\lambda(\alpha^r)(U-1), \lambda(\alpha^{rq})(U-1) < 1$. It is thus invertible over $\tilde{\mathcal{R}}_S^{\mathrm{int}}$, so $\mathbf{v}_1, \dots, \mathbf{v}_n$ form a basis of $M_0 \otimes_{\tilde{\mathcal{R}}_R^{\mathrm{int}}} \tilde{\mathcal{R}}_S^{\mathrm{int}}$ as desired. □

Theorem 7.3.7. — *Let M be a φ^a-module over $\tilde{\mathcal{R}}_R$ of nowhere zero rank. Choose $\beta \in \mathcal{M}(R)$ and choose $c, d \in \mathbb{Z}$ with d a positive multiple of a and $c/d = \mu(M, \beta)$. Suppose that β belongs to the pure locus of M. Then any (c,d)-pure model of $M \otimes_{\tilde{\mathcal{R}}_R} \tilde{\mathcal{R}}_{\mathcal{H}(\beta)}$ extends to a free local (c,d)-pure model of M at β.*

Proof. — We may assume $d = a$, so $q = p^d$. Choose a (c,d)-pure model $M_{0,\beta}$ of $M \otimes_{\tilde{\mathcal{R}}_R} \tilde{\mathcal{R}}_{\mathcal{H}(\beta)}$. By Proposition 4.2.15, for L a completed algebraic closure of $\mathcal{H}(\beta)$, for some choice of r, there exists a basis $\mathbf{e}_1, \dots, \mathbf{e}_n$ of $M_r \otimes_{\tilde{\mathcal{R}}_R^r} \tilde{\mathcal{R}}_L^{[r/q,r]}$ on which $p^c \varphi^d$ acts *via* the identity matrix. We may also ensure that $\mathbf{e}_1, \dots, \mathbf{e}_n$ also form a basis of $M_{0,\beta} \otimes_{\tilde{\mathcal{R}}_{\mathcal{H}(\beta)}^{\mathrm{int}}} \tilde{\mathcal{R}}_L^{\mathrm{int}}$.

By Lemma 2.2.13, any elements $\mathbf{e}'_1, \dots, \mathbf{e}'_n$ of $M_r \otimes_{\tilde{\mathcal{R}}_R^r} \tilde{\mathcal{R}}_L^{[r/q,r]}$ which are sufficiently close to $\mathbf{e}_1, \dots, \mathbf{e}_n$ also generate $M_r \otimes_{\tilde{\mathcal{R}}_R^r} \tilde{\mathcal{R}}_L^{[r/q,r]}$. Since the separable closure of $\mathcal{H}(\beta)$ in L is dense, we can take $\mathbf{e}'_1, \dots, \mathbf{e}'_n$ to be generators of $M_r \otimes_{\tilde{\mathcal{R}}_R^r} \tilde{\mathcal{R}}_E^{[r/q,r]}$ for some finite Galois extension E of $\mathcal{H}(\beta)$ such that $\mathbf{e}'_j = \sum_i C_{ij} \mathbf{e}_i$ for some invertible matrix C over $\tilde{\mathcal{R}}_E^{[r/q,r]}$ satisfying $\lambda(\beta^{r/q})(C-1), \lambda(\beta^r)(C-1) < 1$.

Since R_β is a henselian local ring (by Lemma 2.4.17), by Theorem 1.2.8 we can find a rational localization $R \to R'$ encircling β and a faithfully finite étale R'-algebra S such that S is Galois over R', S admits a unique extension γ of β, and such extension has residue field E. By Lemma 5.1.7, for a suitable choice of R', we can take $\mathbf{e}'_1, \dots, \mathbf{e}'_n$ to be a basis of $M_r \otimes_{\tilde{\mathcal{R}}_R^r} \tilde{\mathcal{R}}_S^{[r/q,r]}$ on which the action of $p^c \varphi^d$ is *via* a matrix F for which $\lambda(\alpha^{r/q})(F-1) < 1$.

In this setting, Lemma 7.1.2 produces a basis $\mathbf{e}''_1, \dots, \mathbf{e}''_n$ of $M \otimes_{\tilde{\mathcal{R}}_R} \tilde{\mathcal{R}}_S$ on which $p^c \varphi^d$ acts *via* an invertible matrix over $\tilde{\mathcal{R}}_S^{\mathrm{int},r/q}$ congruent to 1 modulo p. More precisely, we have $\mathbf{e}''_j = \sum_i U_{ij} \mathbf{e}'_i$ for some invertible matrix U over $\tilde{\mathcal{R}}_S^{[r/q,r]}$ for which $\lambda(\alpha^{r/q})(U-1), \lambda(\alpha^r)(U-1) < 1$. Starting from this new basis and then applying Proposition 7.3.6, we obtain, for some S-algebra S' which is the completed direct limit of faithfully finite étale subalgebras (and which maps to L), a basis $\mathbf{e}_1''', \dots, \mathbf{e}_n'''$ of $M \otimes_{\tilde{\mathcal{R}}_R} \tilde{\mathcal{R}}_{S'}$ fixed by $p^c \varphi^d$. More precisely, we have $\mathbf{e}_j''' = \sum_i V_{ij} \mathbf{e}_i''$ for some invertible matrix V over $\tilde{\mathcal{R}}_{S'}^{\mathrm{int},r}$ for which $\lambda(\alpha^r)(V-1) < 1$ (this bound following from the proof of Proposition 7.3.6). We may also choose S' to have an automorphism lifting each element of $G = \mathrm{Gal}(E/\mathcal{H}(\beta))$.

We thus have $\mathbf{e}_j''' = \sum_i (CUV)_{ij} \mathbf{e}_i$, and the matrix CUV over $\tilde{\mathcal{R}}_L^{[r/q,r]}$ satisfies $\lambda(\beta^{r/q})(CUV-1), \lambda(\beta^r)(CUV-1) < 1$. However, both $\mathbf{e}_i$ and $\mathbf{e}_j'''$ are fixed by $p^c \varphi^d$, so the entries of CUV must be fixed by φ^d. By Lemma 4.2.10, CUV has entries in the field $W(\mathbb{F}_q)[p^{-1}]$. Since $\lambda(\beta^{r/q})(CUV-1) < 1$, we conclude that $CUV-1$ has entries in $pW(\mathbb{F}_q)$. In particular, $\mathbf{e}_1'', \dots, \mathbf{e}_n''$ form a basis of $M_{0,\beta} \otimes_{\tilde{\mathcal{R}}^{\mathrm{int}}_{\mathcal{H}(\beta)}} \tilde{\mathcal{R}}_L^{\mathrm{int}}$ and hence also a basis of $M_{0,\beta} \otimes_{\tilde{\mathcal{R}}^{\mathrm{int}}_{\mathcal{H}(\beta)}} \tilde{\mathcal{R}}_E^{\mathrm{int}}$.

This last result implies that the action of any element $\tau \in G$ on S induces an automorphism of the $\tilde{\mathcal{R}}_E^{\mathrm{int}}$-span of $\mathbf{e}_1'', \dots, \mathbf{e}_n''$. For each τ, choose a lift of τ to S' and define the matrix T_τ over $\tilde{\mathcal{R}}_{S'}$ by $\tau(\mathbf{e}_j''') = \sum_i (T_\tau)_{ij} \mathbf{e}_i'''$. Again, since $\mathbf{e}_i'''$ and $\tau(\mathbf{e}_j''')$ are fixed by $p^c \varphi^d$, the entries of T_τ are forced to belong to the φ^d-fixed subring of $\tilde{\mathcal{R}}_{S'}^{\mathrm{int}}$, which Corollary 5.2.4 identifies as $W((S')^{\overline{\varphi}^d})[p^{-1}]$. By construction, the images of these entries in $\tilde{\mathcal{R}}_L$ belong to $\tilde{\mathcal{R}}_L^{\mathrm{int}}$, and hence to $\mathfrak{o}_L^{\varphi^d}$. By Remark 5.1.5 and Remark 5.0.2, the condition of an element of $W((S')^{\overline{\varphi}^d})[p^{-1}]$ belonging to $W(S')$ is closed *and open* on $\mathcal{M}(S')$. By Remark 2.3.15 (b) and Lemma 2.4.17, the natural map $p : \mathcal{M}(S') \to \mathcal{M}(R')$ is open. We thus can force the entries of T_τ into $W((S')^{\overline{\varphi}^d})$ by shrinking $\mathcal{M}(R')$ again (in a manner dependent on τ). (Note that this last step fails if we try to argue directly with the $\mathbf{e}_i''$ rather than with the $\mathbf{e}_i'''$.)

After the resulting shrinking of $\mathcal{M}(R')$, the $\tilde{\mathcal{R}}_S^{\mathrm{int}}$-span of $\mathbf{e}_1'', \dots, \mathbf{e}_n''$ admits an action of G, so we can apply faithfully flat descent (Theorem 1.3.4) to descend it to a local (c,d)-pure model of M at β. (We are here using Proposition 5.5.3 (b) to deduce that $\tilde{\mathcal{R}}_S^{\mathrm{int}}$ is faithfully finite étale over $\tilde{\mathcal{R}}_{R'}^{\mathrm{int}}$.) This local model is locally free by Theorem 1.3.5; we obtain a free local model at β by applying Lemma 7.3.3. □

Corollary 7.3.8. — *For any φ^a-module M over $\tilde{\mathcal{R}}_R$, the pure locus and étale locus of M are open.*

Proof. — The openness of the pure locus is immediate from Theorem 7.3.7. The étale locus is open because it is the intersection of the pure locus with the closed and open (by Lemma 7.2.2) subset of $\mathcal{M}(R)$ on which the degree of M is zero. □

Corollary 7.3.9. — *For any φ^a-module M over $\tilde{\mathcal{R}}_R$, M is étale (resp. pure) if and only if M is pointwise étale (resp. pointwise pure).*

Corollary 7.3.10. — *For any φ^a-module M over $\tilde{\mathcal{R}}_R$ and any $\beta \in \mathcal{M}(R)$, M is pure at β if and only if M admits a (not necessarily locally free) local pure model at β.*

Proof. — If M admits a local pure model M_0 at β, then M_0 also generates a pure model of $M \otimes_{\tilde{\mathcal{R}}_R} \tilde{\mathcal{R}}_{\mathcal{H}(\beta)}$, which is necessarily free because $\tilde{\mathcal{R}}^{\mathrm{int}}_{\mathcal{H}(\beta)}$ is a principal ideal domain. Hence β belongs to the pure locus of M, which by Theorem 7.3.7 implies that M is pure at β. □

Remark 7.3.11. — Let M be a φ^a-module over $\tilde{\mathcal{R}}^{\mathrm{bd}}_R$. Whereas Lemma 7.3.3 implies that M is pure if and only if $M \otimes_{\tilde{\mathcal{R}}^{\mathrm{bd}}_R} \tilde{\mathcal{E}}_R$ is pure, it is not the case that purity of M can be deduced from purity of $M \otimes_{\tilde{\mathcal{R}}^{\mathrm{bd}}_R} \tilde{\mathcal{R}}_R$.

7.4. Slope filtrations in geometric families

At this point, it is natural to discuss generalizations to the relative case of the existence of slope filtrations for Frobenius modules over the Robba ring (Theorem 4.1.10) or the extended Robba ring (Theorem 4.2.13). We do not have in mind an explicit use for these in p-adic Hodge theory, but we expect them to become relevant in the same way that slope theory over the Robba ring appears in the work of Colmez on trianguline representations [**26**].

We first give a brief review of the formalism of slope filtrations and slope polygons. See [**83**, §3.5] for a more thorough discussion, keeping in mind the change in sign convention. (To compensate for that change, we have also swapped the order of slopes in the slope polygon in order to preserve the convex shape of the polygon.)

Definition 7.4.1. — Suppose that $R = L$ is an analytic field, and let M be a φ^a-module over $\tilde{\mathcal{R}}_L$. Let $0 = M_0 \subset \cdots \subset M_l = M$ be the filtration provided by Theorem 4.2.13, in which $M_1/M_0, \dots, M_l/M_{l-1}$ are pure and $\mu(M_1/M_0) > \cdots > \mu(M_l/M_{l-1})$. Define the *slope polygon* of M to be the polygonal line starting at $(0,0)$ and consisting of, for $i = l, \dots, 1$ in order, a segment of horizontal width $\mathrm{rank}(M_i/M_{i-1})$ and slope $\mu(M_i/M_{i-1})$. Note that the right endpoint of the polygon is $(\mathrm{rank}(M), \deg(M))$.

For M a φ^a-module over $\tilde{\mathcal{R}}^{\mathrm{bd}}_L$, there are two natural ways to associate a slope polygon to M. One is to first extend scalars to $\tilde{\mathcal{R}}_L$ and use the definition given in the previous paragraph; this gives the *special slope polygon* of M. The other is to first extend scalars to $\tilde{\mathcal{E}}_L$, identify the latter with $\tilde{\mathcal{R}}_L$ for the trivial norm on L, then invoke the previous paragraph. This gives the *generic slope polygon* of M. (It is equivalent to define the generic slope polygon using the usual Dieudonné-Manin definition of slopes; see for instance [**85**, Chapter 14].)

Remark 7.4.2. — For $R = L$ an analytic field and L' a complete extension of L, passing from L to L' does not change slope polygons, by virtue of Corollary 4.2.14 and the uniqueness of the slope filtration in Theorem 4.2.13.

Proposition 7.4.3. — *Let $R = L$ be an analytic field, and let M be a φ^a-module over $\tilde{\mathcal{R}}^{\mathrm{bd}}_L$.*

(a) *The special slope polygon lies on or above the generic slope polygon, with the same endpoints.*

(b) *If the two polygons coincide, then the slope filtration of* $M \otimes_{\tilde{\mathcal{R}}_L^{\mathrm{bd}}} \tilde{\mathcal{R}}_L$ *(Theorem 4.2.13) descends to* M*.*

Proof. — For (a), see [**83**, Proposition 5.5.1]. For (b), see [**83**, Theorem 5.5.2]. □

Lemma 7.4.4. — *Let* M *be a* φ^a*-module over* $\tilde{\mathcal{R}}_R^{\mathrm{bd}}$ *admitting a basis on which* φ^a *acts via a matrix of the form* AD*, where* D *is a diagonal matrix with diagonal entries in* $p^{\mathbb{Z}}$*, and* A *is a square matrix such that* $A-1$ *has entries in* $p\tilde{\mathcal{R}}_R^{\mathrm{int}}$*. Then there exist an* R*-algebra* S *which is the union of faithfully finite étale* R*-subalgebras and an invertible matrix* U *over* $W(S)$ *congruent to* 1 *modulo* p*, such that* $U^{-1}AD\varphi^a(U) = D$*. In particular, for each* $\beta \in \mathcal{M}(R)$*, the generic slopes of* $M \otimes_{\tilde{\mathcal{R}}_R^{\mathrm{bd}}} \tilde{\mathcal{R}}_{\mathcal{H}(\beta)}^{\mathrm{bd}}$ *are the negatives of the* p*-adic valuations of the entries of* D*, divided by* a*.*

Proof. — We can proceed as in [**82**, Lemma 5.9]. Note that although the analogous statement in our setup would be taking S to be the completed union of faithfully finite étale R-subalgebras, the argument of [**82**, Lemma 5.9] actually implies that one can take S to be the union of faithfully finite étale R-subalgebras. This refinement will be useful for Theorem 7.4.9. □

Theorem 7.4.5. — *For any* φ^a*-module* M *over* $\tilde{\mathcal{R}}_R$*, the function mapping* $\beta \in \mathcal{M}(R)$ *to the slope polygon of* $M \otimes_{\tilde{\mathcal{R}}_R} \tilde{\mathcal{R}}_{\mathcal{H}(\beta)}$ *is lower semicontinuous. In other words, if* $\operatorname{rank}(M)$ *is constant (which is true locally by Lemma 7.2.2), for any* $x \in [0, \operatorname{rank}(M)]$*, the* y*-coordinate of the point of the slope polygon of* $M \otimes_{\tilde{\mathcal{R}}_R} \tilde{\mathcal{R}}_{\mathcal{H}(\beta)}$ *is a lower semicontinuous function of* β*. (Note that this function is locally constant for* $x = \operatorname{rank}(M)$*, by Lemma 7.2.2 again.)*

Proof. — Choose $\beta \in \mathcal{M}(R)$. Let L be a completed algebraic closure of $\mathcal{H}(\beta)$. By Proposition 4.2.16, for some positive multiple d of a, $M \otimes_{\tilde{\mathcal{R}}_R} \tilde{\mathcal{R}}_L$ admits a basis on which φ^d acts *via* a diagonal matrix D with entries in $p^{\mathbb{Z}}$. We now proceed as in Theorem 7.3.7, by applying Lemma 7.1.2 to a suitably good approximation of this basis. As a result, we obtain a rational localization $R \to R'$ encircling β, a faithfully finite étale R'-algebra S, and a basis $\mathbf{v}_1, \dots, \mathbf{v}_n$ of $M \otimes_{\tilde{\mathcal{R}}_R} \tilde{\mathcal{R}}_S$ on which φ^d acts *via* an invertible matrix over $\tilde{\mathcal{R}}_S^{\mathrm{bd}}$ of the form FD, where $F-1$ has entries in $p\tilde{\mathcal{R}}_S^{\mathrm{int}}$. Let N be the φ^d-module over $\tilde{\mathcal{R}}_S^{\mathrm{bd}}$ spanned by $\mathbf{v}_1, \dots, \mathbf{v}_n$.

By Remark 7.4.2, the slope polygon of M at a given point of $\mathcal{M}(R')$ is the same as at any point of $\mathcal{M}(S)$ restricting to the given point. For one, this means that the negatives of the p-adic valuations of the diagonal entries of D give the slopes in the slope polygon of M at β. By Lemma 7.4.4, this polygon also computes the generic slope polygon of N at each $\gamma \in \mathcal{M}(S)$. By Proposition 7.4.3 (a), we conclude that the special slope polygon of N at each $\gamma \in \mathcal{M}(S)$, or in other words the slope polygon of M at γ, lies on or above the slope polygon of M at β. By Remark 7.4.2 again, this implies that the slope polygon is lower semicontinuous as a function on $\mathcal{M}(R)$, as desired. □

In addition to semicontinuity, we have the following boundedness property for the slope polygon.

Proposition 7.4.6. — *For any φ^a-module M over $\tilde{\mathcal{R}}_R$, the function mapping $\beta \in \mathcal{M}(R)$ to the slope polygon of $M \otimes_{\tilde{\mathcal{R}}_R} \tilde{\mathcal{R}}_{\mathcal{H}(\beta)}$ is bounded above and below.*

Proof. — Choose N as in Proposition 6.2.4; we then obtain a surjection $\tilde{\mathcal{R}}_R^m(-N) \to M$ of φ^a-modules for some nonnegative integer m. For each $\beta \in \mathcal{M}(R)$, for s the smallest slope in the slope polygon of M at β, $M \otimes_{\tilde{\mathcal{R}}_R} \tilde{\mathcal{R}}_{\mathcal{H}(\beta)}$ surjects onto a nonzero φ^a-module over $\tilde{\mathcal{R}}_{\mathcal{H}(\beta)}$ of slope s, as then does $\tilde{\mathcal{R}}_R^m(-N)$. This forces $-N/a \leqslant s$ by (the easy direction of) Theorem 4.2.12; consequently, all of the slopes of M at β are bounded below by $-N/a$. Since the sum of the slopes is a continuous (by Lemma 7.2.2) and hence bounded function on $\mathcal{M}(R)$, the slopes of M at β are also bounded above. This proves the claim. □

Corollary 7.4.7. — *Let M be a φ^a-module over $\tilde{\mathcal{R}}_R$. Then there exists an open dense subset U of $\mathcal{M}(R)$ on which the slope polygon of M is locally constant.*

Proof. — It suffices to check this property locally around some $\beta \in \mathcal{M}(R)$. By Lemma 7.2.2, we may assume that M has constant rank and degree over $\mathcal{M}(R)$.

By Proposition 7.4.6, the slope polygons of M are limited to a finite set S. Let T be the subset of S consisting of those polygons which occur in every neighborhood of β; this set is nonempty because it contains the slope polygon at β. Note that we can find a neighborhood U of β on which no polygon outside T occurs, by eliminating elements of $S \setminus T$ one at a time.

Let P be a maximal polygon in T, *i.e.*, one which does not lie on or below any other element of T. By lower semicontinuity (Theorem 7.4.5), for every neighborhood V of β within U, there is a nonempty open subset on which the slope polygon is identically equal to P. This proves the claim. □

If the slope polygon does indeed vary, it is unclear whether one can expect to construct a slope filtration. One does get a result in case the polygon is locally constant, or more generally if one of its vertices is locally constant.

Lemma 7.4.8. — *Let A be an $n \times n$ matrix over $\tilde{\mathcal{R}}_R^{\mathrm{int}}$ which is invertible over $\tilde{\mathcal{R}}_R^{\mathrm{bd}}$, fix $x_1, \dots, x_n \in \tilde{\mathcal{R}}_R^{\mathrm{bd}}$, and choose $y_1, \dots, y_n \in \tilde{\mathcal{E}}_R$ so that*

$$y_i - x_i = \sum_j A_{ij} \varphi^a(y_j) \qquad (i = 1, \dots, n). \tag{7.4.8.1}$$

Then $y_i \in \tilde{\mathcal{R}}_R^{\mathrm{bd}}$ for $i = 1, \dots, n$ if and only if for each $\beta \in \mathcal{M}(R)$, the image of y_i in $\tilde{\mathcal{E}}_{\mathcal{H}(\beta)}$ belongs to $\tilde{\mathcal{R}}_{\mathcal{H}(\beta)}^{\mathrm{bd}}$ for $i = 1, \dots, n$.

Proof. — We may assume without loss of generality that $x_1, \dots, x_n \in \tilde{\mathcal{R}}_R^{\mathrm{int}}$ and $y_1, \dots, y_n \in W(R)$. Fix $r > 0$ for which $x_i, A_{ij} \in \tilde{\mathcal{R}}_R^{\mathrm{int},r}$ and $(A^{-1})_{ij} \in \tilde{\mathcal{R}}_R^{\mathrm{bd},r}$, and put $x'_i = \sum_j (A^{-1})_{ij} x_j$.

Suppose that $y_1, \dots, y_n \in \tilde{\mathcal{R}}_R^{\mathrm{int}}$. We may then choose some $s > 0$ so that $y_1, \dots, y_n \in \tilde{\mathcal{R}}_R^{\mathrm{int},s}$. By (7.4.8.1), we must also have $y_1, \dots, y_n \in \tilde{\mathcal{R}}_R^{\mathrm{int},s'}$ for $s' = \min\{rp^a, sp^a\}$; it follows that $y_1, \dots, y_n \in \tilde{\mathcal{R}}_R^{\mathrm{int},rp^a}$. In particular,

$$\lambda(\alpha^r)(y_i) \leqslant \lambda(\alpha^{rp^a})(y_i)^{p^{-a}} = \lambda(\alpha^r)(\varphi^a(y_i))^{p^{-a}}.$$

Consequently, if $\lambda(\alpha^r)(\varphi^a(y_i)) > \lambda(\alpha^r)(A^{-1})^{1/(1-p^{-a})}$, then

$$\lambda(\alpha^r)(\varphi^a(y_i)) > \lambda(\alpha^r)(A^{-1})\lambda(\alpha^r)(y_i)$$

and rewriting (7.4.8.1) as $\varphi^a(y_i) + x'_i = \sum_j (A^{-1})_{ij} y_j$ yields

$$\max_i \{\lambda(\alpha^r)(\varphi^a(y_i))\} = \lambda(\alpha^r)(x').$$

To summarize, if $y_1, \dots, y_n \in \tilde{\mathcal{R}}_R^{\mathrm{int}}$, then $y_1, \dots, y_n \in \tilde{\mathcal{R}}_R^{\mathrm{int},rp^a}$ and

$$\lambda(\alpha^{rp^a})(y_i) \leqslant \max\{\lambda(\alpha^r)(x'), \lambda(\alpha^r)(A^{-1})^{1/(1-p^{-a})}\}. \tag{7.4.8.2}$$

In particular, if the images of $y_1, \dots, y_n$ in $W(\mathcal{H}(\beta))$ belong to $\tilde{\mathcal{R}}_{\mathcal{H}(\beta)}^{\mathrm{int}}$ for all $\beta \in \mathcal{M}(R)$, then these images belong to $\tilde{\mathcal{R}}_{\mathcal{H}(\beta)}^{\mathrm{int},rp^a}$ and $\lambda(\beta^{rp^a})(y_i)$ is bounded uniformly over β. This yields the desired result. □

Theorem 7.4.9. — *Let M be a φ^a-module over $\tilde{\mathcal{R}}_R$ of constant rank n. Let $\mu_1(M, \beta) \geqslant \cdots \geqslant \mu_n(M, \beta)$ be the slopes of M at $\beta \in \mathcal{M}(R)$ listed with multiplicity. Suppose that for some $m \in \{1, \dots, n-1\}$, for all $\beta \in \mathcal{M}(R)$, $\mu_m(M, \beta) > \mu_{m+1}(M, \beta)$ and $\mu_1(M, \beta) + \cdots + \mu_m(M, \beta)$ is equal to a constant value. Then there exists a unique φ^a-submodule N of M of rank m such that M/N is a φ^a-module and for each $\beta \in \mathcal{M}(R)$, the slopes of N at β are $\mu_1(M, \beta), \dots, \mu_m(M, \beta)$ while the slopes of M/N at β are $\mu_{m+1}(M, \beta), \dots, \mu_n(M, \beta)$.*

Proof. — It suffices to check the claim locally around a point $\beta \in \mathcal{M}(R)$, since φ^a-modules can be glued by Corollary 6.3.13. Set notation as in the proof of Theorem 7.4.5; using faithfully flat descent, we may reduce to the case $R = R' = S$. Let N be the $\tilde{\mathcal{R}}_R^{\mathrm{bd}}$-span of $\mathbf{v}_1, \dots, \mathbf{v}_n$, and let $N_0 \subset N$ be the $\tilde{\mathcal{R}}_R^{\mathrm{int}}$-span of $\mathbf{v}_1, \dots, \mathbf{v}_n$.

By Lemma 7.4.4, for some S which is the union of faithfully finite étale R-subalgebras $\{S_i\}_{i \in I}$, we may split $N_0 \otimes_{\tilde{\mathcal{R}}_R^{\mathrm{int}}} W(S)$ uniquely as a direct sum of submodules whose base extensions to $\tilde{\mathcal{E}}_R$ are globally pure φ-submodules. From the construction, there is a profinite group G acting on S and each S_i in such a way that $S_i^G = R$.

Each projector in the splitting can be viewed as an element $\mathbf{v}$ of $N_0^\vee \otimes N_0 \otimes_{\tilde{\mathcal{R}}_R^{\mathrm{int}}} W(S)$ which is G-invariant. This forces $\mathbf{v} \in N_0^\vee \otimes N_0 \otimes_{\tilde{\mathcal{R}}_R^{\mathrm{int}}} W(R)$; that is, the projectors are all defined over $W(R)$. In particular, $N \otimes_{\tilde{\mathcal{R}}_R^{\mathrm{bd}}} \tilde{\mathcal{E}}_R$ splits uniquely into globally pure φ-submodules.

We will show that the splitting separating the first m slopes from the others descends to N; for this, we may by the proof of [**83**, Proposition 5.4.5] (plus a descent argument as in the previous paragraph) reduce to the case where N admits a basis $\mathbf{e}_1, \dots, \mathbf{e}_n$ on which φ^a acts *via* a matrix of the form UD, where U is an upper triangular unipotent matrix congruent to 1 modulo p, and $D_{ii} = p^{c_i}$ with $c_1 \geqslant \cdots \geqslant c_n$. By Lemma 7.4.8, we reduce to checking the splitting pointwise on $\mathcal{M}(R)$. But for $R = L$ an analytic field, the splitting in question is given by the proof of [**83**, Theorem 5.5.2]. This completes the proof. □

Corollary 7.4.10. — *Let M be a φ^a-module over $\tilde{\mathcal{R}}_R$ such that the slope polygon function of M is constant on $\mathcal{M}(R)$. Then there exists a unique filtration $0 = M_0 \subset \cdots \subset M_l = M$ of M by φ^a-submodules such that $M_1/M_0, \dots, M_l/M_{l-1}$ are φ^a-modules which are pure of constant slope, and $\mu(M_1/M_0) > \cdots > \mu(M_l/M_{l-1})$.*

Proof. — By Lemma 7.2.2, we may assume that M is of constant rank. We may then induct on the rank using Theorem 7.4.9. □

Corollary 7.4.11. — *Let M be a φ^a-module over $\tilde{\mathcal{R}}_R$ with everywhere negative slopes. Then $H^0_{\varphi^a}(M) = 0$, $H^0_{\varphi^a}(M \otimes_{\tilde{\mathcal{R}}_R} \tilde{\mathcal{R}}_{\mathcal{H}(\beta)}) = 0$ for all $\beta \in \mathcal{M}(R)$, and the map*

$$H^1_{\varphi^a}(M) \longrightarrow \prod_{\beta \in \mathcal{M}(R)} H^1_{\varphi^a}(M \otimes_{\tilde{\mathcal{R}}_R} \tilde{\mathcal{R}}_{\mathcal{H}(\beta)}) \tag{7.4.11.1}$$

is injective.

Proof. — By Theorem 4.2.12, $H^0_{\varphi^a}(M \otimes_{\tilde{\mathcal{R}}_R} \tilde{\mathcal{R}}_{\mathcal{H}(\beta)}) = 0$ for all $\beta \in \mathcal{M}(R)$; since the map

$$H^0_{\varphi^a}(M) \longrightarrow \prod_{\beta \in \mathcal{M}(R)} H^0_{\varphi^a}(M \otimes_{\tilde{\mathcal{R}}_R} \tilde{\mathcal{R}}_{\mathcal{H}(\beta)})$$

is evidently injective, it follows that $H^0_{\varphi^a}(M) = 0$. If $x \in H^1_{\varphi}(M)$ has zero image in $H^1_{\varphi^a}(M \otimes_{\tilde{\mathcal{R}}_R} \tilde{\mathcal{R}}_{\mathcal{H}(\beta)})$ for each $\beta \in \mathcal{M}(R)$, then x defines an extension

$$0 \longrightarrow M \longrightarrow P \longrightarrow \tilde{\mathcal{R}}_R \longrightarrow 0$$

whose base extension to $\tilde{\mathcal{R}}_{\mathcal{H}(\beta)}$ splits for each $\beta \in \mathcal{M}(R)$. This implies that the largest slope of P is identically 0 and the second-largest slope is always negative, so Theorem 7.4.9 implies that $P \cong M \oplus \tilde{\mathcal{R}}_R$. It follows that $x = 0$, completing the proof. □

Remark 7.4.12. — For any given φ^a-module over $\tilde{\mathcal{R}}_R$, the conclusion of Corollary 7.4.11 holds for $M(n)$ for n sufficiently small, since Proposition 7.4.6 ensures that the hypothesis of Corollary 7.4.11 is satisfied. On the other hand, the injectivity of (7.4.11.1) also holds for $M(n)$ for n sufficiently large, as in that case $H^1_{\varphi^a}(M(-n)) = 0$ by Proposition 6.2.2.

Remark 7.4.13. — As noted earlier, there is a generalization of slope theory for φ-modules orthogonal to the one given here, where one continues to work with rings of power series but with coefficients in more general rings (such as affinoid algebras over $\mathbb{Q}_p$). These are called *arithmetic families* in [**90**], where they are distinguished from the *geometric families* arising here. Unfortunately, it seems difficult to achieve any results in the context of arithmetic families as complete as those given here, in no small part because such results would most likely require a heretofore nonexistent slope theory for Frobenius modules over a Robba ring consisting of Laurent series over a *nondiscretely* valued field. One does however get some important information by working in neighborhoods of rigid analytic points; for instance, one can construct global slope filtrations in such neighborhoods [**97**], which is relevant for applications to p-adic automorphic forms *via* the study of eigenvarieties [**98**].

CHAPTER 8

PERFECTOID SPACES

Up to this point, our constructions have generally taken as input a Banach ring or an adic Banach ring. In this sense they are *local*; our next step is to parlay this work into some results of a more global nature, including the relationship between φ-modules and étale local systems. The appropriate category of geometric spaces to use here is the category of *perfectoid spaces*, obtained by glueing together the adic spectra of perfectoid algebras using Huber's formalism of *adic spaces*.

8.1. Some topological properties

We begin by recalling some properties of topological spaces relevant to the study of adic spaces. The key definition of a *spectral space*, and the basic properties of that definition, are due to Hochster [**76**].

Definition 8.1.1. — A topological space X is *sober* if every irreducible closed subset has a unique generic point. This implies that X is T_0: no two distinct points belong to exactly the same closed subsets of X.

A topological space is *quasiseparated* if the intersection of any two quasicompact open subsets is again quasicompact. We will write *qcqs* as an abbreviation for *quasicompact and quasiseparated.*

A *spectral space* is a topological space X which is sober and qcqs and admits a neighborhood basis consisting of quasicompact open subsets.

A *locally spectral space* is a topological space admitting a neighborhood basis consisting of open spectral subspaces. Such a space is spectral if and only if it is qcqs.

A number of equivalent characterizations of spectral spaces can be found in [**76**], including the following.

Lemma 8.1.2. — *A topological space is spectral if and only if it is an inverse limit of finite T_0 spaces.*

Proof. — See [**76**, §13, Proposition 10]. □

Corollary 8.1.3. — *Any inverse limit of spectral spaces is again a spectral space.*

Definition 8.1.4. — For X a spectral space, the *patch topology* on X is the topology generated by quasicompact open subsets for the original topology (which we also call the *spectral topology* to clarify the distinction) and their complements. Beware that one cannot recover the spectral topology from the patch topology alone; for example, there is another spectral topology whose quasicompact open subsets are the complements of the quasicompact open subsets for the original topology [**76**, Proposition 8].

Lemma 8.1.5. — *Let X be a spectral space. Then X is compact for the patch topology.*

Proof. — See [**76**, Theorem 1]. □

Definition 8.1.6. — A map $f : Y \to X$ between locally spectral spaces is *spectral* if it is continuous and for any qcqs open subsets $U \subseteq X, V \subseteq Y$ with $f(V) \subseteq U$, the induced map $V \to U$ is quasicompact (that is, the inverse image of any quasicompact open subset is again quasicompact). Equivalently, the inverse image of any quasicompact open subset is a quasicompact open subset. In particular, any spectral morphism is continuous for the patch topologies.

Remark 8.1.7. — Any scheme is a locally spectral space. Any scheme which is qcqs in the sense of algebraic geometry (*i.e.*, its underlying topological space is quasicompact and the absolute diagonal morphism is quasicompact) is spectral. Any morphism of schemes is spectral.

We have the following refinement of Theorem 2.4.5.

Theorem 8.1.8. — *For any adic Banach ring (A, A^+),* $\mathrm{Spa}(A, A^+)$ *is a spectral space. For any morphism $(A, A^+) \to (B, B^+)$ of adic Banach rings, the map* $\mathrm{Spa}(B, B^+) \to \mathrm{Spa}(A, A^+)$ *is spectral.*

Proof. — See again [**77**, Theorem 3.5 (i)-(ii)]. □

Remark 8.1.9. — Remark 2.4.9 and Corollary 2.5.13 together can be reinterpreted as follows. For (A, A^+) an adic Banach algebra, the image of $\mathcal{M}(A)$ in $\mathrm{Spa}(A, A^+)$ is not necessarily dense for the patch topology; however, if A is an affinoid algebra over an analytic field and $A^+ = A^\circ$, then already $\mathrm{Maxspec}(A)$ is dense in $\mathrm{Spa}(A, A^+)$ for the patch topology.

The comparison of rigid and Berkovich analytic spaces with adic spaces involves the following definition.

Definition 8.1.10. — Let X be a locally spectral topological space. We say X is *taut* if X is quasiseparated and the closure of every quasicompact subset of X is again quasicompact. For example, if X is qcqs, or more generally if X is quasiseparated and admits a locally finite covering by quasicompact open subsets, then X is taut.

A spectral morphism between locally spectral topological spaces is *taut* if the inverse image of every taut open subspace is taut. For example, any qcqs morphism is taut. For more basic properties, see [**79**, Lemma 5.1.3].

Remark 8.1.11. — For k a field, glueing finitely many copies of $\operatorname{Spec} k[t]$ along $\operatorname{Spec} k[t, t^{-1}]$ gives a taut locally spectral space, but glueing infinitely many copies does not. By contrast, for K an analytic field, glueing even two copies of $\operatorname{Spa}(K\{T\}, K\{T\}^\circ)$ along the complement of the origin does not give a taut space, because the complement of the origin is not quasicompact.

8.2. Adic spaces

We now construct spaces out of adic Banach rings, following Huber. Beware that there does not yet seem to be a consensus about terminology concerning adic spaces, so one must check carefully when comparing results across sources.

We begin with an enhancement of the concept of a locally ringed space.

Definition 8.2.1. — We define the category of *locally valuation-ringed spaces* as follows.

- The objects are triples $X = (|X|, \mathcal{O}_X, (v_x)_{x\in X})$ in which $|X|$ is a topological space, $\mathcal{O}_X$ is a sheaf of complete topological rings, and each v_x is a semivaluation on the set-theoretic stalk $\mathcal{O}_{X,x}$.
- The morphisms from $X = (|X|, \mathcal{O}_X, (v_x)_{x\in X})$ to $Y = (|Y|, \mathcal{O}_Y, (v_y)_{y\in Y})$ are pairs (f, φ) where $f : |X| \to |Y|$ is a continuous map and $\varphi : \mathcal{O}_Y \to f_*\mathcal{O}_X$ is a morphism of sheaves of topological rings such that for each $x \in X$, the restriction of v_x along $\varphi_x : \mathcal{O}_{Y,f(x)} \to \mathcal{O}_{X,x}$ is equivalent to $v_{f(x)}$.

We also refer to these spaces for short as *locally v-ringed spaces*.

Definition 8.2.2. — For (A, A^+) a sheafy adic Banach ring, we view $X = \operatorname{Spa}(A, A^+)$ as a locally v-ringed space where $\mathcal{O}_X$ is the structure sheaf on X (Definition 2.4.15) and v_x is the restriction to $\mathcal{O}_{X,x}$ of the canonical valuation on $\mathcal{H}(x)$ (Definition 2.4.11). Any locally v-ringed space of this form is called an *adic affinoid space*. An *adic space* is a locally v-ringed space admitting an open covering by adic affinoid spaces.

A *morphism* of adic spaces is just a morphism of underlying locally v-ringed spaces. Note that any such morphism is automatically an *adic morphism* in the sense of [**78**, §3] because our Banach rings are required to contain topologically nilpotent units; see [**78**, Proposition 3.2]. Note also that if X is an adic space and $U \to X$ is an open immersion of locally ringed spaces, then U naturally acquires the structure of an adic space in such a way that $U \to X$ becomes a morphism of adic spaces. We refer to any such morphism as an *open immersion* of adic spaces.

Definition 8.2.3. — Since the property of being sheafy is not known to be stable under various natural operations (*e.g.*, formation of finite étale extensions), we are forced to associate spaces also to general adic Banach rings. Perhaps the simplest way to do this is to sheafify in the style of Scholze–Weinstein [**116**, Definition 2.1.5]. Let **AdBan** be the category of adic Banach rings. View $\mathbf{AdBan}^{\mathrm{op}}$ as a site in which the coverings are rational coverings. Let $(\mathbf{AdBan}^{\mathrm{op}})\tilde{}$ be the associated topos (consisting of set-valued sheaves on $\mathbf{AdBan}^{\mathrm{op}}$). For $(A, A^+) \in \mathbf{AdBan}$, let $\widetilde{\mathrm{Spa}}(A, A^+) \in (\mathbf{AdBan}^{\mathrm{op}})\tilde{}$ be the sheafification of the presheaf

$$(B, B^+) \longrightarrow \mathrm{Hom}_{\mathbf{AdBan}}((A, A^+), (B, B^+));$$

any such object is called a *preadic affinoid space*. A *rational subspace* of $\widetilde{\mathrm{Spa}}(A, A^+)$ is a morphism of the form $\widetilde{\mathrm{Spa}}(B, B^+) \to \widetilde{\mathrm{Spa}}(A, A^+)$ for some rational localization $(A, A^+) \to (B, B^+)$. An *open immersion* in $(\mathbf{AdBan}^{\mathrm{op}})\tilde{}$ is a morphism $f : \mathcal{F} \to \mathcal{G}$ such that for all $(A, A^+) \in \mathbf{AdBan}$ and all morphisms $\widetilde{\mathrm{Spa}}(A, A^+) \to \mathcal{G}$ in $(\mathbf{AdBan}^{\mathrm{op}})\tilde{}$, there is an open subset $U \subseteq \mathrm{Spa}(A, A^+)$ such that

$$\mathcal{F} \times_{\mathcal{G}} \widetilde{\mathrm{Spa}}(A, A^+) = \varinjlim_{V \subseteq U, V \text{ rational}} \widetilde{\mathrm{Spa}}(\mathcal{O}_{\mathrm{Spa}(A,A^+)}(V), \mathcal{O}^+_{\mathrm{Spa}(A,A^+)}(V)).$$

A *preadic space* (called an *adic space* in [**116**]) is a functor $\mathcal{F} \in (\mathbf{AdBan}^{\mathrm{op}})\tilde{}$ such that

$$\mathcal{F} = \varinjlim_{\widetilde{\mathrm{Spa}}(A,A^+) \to \mathcal{F} \text{ open}} \widetilde{\mathrm{Spa}}(A, A^+).$$

There are natural functors from adic spaces to preadic spaces and from preadic spaces to locally v-ringed spaces, whose composition is the full embedding of adic spaces into locally v-ringed spaces. This allows us to regard adic spaces as a full subcategory of preadic spaces; in particular, when (A, A^+) is sheafy we will freely confuse $\mathrm{Spa}(A, A^+)$ with $\widetilde{\mathrm{Spa}}(A, A^+)$. We may also associate to any preadic space X an underlying topological space $|X|$ in such a way that $\widetilde{\mathrm{Spa}}(A, A^+)$ has underlying topological space $\mathrm{Spa}(A, A^+)$.

Remark 8.2.4. — Beware that a preadic affinoid space can be an adic space without being an adic affinoid space. See Remark 3.6.27.

Remark 8.2.5. — The categories of locally v-ringed spaces and preadic spaces admit fibred products. However, it is unknown whether the fibred product of adic spaces (over an adic space) is again an adic space. A counterexample would necessarily involve nonnoetherian Banach rings thanks to Proposition 2.4.16; on the other hand, the example of perfectoid spaces (§8.3) shows that failure of the noetherian property alone is not sufficient.

Definition 8.2.6. — Let X be a preadic space. We say that X is *quasicompact* if it admits a finite covering by open immersions of preadic affinoid spaces. We say that X is *quasiseparated* if for any two open immersions $U_1 \to X, U_2 \to X$ with U_1, U_2 quasicompact, $U_1 \times_X U_2$ is also quasicompact. We say that X is *taut* if X is quasiseparated and for any open immersion $U \to X$ with U quasicompact, there exists a covering $V_1 \to X, V_2 \to X$ by open immersions such that V_1 is quasicompact, $U \to X$ factors through V_1, and $U \times_X V_2 = \varnothing$. Note that for X an adic space, X is quasicompact, quasiseparated, or taut if and only if $|X|$ has the same property.

Definition 8.2.7. — The assignments $(A, A^+) \mapsto A, (A, A^+) \mapsto A^+$ give rise to sheaves $\mathcal{O}_X, \mathcal{O}_X^+$ of rings on any preadic space X. In case X is an adic space, the sheaf $\mathcal{O}_X$ coincides with the structure sheaf defined previously, while $\mathcal{O}_X^+$ coincides with the subsheaf of $\mathcal{O}_X$ such that for each open subset U of X, a section $f \in \mathcal{O}_X(U)$ belongs to $\mathcal{O}_X^+(U)$ if and only if for each $x \in X$, the image of f in $\mathcal{H}(x)$ belongs to $\mathcal{H}(x)^+$. The stalk $\mathcal{O}_{X,x}^+$ of $\mathcal{O}_X^+$ at $x \in X$ may then be identified with the inverse image of $\mathcal{H}(x)^+$ in $\mathcal{O}_{X,x}$.

For (A, A^+) an adic Banach ring and $X = \widetilde{\operatorname{Spa}}(A, A^+)$, by (2.4.3.1) we have $A^+ = \mathcal{O}_X^+(X)$.

Remark 8.2.8. — For X an adic affinoid space, the sheaf $\mathcal{O}_X^+$ need not be acyclic. However, for perfectoid spaces, $\mathcal{O}_X^+$ is acyclic as a sheaf of almost modules; see Proposition 8.3.2.

Lemma 8.2.9. — *Let (A, A^+) be a sheafy adic Banach ring. Then for each adic space X, the global sections functor induces a bijection between morphisms $X \to \operatorname{Spa}(A, A^+)$ of preadic spaces and morphisms $A \to \mathcal{O}_X(X)$ of topological rings taking A^+ into $\mathcal{O}_X^+(X)$.*

Proof. — See [**78**, Proposition 2.1]. □

Remark 8.2.10. — The statement of Lemma 8.2.9 also holds when X is a preadic space, but in that case it is purely formal. The essential content of the lemma is to define the functor from adic spaces to preadic spaces.

The construction of Definition 2.4.8 extends to preadic and adic spaces as follows. (See also [**79**, §8.1].)

Definition 8.2.11. — Recall that for any adic Banach ring (A, A^+), there is a continuous map $\operatorname{Spa}(A, A^+) \to \mathcal{M}(A)$ constructed in Definition 2.4.8 and a canonical but discontinuous section $\mathcal{M}(A) \to \operatorname{Spa}(A, A^+)$. One way to interpret these constructions is that the first map takes each $x \in \operatorname{Spa}(A, A^+)$ to the unique $y \in \mathcal{M}(A)$ belonging to the intersection of all open neighborhoods of x in $\operatorname{Spa}(A, A^+)$, and that the topology on $\mathcal{M}(A)$ coincides with the quotient topology (not the subspace topology).

Now let X be a preadic spaceLet $\overline{X}$ be the set of $x \in |X|$ for which v_x is equivalent to a real semivaluation. Then for any $x \in X$, the intersection of all open neighborhoods of x in X contains a unique point y of $\overline{X}$ (by the previous paragraph). We thus obtain a set-theoretic map $|X| \to \overline{X}$; we equip $\overline{X}$ with the quotient topology and call it the *real quotient* of X. An open subset of X is *partially proper* if it arises as the inverse image of a (necessarily open) subset of $\overline{X}$.

Lemma 8.2.12. — *Let X be an adic space.*

(a) *If X is quasicompact, then so is $\overline{X}$.*

(b) *If X is taut, then $\overline{X}$ is Hausdorff (and thus is the maximal Hausdorff quotient of X).*

(c) *If X is taut, then any partially proper open subset of X is also taut.*

Proof. — Part (a) is trivial. Part (b) follows from the proof of [**79**, Lemma 8.1.8 (ii)]. To prove (c), let U be a partially proper open subset of X and let V be a quasicompact open subset of U. Let $\overline{V}$ be the image of V in $\overline{X}$; it is a closed subset by (a) and (b). Let W be the inverse image of $\overline{V}$ in $\overline{X}$; then W is the closure of V in X. Since $W \subseteq U$, W is also the closure of V in U; this yields (c). □

Lemma 8.2.12 (b) fails without the taut condition as follows.

Example 8.2.13. — Let K be an analytic field. Let X be the adic space obtained by glueing two copies of the closed unit disc $\mathrm{Spa}(K\{T\}, K\{T\}^\circ)$ along the complement of the origin. Then $\overline{X}$ is obtained by glueing two copies of $\mathcal{M}(K\{T\})$ along the complement of the origin, and thus is not Hausdorff.

Using Lemma 8.2.12, one can formulate the relationship between adic spaces and Berkovich spaces.

Proposition 8.2.14. — *The real quotient functor defines an equivalence of categories between taut adic spaces locally of finite type over an analytic field K (*i.e.*, covered by the adic spectra of affinoid algebras over K) and Hausdorff strictly K-analytic spaces in the sense of Berkovich* [**15**].

Proof. — This follows from Lemma 8.2.12 and [**79**, Proposition 8.3.1]. □

Remark 8.2.15. — In practice, the adic spaces arising in applications are almost always taut. However, nontaut adic spaces do arise in *arithmetic* relative p-adic Hodge theory, where one considers Galois representations not on $\mathbb{Q}_p$-vector spaces but on vector bundles over more general analytic spaces. In that context, it has been shown by us [**90**] and Hellmann [**70**] that the analogue of the étale locus for an arithmetic family of (φ, Γ)-modules is in general a nontaut adic space.

In [**79**, Chapter 1], Huber introduces a large number of properties of morphisms of adic spaces, but only under some noetherian hypotheses which are too restrictive to apply to perfectoid spaces. We thus introduce some *ad hoc* definitions that agree with Huber's definitions when the latter are applicable, as in [**114**, Definition 7.1].

Definition 8.2.16. — Let $\psi : Y \to X$ be a morphism of preadic spaces. We have already defined what it means for ψ to be an *open immersion* (see Definition 8.2.3).

- We say that ψ is *surjective* if for any morphism $Z \to X$, the map $|Y \times_X Z| \to |Z|$ is surjective.

– We say that ψ is *finite étale* if locally on the target, ψ corresponds to a morphism of the form $\widetilde{\operatorname{Spa}}(B, B^+) \to \widetilde{\operatorname{Spa}}(A, A^+)$ where B is a finite étale A-algebra and B^+ is the integral closure of A^+ in B.
– We say that ψ is *étale* if for each $y \in Y$, there exists an open neighborhood U of y in Y such that the restriction of ψ to U factors as an open immersion followed by a finite étale morphism followed by another open immersion.

These properties are evidently stable under base extension.

Lemma 8.2.17. — *The following statements are true.*

(a) *For (A, A^+) an adic Banach ring, the global sections functor induces an equivalence of categories between finite étale morphisms to $\widetilde{\operatorname{Spa}}(A, A^+)$ and* $\mathbf{F\acute{E}t}(A)$.

(b) *Any étale morphism $Y \to X$ of preadic spaces with Y quasicompact factors uniquely as $Y \to Z \to X$ where $Y \to Z$ is surjective, Z is quasicompact, and $Z \to X$ is an open immersion.*

(c) *The properties introduced in Definition 8.2.16 are stable under compositions and fibred products.*

Proof. — Part (a) is immediate from Theorem 2.6.9 and the formal properties of the functor $\widetilde{\operatorname{Spa}}$. To prove (b)-(c), we may use Lemma 2.6.2 and Proposition 2.6.8 to reduce to the case where all of the spaces involved are classical affinoid spaces, for which we may appeal to [**33**, Proposition 3.1.7] for (b) and [**79**, Proposition 1.7.5] for (c). □

Remark 8.2.18. — When applying Lemma 8.2.17 (a), beware that we do not know that a finite étale extension of a sheafy adic Banach algebra is sheafy, or even that a finite étale cover of an adic affinoid space is itself an adic space. In particular, these statements will not follow from Theorem 8.2.22.

Definition 8.2.19. — A family $\{Y_i \to X\}_i$ of morphisms of preadic spaces is a *set-theoretic covering* if the morphism from the disjoint union of the Y_i to X is surjective. Note that if $Y_i \to X$ is étale, we may factor $Y_i \to Z_i \to X$ as in Lemma 8.2.17 (b), and then the original family is a set-theoretic covering if and only if the family $\{Z_i \to X\}$ is.

We may define the *small finite étale site* (resp. the *small étale site*) on a preadic space X over an analytic field as the site $X_{\mathrm{f\acute{e}t}}$ (resp. $X_{\mathrm{\acute{e}t}}$) whose objects consist of finite étale (resp. étale) morphisms $Y \to X$ and whose coverings are set-theoretic coverings.

We define a *stable basis* of $X_{\mathrm{\acute{e}t}}$ to be a basis $\mathcal{B}$ of $X_{\mathrm{\acute{e}t}}$ consisting of adic affinoid spaces such that for any morphism $Y' \to Y$ in $X_{\mathrm{\acute{e}t}}$ which is either finite étale or a rational subdomain embedding, if $Y \in \mathcal{B}$ then $Y' \in \mathcal{B}$. (Note that the basis property also includes stability under formation of fibred products.) We say that X is *stably adic* if there exists a stable basis of $X_{\mathrm{\acute{e}t}}$. For instance, rigid analytic spaces are stably adic; we will see later that perfectoid spaces are also stably adic (Definition 8.3.1).

Many properties of the étale topology may be verified by making the corresponding verifications for the finite étale topology and the adic topology separately. In this process, the role of Tate's reduction argument for the adic topology (Proposition 2.4.20) will be played by the following argument, adapted from de Jong and van der Put [**33**, Proposition 3.2.2].

Proposition 8.2.20. — *Let X be a preadic space, and let $\mathcal{B}$ be a basis of $X_{\text{ét}}$ consisting of preadic affinoid subspaces which is closed under formation of finite étale extensions and rational subdomain embeddings (*e.g.*, a stable basis). Let $\mathcal{P}$ be a property of coverings in $X_{\text{ét}}$ of and by elements of $\mathcal{B}$, and assume that the following conditions hold.*

(a) *Any covering admitting a refinement having property $\mathcal{P}$ also has property $\mathcal{P}$.*

(b) *Any composition of coverings having property $\mathcal{P}$ also has property $\mathcal{P}$.*

(c) *For any $Y \in \mathcal{B}$, any rational covering of Y has property $\mathcal{P}$.*

(d) *For any $Y \in \mathcal{B}$, any faithfully finite étale morphism $Y' \to Y$, viewed as a covering, has property $\mathcal{P}$.*

Then every covering in $X_{\text{ét}}$ of and by elements of $\mathcal{B}$ has property $\mathcal{P}$.

Proof. — We first note that using (a) and Lemma 8.2.17 (b), we may formally extend (c) to any covering for the adic topology. We will use (c) in this stronger form without further comment.

We next establish the following extra condition.

(e) Let $Z \to U$ be a surjective morphism between elements of $\mathcal{B}$ which factors as $Z \to V \to U$ with $V \to U$ finite étale and $Z \to V$ an open immersion. Then $Z \to U$, viewed as a covering, has property $\mathcal{P}$.

We induct on the maximum degree d of $V \to U$. The case $d = 1$ holds because in this case $Z \to U$ is a surjective open immersion and hence an isomorphism (so for instance (d) applies). For $d > 1$, let W be the complement of the diagonal in $V \times_U V$; then the second projection $\mathrm{pr}_2 : W \to V$ is finite étale of maximum degree $d-1$. Put $Z' = (Z \times_U V) \times_{V \times_U V} W$ and $U' = \mathrm{pr}_2(Z')$; these spaces are both quasicompact by Lemma 8.2.17 (b). We may thus find a finite covering $\{U'_i \to U'\}_i$ for the adic topology by elements of $\mathcal{B}$. Put $Z'_i = Z' \times_{U'} U'_i$; the surjective étale morphism $\phi' : Z'_i \to U'_i$ induced by pr_2 factors through $V'_i \to U'_i$ for $V'_i = \mathrm{pr}_2^{-1}(U'_i) \times_{V \times_U V} W$. The latter morphism is finite étale of maximum degree $d - 1$, so the following coverings have property $\mathcal{P}$:

- $\{Z'_i \to U'_i\}$, by the induction hypothesis;
- $\{Z \to V\} \cup \{U'_i \to V\}_i$, by (c);
- $\{V \to U\}$, by (d);
- $\{Z \to U\} \cup \{Z'_i \to U\}_i$, by (b);
- $\{Z \to U\}$, by (a).

This completes the induction and hence the proof of (e).

Given (e), let $\{Y_i \to Y\}_i$ be any covering in $X_{\text{ét}}$ of and by elements of $\mathcal{B}$. For each i, we can find a covering $\{Y_{ij} \to Y_i\}_j$ such that the composition $Y_{ij} \to Y_i \to Y$ factors as $Y_{ij} \to Z_{ij} \to Y$ where Z_{ij} is finite étale and Y_{ij} is an open immersion. By Lemma 8.2.17 (b), we may write the image of $Y_{ij} \to Y$ as a finite union $\{U_{ijk}\}_k$ of elements of $\mathcal{B}$. Put $Y_{ijk} = Y_{ij} \times_Y U_{ijk}$. The following coverings then have property $\mathcal{P}$:

- $\{Y_{ijk} \to U_{ijk}\}$, by (e);
- $\{U_{ijk} \to Y\}$, by (c);
- $\{Y_{ijk} \to Y\}$, by (b);
- $\{Y_i \to Y\}$, by (a).

This completes the proof. □

This gives rise to the following analogue of Proposition 2.4.21.

Proposition 8.2.21. — *For $X, \mathcal{B}$ as in Proposition 8.2.20, let $\mathcal{F}$ be a presheaf of abelian groups on $X_{\text{ét}}$ such that for every $Y = \widetilde{\operatorname{Spa}}(A, A^+) \in \mathcal{B}$ and every covering $\mathfrak{V}$ of one of the following forms:*

(a) *a simple Laurent covering, or*

(b) *a faithfully finite étale morphism;*

we have

$$\check{H}^0(Y, \mathcal{F}; \mathfrak{V}) = \mathcal{F}(Y), \qquad \textit{resp.} \quad \check{H}^i(Y, \mathcal{F}; \mathfrak{V}) = \begin{cases} \mathcal{F}(Y) & i = 0 \\ 0 & i > 0. \end{cases}$$

Then for every covering $\mathfrak{V}$ of $Y \in \mathcal{B}$ by elements of $\mathcal{B}$,

$$H^0(Y, \mathcal{F}) = \check{H}^0(Y, \mathcal{F}; \mathfrak{V}) = \mathcal{F}(Y), \quad \textit{resp.} \ H^i(Y, \mathcal{F}) = \check{H}^i(Y, \mathcal{F}; \mathfrak{V}) = \begin{cases} \mathcal{F}(Y) & i = 0 \\ 0 & i > 0. \end{cases}$$

In particular, on Y, $\mathcal{F}$ takes the same value as its sheafification.

Proof. — As in the proof of Proposition 2.4.21, using Proposition 2.4.20 and Proposition 8.2.20 we may successively verify that:

- $\mathcal{F}(Y) \to \check{H}^0(Y, \mathcal{F}; \mathfrak{V})$ is injective;
- $\mathcal{F}(Y) \to \check{H}^0(Y, \mathcal{F}; \mathfrak{V})$ is bijective;
- in the second situation, $\mathfrak{V}$ is universally Čech-acyclic (its pullback along any morphism $Y' \to Y$ of elements of $\mathcal{B}$ is Čech-acyclic);

and then check the claims for $H^i(Y, \mathcal{F})$ by standard homological algebra. □

For vector bundles on adic spaces, one has analogues of the theorems of Tate and Kiehl. Recall that we cannot handle general coherent sheaves because rational localizations are in general not flat. (We will return to this issue in a subsequent paper.)

Theorem 8.2.22. — *Let $X = \operatorname{Spa}(A, A^+)$ be an adic affinoid space.*

(a) *For any finite projective A-module M, the presheaf $\tilde{M}$ with $\tilde{M}(U) = M \otimes_A \mathcal{O}_X(U)$ is an acyclic sheaf for the adic topology and the finite étale topology.*

(b) *The categories of finite projective A-modules, finite locally free $\mathcal{O}_X$-modules, and finite locally free $\mathcal{O}_{X_{\text{fét}}}$-modules are equivalent.*

(c) *Suppose that $X_{ét}$ admits a stable basis $\mathcal{B}$. Then for any finite projective A-module M, the presheaf $\tilde{M}$ on $X_{ét}$ defined as in (a) is an acyclic sheaf on $\mathcal{B}$, meaning that for $Y \in \mathcal{B}$ we have*

$$H^i(Y, \tilde{M}) = \begin{cases} \tilde{M}(Y) & i = 0 \\ 0 & i > 0. \end{cases}$$

(d) *Suppose that X is stably adic. Then the categories of finite projective A-modules and finite locally free $\mathcal{O}_{X_{ét}}$-modules are equivalent.*

Proof. — To prove (a), note that the adic case follows from Theorem 2.4.23 and the finite étale case follows from Lemma 8.2.17 (a). To prove (b), note that the equivalence between the first and second categories follows from Theorem 2.7.7, while the equivalence between the first and third categories follows from faithfully flat descent for finite étale morphisms (Theorem 1.3.4 and Theorem 1.3.5). Given (a), we may deduce (c) using Proposition 8.2.21. Given (b), we may deduce (d) using Proposition 8.2.20. □

8.3. Perfectoid spaces

We next globalize the theory of perfectoid algebras to obtain perfectoid spaces, following Scholze [**114**] (plus some minor modifications to avoid having to work over a perfectoid field). We also introduce the related notions of *preperfectoid spaces* and *relatively perfectoid spaces*, following Scholze–Weinstein [**116**].

Definition 8.3.1. — A *perfect uniform/perfectoid/preperfectoid/strongly preperfectoid/relatively perfectoid affinoid space* is a preadic affinoid space of the form $\widetilde{\mathrm{Spa}}(A, A^+)$ where (A, A^+) is a perfect uniform/perfectoid/preperfectoid/strongly preperfectoid/relatively perfectoid adic Banach algebra. Any such space is in fact an adic affinoid space by Theorem 3.1.13 (in the perfect case), Theorem 3.6.15 (in the perfectoid case), and Theorem 3.7.4 (in the remaining cases).

A *perfect/perfectoid/preperfectoid/strongly preperfectoid/relatively perfectoid space* is a preadic space covered by open subspaces which are perfect/perfectoid/preperfectoid/strongly preperfectoid/relatively perfectoid affinoid spaces. Any such space is in fact an adic space, and even a stably adic space: thanks to Theorem 3.6.14 and Theorem 3.6.21, the subspaces which are themselves perfect/perfectoid/preperfectoid/strongly preperfectoid/relatively perfectoid form a basis for the étale topology.

Proposition 8.3.2. — *Let $X = \mathrm{Spa}(A, A^+)$ be a perfect uniform or perfectoid affinoid space, and equip A with the spectral norm.*

(a) *We have $H^0(X, \mathcal{O}) = H^0(X_{ét}, \mathcal{O}) = A$.*

(b) *For $i > 0$, $H^i(X, \mathcal{O}) = H^i(X_{ét}, \mathcal{O}) = 0$.*

(c) *For $i > 0$, the groups $H^i(X, \mathcal{O}^+_X), H^i(X_{ét}, \mathcal{O}^+_X)$ are annihilated by $\mathfrak{m}_A$ (*i.e.*, they are* almost zero*).*

Proof. — The first two assertions follow from Theorem 2.8.10, Theorem 2.4.23, and Theorem 8.2.22. The third assertion follows from the first two assertions plus Remark 3.1.6 (in the perfect uniform case) or Proposition 3.6.9 (in the perfectoid case). □

Remark 8.3.3. — By Proposition 3.1.16, any adic affinoid space which is also a perfect uniform space is in fact a perfect uniform affinoid space. The corresponding statement for perfectoid spaces is unknown; see Remark 3.6.27.

Definition 8.3.4. — For X a perfect adic space, we may define sheaves of rings $*_X$ for

$$* = \tilde{\mathcal{E}}^{\mathrm{int}}, \tilde{\mathcal{E}}, \tilde{\mathcal{R}}^{\mathrm{int},r}, \tilde{\mathcal{R}}^{\mathrm{int},+}, \tilde{\mathcal{R}}^{\mathrm{int}}, \tilde{\mathcal{R}}^{\mathrm{bd},r}, \tilde{\mathcal{R}}^{\mathrm{bd},+}, \tilde{\mathcal{R}}^{\mathrm{bd}}, \tilde{\mathcal{R}}^{[s,r]}, \tilde{\mathcal{R}}^{r}, \tilde{\mathcal{R}}^{+}, \tilde{\mathcal{R}},$$

by glueing together the presheaves defined in Definition 5.3.1, which are sheaves by Theorem 5.3.3. We may also view these as presheaves on $X_{\mathrm{ét}}$; by Proposition 8.2.21, sheafifying these presheaves does not change their values on any perfect adic affinoid space.

We globalize the perfectoid correspondence as follows.

Theorem 8.3.5. — *There is an equivalence of categories between perfectoid adic spaces and pairs $(X, \mathcal{I})$ in which X is a perfect adic space and $\mathcal{I}$ is an ideal subsheaf of $\tilde{\mathcal{R}}_X^{\mathrm{int},+}$ which is locally generated by a single element which is primitive of degree* 1. *(In the latter category, morphisms have the form $(Y, \mathcal{J}) \to (X, \mathcal{I})$ where $Y \to X$ is a morphism of perfect uniform adic spaces and $\mathcal{J}$ is isomorphic to the pullback of $\mathcal{I}$.) This equivalence is compatible with rational localizations, fibred products, and étale morphisms; moreover, corresponding spaces are functorially homeomorphic.*

Proof. — This follows by combining Theorem 3.6.5, Theorem 3.6.14, and Theorem 3.6.21. □

This recovers Scholze's tilting correspondence from [**114**].

Corollary 8.3.6. — *Suppose that K is a perfectoid analytic field of characteristic* 0, *and let K' be the corresponding perfect analytic field of characteristic p from Theorem 3.5.3 (*i.e.*, the* tilt *of K in the sense of Scholze). Then there is a canonical equivalence of categories between perfectoid adic spaces over K and perfect uniform adic spaces over K' which is compatible with fibred products. Moreover, corresponding spaces have homeomorphic underlying topological spaces and isomorphic étale topoi.*

8.4. Étale local systems on adic spaces

We now wish to define and study étale local systems on adic spaces. For this, we must clarify the distinction between étale local systems on the Zariski spectrum and the adic spectrum of an adic Banach ring. We begin with a refinement of Lemma 2.4.17.

Lemma 8.4.1. — *Let (A, A^+) be an adic Banach ring. Let $\mathrm{Spec}(A') \to \mathrm{Spec}(A)$ be a surjective étale morphism of schemes. Then for any $\alpha \in \mathcal{M}(A)$, there exists a rational localization $(A, A^+) \to (B, B^+)$ encircling α such that $A' \otimes_A B$ splits as a direct sum of subrings, at least one of which is faithfully finite étale over B.*

Proof. — By hypothesis, we can choose some prime ideal $\mathfrak{q}$ of A' above $\mathfrak{p}_\alpha$. The norm on $\kappa(\mathfrak{p}_\alpha)$ induced from $\mathcal{H}(\alpha)$ then extends to $\kappa(\mathfrak{q})$ and thus defines a point $\beta \in \mathcal{M}(A')$. By the Jacobian criterion, we can write $A'_{\mathfrak{q}}$ as a complete intersection $A_{\mathfrak{p}_\alpha}[x_1, \dots, x_n]/(f_1, \dots, f_n)$ with invertible Jacobian determinant J. We may then choose a rational localization $(A, A^+) \to (B, B^+)$ encircling α so that $f_1, \dots, f_n$ have coefficients in B and that J is invertible in

$$S[x_1, \dots, x_n]/(f_1, \dots, f_n).$$

This yields the desired result. □

Lemma 8.4.2. — *Let (A, A^+) be an adic Banach ring. Let V be an étale $\mathbb{Q}_p$-local system over* $\mathrm{Spec}(A)$. *Then for any $\alpha \in \mathcal{M}(A)$, there exists a rational localization $(A, A^+) \to (B, B^+)$ encircling α such that $V \times_{\mathrm{Spec}(A)} \mathrm{Spec}(B)$ is an isogeny $\mathbb{Z}_p$-local system on* $\mathrm{Spec}(B)$.

Proof. — This follows by combining Lemma 1.4.8 with Lemma 8.4.1. □

Definition 8.4.3. — Let X be a preadic space. For each covering of X by preadic affinoid spaces $U_i = \widetilde{\mathrm{Spa}}(A_i, A_i^+)$, construct the categories of descent data for étale $\mathbb{Z}_p$-local systems and etale $\mathbb{Q}_p$-local systems for the covering: that is, specify a local system V_i over $\mathrm{Spec}(A_i)$ for each i, then cover each intersection $U_i \cap U_j$ with preadic affinoid spaces $\widetilde{\mathrm{Spa}}(B_k, B_k^+)$ and define isomorphisms of the restrictions of V_i and V_j to B_k satisfying the cocycle condition. Then form the 2-limit (as in Remark 1.2.9) over all covering families; we call the resulting categories the categories of *étale $\mathbb{Z}_p$-local systems over X* and *étale $\mathbb{Q}_p$-local systems over X*. When $X = \mathrm{Spa}(A, A^\circ)$ for A an affinoid algebra over an analytic field, these categories are equivalent to the corresponding categories defined by de Jong [**32**] in terms of étale covering spaces.

Remark 8.4.4. — Thanks to the local factorization of étale morphisms, one gets the same categories of local systems if one takes coverings in the étale topology rather than the adic topology.

Remark 8.4.5. — The natural functor from étale $\mathbb{Z}_p$-local systems over $\mathrm{Spec}(A)$ to étale $\mathbb{Z}_p$-local systems over $\widetilde{\mathrm{Spa}}(A, A^+)$ is an equivalence of categories, since $\mathbb{Z}_p$-local systems are determined by finite étale algebras (Remark 1.4.3) and these glue over covering families of rational localizations (Theorem 2.6.9). The corresponding statement for $\mathbb{Q}_p$-local systems is false; see Remark 8.4.8.

We have the following variant of Lemma 8.4.2, with essentially the same proof.

Proposition 8.4.6. — *Let (A, A^+) be an adic Banach ring. Let V be an étale $\mathbb{Q}_p$-local system over $\widetilde{\mathrm{Spa}}(A, A^+)$. Then for any $\alpha \in \mathcal{M}(A)$, there exists a rational localization $(A, A^+) \to (B, B^+)$ encircling α such that the restriction of V to $\widetilde{\mathrm{Spa}}(B, B^+)$ is an isogeny $\mathbb{Z}_p$-local system.*

Proof. — By Remark 1.4.7 and Lemma 8.2.17 (a), given a preadic space X and an object $T \in \mathbb{Z}_p\text{-}\mathbf{Loc}(X)$, the functor taking a morphism $Y \to X$ to the pairs (T', ι) in which $T' \in \mathbb{Z}_p\text{-}\mathbf{Loc}(Y)$ and $\iota : T_Y \otimes \mathbb{Q}_p \to T' \otimes \mathbb{Q}_p$ is an isomorphism such that $p^m \iota \in \mathrm{Mor}(T_Y, T')$, $p^m \iota^{-1} \in \mathrm{Mor}(T', T_Y)$ is representable by a finite étale morphism $L_m(T) \to X$. By this construction plus Theorem 1.2.8 and Lemma 2.4.17, we may find a rational localization $(A, A^+) \to (B, B^+)$ encircling α, a faithfully finite étale morphism $(B, B^+) \to (C, C^+)$, and a $\mathbb{Z}_p$-lattice in V over $\widetilde{\mathrm{Spa}}(C, C^+)$ admitting a descent datum relative to $\widetilde{\mathrm{Spa}}(B, B^+)$. We may thus invoke Lemma 1.4.8 to conclude. □

Corollary 8.4.7. — *Let (A, A^+) be an adic Banach ring. Then any étale $\mathbb{Q}_p$-local system over $\widetilde{\mathrm{Spa}}(A, A^+)$ can be realized as a descent datum for isogeny $\mathbb{Z}_p$-local systems for some strong covering family of rational localizations.*

Proof. — This is immediate from Proposition 8.4.6 and the compactness of $\mathcal{M}(A)$. □

Remark 8.4.8. — The natural functor from étale $\mathbb{Q}_p$-local systems over $\mathrm{Spec}(A)$ to étale $\mathbb{Q}_p$-local systems over $\widetilde{\mathrm{Spa}}(A, A^+)$ is fully faithful, but it need not be essentially surjective even when A is a reduced affinoid algebra over an analytic field, as observed by de Jong. For instance, suppose that A is an integral affinoid algebra over an analytic field. Then on one hand $\mathrm{Spec}(A)$ is normal and noetherian, so $\mathbb{Z}_p\text{-}\mathbf{ILoc}(\mathrm{Spec}(A)) = \mathbb{Q}_p\text{-}\mathbf{Loc}(\mathrm{Spec}(A))$ by Remark 1.4.9. On the other hand, étale $\mathbb{Q}_p$-local systems over $\mathrm{Spa}(A, A^+)$ correspond to continuous representations of the étale fundamental group of $\mathcal{M}(A)$ (as defined in [**32**, §2.6]) on finite-dimensional $\mathbb{Q}_p$-vector spaces, and such representations can fail to have compact image. Typical examples arise from instances of p-adic uniformization, such as the Tate uniformization of an elliptic curve of split multiplicative reduction; see Example 8.5.17. More examples of this sort arise from Rapoport-Zink period morphisms; see [**32**, §7].

Remark 8.4.9. — For any adic Banach ring (A, A^+) and any isogeny $\mathbb{Z}_p$-local systems V_1, V_2 on $\mathrm{Spec}(A)$, any extension $0 \to V_1 \to V \to V_2 \to 0$ in the category of étale $\mathbb{Q}_p$-local systems on $\widetilde{\mathrm{Spa}}(A, A^+)$ descends to an extension of isogeny $\mathbb{Z}_p$-local systems on $\mathrm{Spec}(A)$. To see this, start with identifications $V_i = T_i \otimes_{\mathbb{Z}_p} \mathbb{Q}_p$ for some étale $\mathbb{Z}_p$-local systems on $\mathrm{Spec}(A)$. The extension $0 \to V_1 \to V \to V_2 \to 0$ then corresponds étale locally on $\widetilde{\mathrm{Spa}}(A, A^+)$ to a class in $\mathrm{Ext}(T_2, T_1) \otimes_{\mathbb{Z}_p} \mathbb{Q}_p$; by Corollary 8.4.7, we can find a strong covering family $\{(A, A^+) \to (B_i, B_i^+)\}_i$ of rational localizations such that the extension corresponds to a class x_i in $\mathrm{Ext}(T_2, T_1) \otimes_{\mathbb{Z}_p} \mathbb{Q}_p$ over $\mathrm{Spec}(B_i)$. Put $B_{ij} = B_i \widehat{\otimes}_A B_j$; then $x_i - x_j$ vanishes as an element of $\mathrm{Ext}(T_2, T_1) \otimes_{\mathbb{Z}_p} \mathbb{Q}_p$ over

$\mathrm{Spec}(B_{ij})$. We may rescale T_1 by a power of p first to force each x_i into $\mathrm{Ext}(T_2, T_1)$ over $\mathrm{Spec}(B_i)$, then to force $x_i - x_j$ to vanish in $\mathrm{Ext}(T_2, T_1)$ over $\mathrm{Spec}(B_{ij})$. The extensions then glue to define an extension of étale $\mathbb{Z}_p$-local systems on $\widetilde{\mathrm{Spa}(A, A^+)}$. We may now invoke Remark 8.4.5 to conclude.

8.5. φ-modules and local systems

We now relate pure and étale φ-modules to étale local systems on perfectoid adic spaces.

Hypothesis 8.5.1. — Throughout §8.5, let d be a positive integer. Write $\mathbb{Q}_{p^d}$ for the finite unramified extension of $\mathbb{Q}_p$ of degree d and $\mathbb{Z}_{p^d}$ for the valuation subring of $\mathbb{Q}_{p^d}$.

We begin with the case of $\mathbb{Z}_p$-local systems.

Lemma 8.5.2. — *Let R be a perfect Banach algebra over $\mathbb{F}_p$ with spectral norm α. For any $c > 1$, any positive integer d, and any $x \in R$, there exists $y \in R$ with $\alpha(x - y + y^{p^d}) < c$.*

Proof. — If $\alpha(x) \leqslant 1$, we may take $y = 0$. Otherwise, we may take $y = -(x^{p^{-d}} + \cdots + x^{p^{-md}})$ for any positive integer m which is large enough that $\alpha(x)^{p^{-md}} < c$, as then $x - y + y^{p^d} = x^{p^{-md}}$. □

Theorem 8.5.3. — *For (R, R^+) a perfect uniform adic Banach algebra over $\mathbb{F}_{p^d}$, the following categories are equivalent.*

(a) *The category of étale $\mathbb{Z}_{p^d}$-local systems over* $\mathrm{Spec}(R)$ *(or equivalently* $\mathrm{Spa}(R, R^+)$*).*

(b) *The category of étale $\mathbb{Z}_{p^d}$-local systems over* $\mathrm{Spec}(R_0)$ *(or equivalently* $\mathrm{Spa}(R_0, R_0^+)$*) for any complete subring R_0 of R whose completed direct perfection is equal to R, taking $R_0^+ = R_0 \cap R^+$.*

(c) *The category of étale $\mathbb{Z}_{p^d}$-local systems over* $\mathrm{Spec}(A)$ *(or equivalently* $\mathrm{Spa}(A, A^+)$*) for (A, A^+) corresponding to (R, R^+) as in Theorem 3.6.5.*

(d) *The category of φ^d-modules over $\tilde{\mathcal{E}}_R^{\mathrm{int}}$.*

(e) *The category of φ^d-modules over $\tilde{\mathcal{R}}_R^{\mathrm{int}}$.*

More precisely, the functor from (e) to (d) is base extension.

Proof. — The equivalences between (a) and (b) and between (a) and (c) follow from Remark 1.4.3 combined with Theorem 3.1.15 and Theorem 3.6.21, respectively. The equivalence between (a) and (d) follows immediately from Proposition 3.2.7.

We next check that the base extension functor from (e) to (d) is fully faithful. As in Remark 4.3.4, it suffices to check that for M a φ^d-module over $\tilde{\mathcal{R}}_R^{\mathrm{int}}$, any φ^d-stable element $\mathbf{v} \in M \otimes_{\tilde{\mathcal{R}}_R^{\mathrm{int}}} W(R)$ belongs to M. This claim may be checked locally on $\mathcal{M}(R)$, so we may assume that M is free over $\tilde{\mathcal{R}}_R^{\mathrm{int}}$. By Proposition 7.3.6, we can choose an R-algebra S which is a completed direct limit of faithfully finite étale R-subalgebras, in such a way that $M \otimes_{\tilde{\mathcal{R}}_R^{\mathrm{int}}} \tilde{\mathcal{R}}_S^{\mathrm{int}}$ admits a φ^d-invariant basis $\mathbf{e}_1, \dots, \mathbf{e}_n$.

If we write $\mathbf{v} = \sum_i x_i \mathbf{e}_i$ with $x_i \in W(S)$, then $x_i \in W(S)^{\varphi^d} = W(S^{\overline{\varphi}^d})$, and the latter ring is contained in $\tilde{\mathcal{R}}_S^{\mathrm{int}}$ by Remark 5.1.5. Since $W(R) \cap \tilde{\mathcal{R}}_S^{\mathrm{int}} = \tilde{\mathcal{R}}_R^{\mathrm{int}}$, it follows that $\mathbf{v} \in M$ as desired.

We finally check that the functor from (e) to (d) is essentially surjective. Let M be a φ^d-module over $W(R)$. By faithfully flat descent (Theorem 1.3.4 and Theorem 1.3.5), to check that M arises by base extension from $\tilde{\mathcal{R}}_R^{\mathrm{int}}$, it suffices to do so after replacing R with a faithfully finite étale extension. Since (a) and (d) are equivalent, we may reduce to the case where M admits a basis $\mathbf{e}_1, \dots, \mathbf{e}_n$ on which φ^d acts *via* a matrix F for which $F - 1$ has entries in $pW(R)$.

Let α be the spectral norm on R. We define matrices F_n, G_n for each positive integer n such that $F_1 = F$, $G_1 = 1$, $F_n - 1$ has entries in $pW(R)$, G_n has entries in $\tilde{\mathcal{R}}_R^{\mathrm{int},1}$, $\lambda(\alpha)(G_n - 1) < 1$, and $X_n = p^{-n}(F_n - G_n)$ has entries in $W(R)$. Namely, given F_n and G_n, apply Lemma 8.5.2 to construct a matrix $\overline{Y}_n$ over R so that $\alpha(\overline{X}_n - \overline{Y}_n + \overline{\varphi}^d(\overline{Y}_n)) < p^{n/2}$. Then put $U_n = 1 + p^n[\overline{Y}_n]$ (the entries of $[\overline{Y}_n]$ are Teichmüller liftings of corresponding entries of $\overline{Y}_n$), $F_{n+1} = U_n^{-1} F_n \varphi^d(U_n)$, $G_{n+1} = G_n + p^n[\overline{X}_n - \overline{Y}_n + \overline{\varphi}^d(\overline{Y}_n)]$. The product $U_1 U_2 \cdots$ converges to a matrix U so that $U^{-1} F \varphi^d(U)$ is equal to the p-adic limit of the G_n, which is invertible over $\tilde{\mathcal{R}}_R^{\mathrm{int},r}$ for any $r \in (0,1)$. Consequently, the $\tilde{\mathcal{R}}_R^{\mathrm{int}}$-span of the vectors $\mathbf{v}_1, \dots, \mathbf{v}_n$ defined by $\mathbf{v}_j = \sum_i U_{ij} \mathbf{e}_i$ gives a φ^d-module N over $\tilde{\mathcal{R}}_R^{\mathrm{int}}$ for which $N \otimes_{\tilde{\mathcal{R}}_R^{\mathrm{int}}} W(R) \cong M$. □

Remark 8.5.4. — In Theorem 8.5.3, the equivalence between étale $\mathbb{Z}_p$-local systems over R and φ-modules over $W(R)$ can also be interpreted as a form of nonabelian Artin-Schreier theory, using *Lang torsors*. For example, see [**100**, Proposition 4.12] for a derivation of ordinary Artin-Schreier theory in this framework.

Thanks to Theorem 8.3.5, Theorem 8.5.3 immediately globalizes as follows.

Theorem 8.5.5. — *Let X be a perfectoid adic space over $\mathbb{Q}_{p^d}$ and let X' be the corresponding perfect uniform adic space over $\mathbb{F}_{p^d}$. Then the following categories are equivalent.*

(a) *The category of étale $\mathbb{Z}_{p^d}$-local systems over X.*

(b) *The category of étale $\mathbb{Z}_{p^d}$-local systems over X'.*

(c) *The category of étale $\mathbb{Z}_{p^d}$-local systems over X'_0 for any adic space X'_0 whose inverse perfection (*i.e., *its inverse limit along absolute Frobenius) is isomorphic to X'.*

(d) *The category of φ^d-modules over $\tilde{\mathcal{E}}_{X'}^{\mathrm{int}}$.*

(e) *The category of φ^d-modules over $\tilde{\mathcal{R}}_{X'}^{\mathrm{int}}$.*

We next consider isogeny $\mathbb{Z}_p$-local systems.

Theorem 8.5.6. — *For (R, R^+) a perfect uniform adic Banach algebra over $\mathbb{F}_{p^d}$, the following categories are equivalent.*

(a) *The category of isogeny $\mathbb{Z}_{p^d}$-local systems over* Spec(R) *(or equivalently* Spa(R, R^+)*).*

(b) *The category of isogeny* $\mathbb{Z}_{p^d}$*-local systems over* $\operatorname{Spec}(R_0)$ *(or equivalently* $\operatorname{Spa}(R_0, R_0^+)$*) for any complete subring* R_0 *of* R *whose completed direct perfection is equal to* R*, taking* $R_0^+ = R_0 \cap R^+$.

(c) *The category of isogeny* $\mathbb{Z}_{p^d}$*-local systems over* $\operatorname{Spec}(A)$ *(or equivalently* $\operatorname{Spa}(A, A^+)$*) for* (A, A^+) *corresponding to* (R, R^+) *as in Theorem 3.6.5.*

(d) *The category of globally étale* φ^d*-modules over* $\tilde{\mathcal{E}}_R$.

(e) *The category of globally étale* φ^d*-modules over* $\tilde{\mathcal{R}}_R^{\mathrm{bd}}$.

(f) *The category of globally étale* φ^d*-modules over* $\tilde{\mathcal{R}}_R$.

More precisely, the functors from (e) to (d) and (f) are base extensions.

Proof. — The equivalences between (a) and (b) and between (a) and (c) again follow from Remark 1.4.3 combined with Theorem 3.1.15 and Theorem 3.6.21, respectively.

The functor from (a) to (e) is constructed as follows. Let V be an isogeny $\mathbb{Z}_{p^d}$-local system over $\operatorname{Spec}(R)$; we may write $V = T \otimes_{\mathbb{Z}_p} \mathbb{Q}_p$ for some $\mathbb{Z}_{p^d}$-local system T on $\operatorname{Spec}(R)$. The latter corresponds to a φ^d-module $M_{0,R}$ over $\tilde{\mathcal{R}}_R^{\mathrm{int}}$ by Theorem 8.5.3. The assignment $V \to M_{0,R} \otimes_{\tilde{\mathcal{R}}_R^{\mathrm{int}}} \tilde{\mathcal{R}}_R^{\mathrm{bd}}$ then defines a fully faithful functor by Corollary 5.2.4; by the same reasoning, the resulting functors from (a) to (d) and from (a) to (f) are fully faithful.

To construct the functor from (e) back to (a), given a φ^d-module M over $\tilde{\mathcal{R}}_R^{\mathrm{bd}}$ admitting a locally free étale model M_0, apply Theorem 8.5.3 to convert M_0 into a $\mathbb{Z}_{p^d}$-local system T on $\operatorname{Spec}(R)$. The assignment $M \to T \otimes_{\mathbb{Z}_p} \mathbb{Q}_p$ defines a quasi-inverse to the functor from (a) to (e). By similar reasoning, the functors from (a) to (d) and from (a) to (f) are equivalences of categories. □

Definition 8.5.7. — For $c, d \in \mathbb{Z}$ with $d > 0$, define an *isogeny* (c,d)*-*$\mathbb{Z}_p$*-local system* on a scheme X to be an isogeny $\mathbb{Z}_{p^d}$-local system V on X equipped with a semilinear action of the Frobenius automorphism τ of $\mathbb{Q}_{p^d}$ on sections, such that $p^c\tau^d$ acts as the identity. For $c', d' \in \mathbb{Z}$ with $d' > 0$ and $c'/d' = c/d$, the categories of isogeny (c,d)-$\mathbb{Z}_p$-local systems and isogeny (c',d')-$\mathbb{Z}_p$-local systems on any scheme are naturally equivalent: this reduces to the case where $c' = ce, d' = de$ for some positive integer e, in which case the claim is an easy exercise using Hilbert's Theorem 90. (See [**11**, Théorème 3.2.3] for a similar construction.) We may similarly define *étale* (c,d)*-*$\mathbb{Q}_p$*-local systems.*

Theorem 8.5.8. — *For* (R, R^+) *a perfect uniform adic Banach algebra over* $\mathbb{F}_{p^d}$ *and* $c \in \mathbb{Z}$*, the following categories are equivalent.*

(a) *The category of isogeny* (c,d)*-*$\mathbb{Z}_p$*-local systems over* $\operatorname{Spec}(R)$.

(b) *The category of isogeny* (c,d)*-*$\mathbb{Z}_p$*-local systems over* $\operatorname{Spec}(R_0)$ *for any subring* R_0 *of* R *whose completed direct perfection is equal to* R.

(c) *The category of isogeny* (c,d)-$\mathbb{Z}_p$*-local systems over* $\mathrm{Spec}(A)$ *for* $A = \tilde{\mathcal{R}}_R^{\mathrm{int},1}/(z)$ *for any* $z \in W(R^+)$ *which is primitive of degree* 1.

(d) *The category of globally* (c,d)*-pure* φ*-modules over* $\tilde{\mathcal{E}}_R$.

(e) *The category of globally* (c,d)*-pure* φ*-modules over* $\tilde{\mathcal{R}}_R^{\mathrm{bd}}$.

(f) *The category of globally* (c,d)*-pure* φ*-modules over* $\tilde{\mathcal{R}}_R$.

More precisely, the functors from (e) to (d) and (f) are base extensions.

Proof. — This is immediate from Theorem 8.5.6. □

We finally pass to $\mathbb{Q}_p$-local systems on adic spaces.

Definition 8.5.9. — Let X be a perfect uniform adic space over $\mathbb{F}_p$ and let M be a φ-module over one of $\tilde{\mathcal{E}}_X$, $\tilde{\mathcal{R}}_X^{\mathrm{bd}}$, $\tilde{\mathcal{R}}_X$. For $x \in X$, we say that M is (c,d)*-pure at* x if M is (c,d)-pure at the rank 1 seminorm induced by x. We say M is (c,d)*-pure* if it is (c,d)-pure at each $x \in X$. Similarly, we define the *slope polygon* of M at x by passing to the induced rank 1 seminorm; this leads to corresponding definitions of the *pure locus* and *étale locus* of M. By definition, these sets are pullback from subsets of the real quotient $\overline{X}$.

Remark 8.5.10. — Let (R, R^+) be a perfect uniform Banach algebra over $\mathbb{F}_p$ and put $X = \mathrm{Spa}(R, R^+)$. For $* = \tilde{\mathcal{E}}, \tilde{\mathcal{R}}^{\mathrm{bd}}, \tilde{\mathcal{R}}$, there is a natural functor from φ^d-modules over $*_R$ to φ^d-modules over $*_X$; we sometimes refer to the latter as *local* φ^d*-modules* over $*_R$. These functors are fully faithful by Theorem 5.3.3. The functors for $\tilde{\mathcal{E}}$ and $\tilde{\mathcal{R}}^{\mathrm{bd}}$ are not equivalences of categories (see Example 8.5.17), but the functor for $\tilde{\mathcal{R}}$ is an equivalence of categories by Corollary 6.3.13. In any case, thanks to the local nature of the pure and étale conditions, one sees easily that a φ^d-module over $*_R$ is pure or étale if and only if the corresponding φ^d-module over $*_X$ has this property.

Lemma 8.5.11. — *Let* X *be a perfect uniform adic space over* $\mathbb{F}_{p^d}$ *and let* M *be a* φ^d*-module over* $\tilde{\mathcal{R}}_X$. *Then the pure locus and the étale locus of* $\tilde{\mathcal{R}}_X$ *are open and partially proper. In particular, by Lemma 8.2.12, if* X *is taut, then so are the pure locus and the étale locus.*

Proof. — This is immediate from Corollary 7.3.8. □

Theorem 8.5.12. — *Let* X *be a perfectoid adic space over* $\mathbb{Q}_{p^d}$ *and let* X' *be the corresponding perfect uniform adic space over* $\mathbb{F}_{p^d}$. *Then the following categories are equivalent.*

(a) *The category of étale* (c,d)-$\mathbb{Q}_p$*-local systems over* X.

(b) *The category of étale* (c,d)-$\mathbb{Q}_p$*-local systems over* X'.

(c) *The category of étale* (c,d)-$\mathbb{Q}_p$*-local systems over* X'_0 *for any adic space* X'_0 *whose inverse perfection is isomorphic to* X'.

(d) *The category of (c,d)-pure φ-modules over $\tilde{\mathcal{E}}_{X'}$.*

(e) *The category of (c,d)-pure φ-modules over $\tilde{\mathcal{R}}^{\mathrm{bd}}_{X'}$.*

(f) *The category of (c,d)-pure φ-modules over $\tilde{\mathcal{R}}_{X'}$.*

Proof. — This is immediate from Theorem 8.3.5 and Theorem 8.5.8. □

Corollary 8.5.13. — *Let X be a perfect uniform adic space over $\mathbb{F}_{p^d}$. Then a φ^d-module over $\tilde{\mathcal{E}}_X$, $\tilde{\mathcal{R}}^{\mathrm{bd}}_X$, $\tilde{\mathcal{R}}_X$ is pure of slope s at some $x \in X$ if and only if it is (c',d')-pure at x for every (not just one) pair c', d' of integers for which d' is a positive multiple of d and $c'/d' = s$.*

Proof. — This follows from Theorem 8.5.8 plus the corresponding equivalence on the side of local systems (Definition 8.5.7). □

Corollary 8.5.14. — *Let M be a φ^d-module over $\tilde{\mathcal{E}}_R$, $\tilde{\mathcal{R}}^{\mathrm{bd}}_R$, or $\tilde{\mathcal{R}}_R$. If M is pointwise pure, then M is pure.*

Proof. — Over $\tilde{\mathcal{R}}_R$ this is immediate from Corollary 7.3.9, so from now on we assume that M is a φ^d-module over $\tilde{\mathcal{E}}_R$ (resp. $\tilde{\mathcal{R}}^{\mathrm{bd}}_R$). Choose $\beta \in \mathcal{M}(R)$, put $n = \mathrm{rank}(M,\beta)$, and choose $c', d' \in \mathbb{Z}$ with d' a positive multiple of d and $c'/d' = \mu(M,\beta)$. Put $M_\beta = M \otimes_{\tilde{\mathcal{E}}_R} \tilde{\mathcal{E}}_{\mathcal{H}(\beta)}$ (resp. $M_\beta = M \otimes_{\tilde{\mathcal{R}}^{\mathrm{bd}}_R} \tilde{\mathcal{R}}^{\mathrm{bd}}_{\mathcal{H}(\beta)}$).

Suppose first that M_β admits a *cyclic vector*, *i.e.*, an element $\mathbf{e}$ such that $(p^{c'}\varphi^{d'})^i(\mathbf{e})$ for $i = 0,\dots,n-1$ are linearly independent. Then any element of M_β sufficiently close to $\mathbf{e}$ for the weak topology (resp. the LF topology) is also a cyclic vector, so we may choose $\mathbf{e} \in M_S$ for $R \to S$ a rational localization encircling β and $M_S = M \otimes_{\tilde{\mathcal{E}}_R} \tilde{\mathcal{E}}_S$ (resp. $M_S = M \otimes_{\tilde{\mathcal{R}}^{\mathrm{bd}}_R} \tilde{\mathcal{R}}^{\mathrm{bd}}_S$). For a suitable choice of S, $(p^{c'}\varphi^{d'})^i(\mathbf{e})$ for $i = 0,\dots,n-1$ form a basis of M_S. For such S, the $W(S)$-submodule (resp. $\tilde{\mathcal{R}}^{\mathrm{int}}_S$-submodule) of M spanned by this basis forms a free pure model by [**83**, Lemma 5.2.4].

To handle the general case, write R as a Banach algebra over a perfect analytic field L. Choose $\overline{x} \in L^\times$ of norm less than 1. Let $\mathbf{e}_1,\dots,\mathbf{e}_n$ be a basis of M_β. Then for some $r_1,\dots,r_n \in \mathbb{Z}[p^{-1}]$, $[\overline{x}]^{r_1}\mathbf{e}_1 + \cdots + [\overline{x}]^{r_n}\mathbf{e}_n$ is a cyclic vector of M_β, so the previous paragraph shows that M is pure. □

Corollary 8.5.15. — *Let $(R,R^+) \to (S,S^+)$ be a bounded homomorphism of perfect uniform adic Banach algebras over $\mathbb{F}_{p^d}$ for which $\mathrm{Spa}(S,S^+) \to \mathrm{Spa}(R,R^+)$ is surjective. Let M be a local φ^d-module over $\tilde{\mathcal{E}}_R$ (resp. $\tilde{\mathcal{R}}^{\mathrm{bd}}_R$, $\tilde{\mathcal{R}}_R$). Then M is pure if and only if $M \otimes_{\tilde{\mathcal{E}}_R} \tilde{\mathcal{E}}_S$ (resp. $M \otimes_{\tilde{\mathcal{R}}^{\mathrm{bd}}_R} \tilde{\mathcal{R}}^{\mathrm{bd}}_S$, $M \otimes_{\tilde{\mathcal{R}}_R} \tilde{\mathcal{R}}_S$) is pure.*

Proof. — By Corollary 4.2.14, M is pointwise pure. By Corollary 8.5.14, M is pure. □

Corollary 8.5.16. — *Let $Y \to X$ be a surjective morphism of perfectoid adic spaces. Let M be a φ^d-module over $\tilde{\mathcal{R}}_X$. Then M is pure (resp. étale) if and only if the pullback of M to $\tilde{\mathcal{R}}_Y$ is pure (resp. étale).*

Here are some examples to illustrate the difference between φ-modules and local φ-modules, and between globally étale and étale φ-modules.

Example 8.5.17. — Put $K = \mathbb{F}_p((q))$ for an arbitrary normalization $|q| = \omega < 1$ of the q-adic norm, and define the strictly affinoid algebras

$$B = K\{\omega^2/T, T, U/\omega^{-2}\}/(U(T-q)-1), \quad B_1 = K\{\omega^2/T, T/\omega^2\}, \quad B_2 = K\{1/T, T\}.$$

over K. In words, $\mathrm{Spa}(B, B^\circ)$ is the annulus $\omega^2 \leqslant |T| \leqslant 1$ minus the open disc $|T-q| < \omega^2$, and $\mathrm{Spa}(B_1, B_1^\circ)$ and $\mathrm{Spa}(B_2, B_2^\circ)$ are the boundary circles $|T| = \omega^2$ and $|T| = 1$, respectively, within $\mathrm{Spa}(B, B^\circ)$. Let $\sigma_q : B_2 \to B_1$ be the substitution $T \mapsto q^2T$. If we quotient $\mathrm{Spa}(B, B^\circ)$ by the identification $\mathrm{Spa}(B_2, B_2^\circ) \cong \sigma_q^* \mathrm{Spa}(B_1, B_1^\circ)$, we obtain a strictly affinoid subspace $\mathrm{Spa}(A, A^\circ)$ of the Tate curve over X for the parameter q^2. The latter is the analytification of a smooth projective curve over K of genus 1; see for instance [**121**, Theorem V.3.1] for explicit equations.

We may construct an étale $\mathbb{Q}_p$-local system V on $\mathrm{Spa}(A, A^\circ)$ as follows. Let $\tilde{V}$ be the trivial $\mathbb{Q}_p$-local system on $\mathrm{Spa}(B, B^\circ)$, equipped with the distinguished generator 1. Let $\tilde{V}_1, \tilde{V}_2$ be the restrictions of $\tilde{V}$ to $\mathrm{Spa}(B_1, B_1^\circ), \mathrm{Spa}(B_2, B_2^\circ)$, respectively. To specify V, it suffices to specify an isomorphism $\tilde{V}_1 \cong \sigma_q^* \tilde{V}_2$; we choose the isomorphism matching $1 \in \tilde{V}_1$ with $p \in \sigma_q^* \tilde{V}_2$.

Let R, S, S_1, S_2 be the completed perfections of A, B, B_1, B_2, respectively. We claim that V cannot correspond to an étale φ-module M over $\tilde{\mathcal{E}}_R$ (or over the subring $\tilde{\mathcal{R}}_R^{\mathrm{bd}}$ thereof). To check this, suppose the contrary, and choose any nonzero element $\mathbf{v} \in M$. The pullback of M to $\tilde{\mathcal{E}}_S$ can be identified with the trivial φ-module $\tilde{\mathcal{E}}_S$ itself, and $\mathbf{v}$ must correspond to an element $x \in \tilde{\mathcal{E}}_S$. Let x_1, x_2 be the images of x in $\tilde{\mathcal{E}}_{S_1}, \tilde{\mathcal{E}}_{S_2}$, respectively. We must then have

$$x_2 = p\sigma_q(x_1) \in \tilde{\mathcal{E}}_{S_2} = W(S_2)[p^{-1}].$$

However, this is impossible: the maps $S \to S_1, S_2$ are injective, so an element of $W(S)[p^{-1}]$ which maps to $W(S_1)$ or $W(S_2)$ must itself belong to $W(S)$. Thus if $x \in p^m W(S)$ for some $m \in \mathbb{Z}$, then also $x \in p^{m+1}W(S)$, which cannot hold for all m if $x \neq 0$.

By contrast, by Theorem 8.5.12, V does correspond to an étale φ-module over $\tilde{\mathcal{R}}_R$, and to étale φ-modules over $\tilde{\mathcal{E}}_X$ and $\tilde{\mathcal{R}}_X^{\mathrm{bd}}$. By the previous paragraph, however, the latter do not descend to étale φ-modules over $\tilde{\mathcal{E}}_R$ or $\tilde{\mathcal{R}}_R^{\mathrm{bd}}$; in particular, we obtain an obstruction to glueing finite projective modules over these rings as indicated in Remark 5.3.7. This lack of descent in turn provides an example of an étale φ-module over $\tilde{\mathcal{R}}_R$ which does not admit a (not necessarily finite locally free) étale model.

Example 8.5.18. — Let K be an algebraically closed analytic field of characteristic $p \neq 2$. Let R be the completed perfection of $K\{X, Y\}/(y^2 - (x^2-1)^2)$. Put $X = \mathrm{Spa}(R, R^\circ)$. Then X admits a étale cover consisting of a doubly infinite chain of copies of $\mathrm{Spa}(K\{X\}, K\{X\}^\circ)$; as in Example 8.5.17, this gives rise to an étale $\mathbb{Q}_p$-local system V on X which is not an isogeny $\mathbb{Z}_p$-local system. By Theorems 8.5.8 and 8.5.12, V corresponds to étale φ-modules over $\tilde{\mathcal{E}}_X$, $\tilde{\mathcal{E}}_X^{\mathrm{bd}}, \tilde{\mathcal{R}}_X$ which are not globally étale.

Remark 8.5.19. — If A is a connected affinoid algebra over an analytic field K, then an *étale fundamental group* of $\mathcal{M}(A)$ has been defined by de Jong [**32**]; its continuous representations on finite-dimensional $\mathbb{Q}_p$-vector spaces correspond precisely to étale $\mathbb{Q}_p$-local systems on $\mathrm{Spa}(A, A^\circ)$ in our sense [**32**, Lemma 2.6]. It should be possible to show using Theorem 3.3.7 (c) and Theorem 3.6.21 that $\mathcal{M}(A)$ and $\mathcal{M}(\tilde{\mathcal{R}}_R^{\mathrm{int},1}/(z))$ have the same étale fundamental group; the only serious issue is that $\tilde{\mathcal{R}}_R^{\mathrm{int},1}/(z)$ need not be an affinoid algebra over an analytic field, so some work is needed to define the étale fundamental group and check some basic properties.

8.6. A bit of cohomology

We next relate the cohomology of $\mathbb{Z}_p$-local systems and isogeny $\mathbb{Z}_p$-local systems with φ-modules; again, this comes down to nonabelian Artin-Schreier-Witt theory. The corresponding statements for adic spaces require additional work even to assert what is meant by étale cohomology; this work is carried out in the next section.

Hypothesis 8.6.1. — Throughout §8.6, retain Hypothesis 8.5.1, and in addition let R be a perfect uniform Banach algebra over $\mathbb{F}_{p^d}$.

Theorem 8.6.2. — *Let T be an étale $\mathbb{Z}_{p^d}$-local system on* $\mathrm{Spec}(R)$. *Let M be the φ^d-module over $\tilde{\mathcal{R}}_R^{\mathrm{int}}$ or $W(R)$ corresponding to T* via *Theorem 8.5.3. Then there are natural (in T and R) bijections $H^i_{\mathrm{\acute{e}t}}(X,T) \cong H^i_{\varphi^d}(M)$ for all $i \geqslant 0$.*

Proof. — Suppose first that M is defined over $W(R)$. For each positive integer n, view T/p^nT as a locally constant étale sheaf on $\mathrm{Spec}(R)$, and let $\tilde{M}_n$ be the étale sheaf on $\mathrm{Spec}(R)$ corresponding to the quasicoherent sheaf on $\mathrm{Spec}(W(R)/(p^n))$ with global sections M/p^nM. We then have an exact sequence

$$0 \longrightarrow T/p^nT \longrightarrow \tilde{M}_n \stackrel{\varphi^d-1}{\longrightarrow} \tilde{M}_n \longrightarrow 0,$$

where exactness at the right is given by Theorem 8.5.3 (or more directly by Proposition 7.3.6). We see by induction on n that $\tilde{M}_n$ is acyclic: it is enough to check that $\ker(\tilde{M}_n \to \tilde{M}_{n-1})$ is acyclic, which it is because it arises from a quasicoherent sheaf on an affine scheme. Taking the long exact sequence in cohomology thus yields the desired result.

Suppose next that M is defined over $\tilde{\mathcal{R}}_R^{\mathrm{int}}$; by the previous paragraph, we need only check the cases $i = 0, 1$. For these, interpret $H^i_{\varphi^d}(M)$ as an extension group as in Definition 1.5.4, then note that any extension of two φ^d-modules is again a φ^d-module. By Theorem 8.5.3, the 0th and 1st extension groups do not change upon base extension to $W(R)$; we may thus deduce the claim from the previous paragraph. □

Lemma 8.6.3. — *Let $0 \to M_1 \to M \to M_2 \to 0$ be a short exact sequence of φ-modules over $\tilde{\mathcal{R}}_R$. If any two of M, M_1, M_2 are (c,d)-pure, then so is the third.*

Proof. — By Corollary 7.3.9, it suffices to treat the case that $R = L$ is an analytic field. In this case, the lemma follows immediately from Theorem 4.2.12. □

We have a similar result for $\mathbb{Q}_p$-local systems. Note that this result can also be formulated in terms of φ-bundles using Proposition 6.3.19.

Theorem 8.6.4. — *Suppose that R is an $\mathbb{F}_{p^d}$-algebra for a positive integer d. Let E be an isogeny $\mathbb{Z}_{p^d}$-local system on* Spec(R). *Let M be the globally étale φ^d-module over $\tilde{\mathcal{R}}_R^{\mathrm{bd}}$, $\tilde{\mathcal{E}}_R$, or $\tilde{\mathcal{R}}_R$ corresponding to E* via *Theorem 8.5.6. Then there are natural (in E and R) bijections $H^i_{\text{ét}}(X,E) \cong H^i_{\varphi^d}(M)$ for all $i \geqslant 0$.*

Proof. — We first treat the case over $\tilde{\mathcal{R}}_R^{\mathrm{bd}}$, the case over $\tilde{\mathcal{E}}_R$ being similar. Let T be an étale $\mathbb{Z}_{p^d}$-local system on X for which $E = T \otimes_{\mathbb{Z}_p} \mathbb{Q}_p$. Let M_0 be the φ^d-module over $\tilde{\mathcal{R}}_R^{\mathrm{int}}$ corresponding to T. By Theorem 8.6.2, we have natural (in T and A) bijections $H^i_{\text{ét}}(X,T) \cong H^i_{\varphi^d}(M_0)$ for all $i \geqslant 0$. By definition, we may identify $H^i_{\text{ét}}(X,E)$ with $H^i_{\text{ét}}(X,T) \otimes_{\mathbb{Z}_p} \mathbb{Q}_p$; in particular, it is zero for $i > 1$. On the other hand, for $i = 0,1$, we may identify $H^i_{\varphi^d}(M)$ with $H^i_{\varphi^d}(M_0) \otimes_{\mathbb{Z}_p} \mathbb{Q}_p$ by identifying M with $M_0 \otimes_{\mathbb{Z}_p} \mathbb{Q}_p = \cup_{n=0}^{\infty} p^{-n} M_0$ and noting that the computation of $H^i_{\varphi^d}$ commutes with direct limits. This proves the claim in this case.

We next treat the case over $\tilde{\mathcal{R}}_R$. Put $M_1 = M_0 \otimes_{\mathbb{Z}_p} \mathbb{Q}_p$, so that we may identify M with $M_1 \otimes_{\tilde{\mathcal{R}}_R^{\mathrm{bd}}} \tilde{\mathcal{R}}_R$. It follows from Theorem 8.5.6 that the natural map $H^0_{\varphi^d}(M_1) \to H^0_{\varphi^d}(M)$ is bijective; hence $H^0_{\varphi^d}(M)$ is naturally isomorphic to $H^0_{\text{ét}}(X,E)$. Recall that by Remark 8.4.9, the extension of two isogeny $\mathbb{Z}_{p^d}$-local systems on Spec(R) in the category of étale $\mathbb{Q}_{p^d}$-local systems on Spa(R,R^+) descends to an extension of isogeny $\mathbb{Z}_{p^d}$-local systems on Spec(R). That is, we may compute $H^1_{\text{ét}}(X,E)$ as an extension group in the category of étale local systems over Spa(R,R^+). We then obtain an isomorphism between this group and $H^1_{\varphi^d}(M)$ by applying Theorem 6.2.9 and noting that any extension of étale φ^d-modules over $\tilde{\mathcal{R}}_R$ is again étale by Lemma 8.6.3. □

Remark 8.6.5. — One can formulate an analogue of Theorem 8.6.4 for isogeny (c,d)-local systems comparing a suitably modified étale cohomology to $H^1_{p^c\varphi^d}$ of the corresponding φ-module. As this statement is a formal consequence of the one given, and we have no particular use for it, we omit further details.

8.7. The relative Fargues-Fontaine curve

We have already seen (Theorem 6.3.12) that for R a perfect Banach algebra over $\mathbb{F}_p$, the φ-modules over $\tilde{\mathcal{R}}_R$ can be described in terms of a vector bundle on a certain scheme Proj(P), which in the case of an analytic field is a Fargues-Fontaine curve. In order to globalize this construction, we must replace the scheme with what amounts to an *analytification* thereof, although the latter construction does not come equipped with a universal property like the one for analytification of schemes of finite type over a field [**62**, Exposé XII].

Hypothesis 8.7.1. — Throughout §8.7, fix a positive integer a and put $q = p^a$. Let (A,A^+) be a perfectoid adic Banach algebra over $\mathbb{Q}_p$ and let (R,R^+) be the perfect uniform adic Banach algebra over $\mathbb{F}_p$ associated to (A, A^+) *via* the perfectoid correspondence (Theorem 3.6.5). Let X be a perfectoid adic space over $\mathbb{Q}_p$ and let X' be the corresponding perfect uniform adic space over $\mathbb{F}_p$ associated to X *via* the global perfectoid correspondence (Theorem 8.3.5).

Remark 8.7.2. — Since (R, R^+) arises from (A, A^+) *via* the perfectoid correspondence, there exists $z \in W(R^+)$ which is primitive of degree 1 and generates the kernel of $\theta : W(R^+) \to A^+$.

Definition 8.7.3. — For $0 < s \leqslant r$, the ring $\tilde{\mathcal{R}}_R^{[s,r]}$ is a relatively perfectoid Banach ring by Theorem 5.3.9. We promote it to an adic Banach ring as in Definition 5.3.10.

Definition 8.7.4. — Define the space

$$U_R = \bigcup_{0<s<r} \operatorname{Spa}(\tilde{\mathcal{R}}_R^{[s,r]}, \tilde{\mathcal{R}}_R^{[s,r],+})$$

whose maximal Hausdorff quotient is the space T_R considered in Proposition 5.4.6. By Theorem 5.3.9, U_R is a relatively perfectoid space. By Proposition 5.4.6, φ^{a*} acts properly discontinuously on U_R, so we may form the orbit space FF_R which again is a relatively perfectoid space.

Using Lemma 5.3.11, we may glue to obtain a relatively perfectoid space $\mathrm{FF}_{X'}$ with the property that for $X' = \operatorname{Spa}(R, R^+)$, we have a natural isomorphism $\mathrm{FF}_{X'} \cong \mathrm{FF}_R$. We also have a natural map $|\mathrm{FF}_{X'}| \to |X'|$ of topological spaces. However, this map cannot arise from a morphism of adic spaces due to the mismatch in characteristics.

We will use the notation FF_A to denote the space FF_R additionally equipped with the morphism $\operatorname{Spa}(A, A^+) \to \mathrm{FF}_R$ induced by the map $\theta : \tilde{\mathcal{R}}_R^{[1,1]} \to A$ (see Lemma 5.5.5). We again glue to obtain a space FF_X which is naturally isomorphic to $\mathrm{FF}_{X'}$ but additionally is equipped with a distinguished morphism $X \to \mathrm{FF}_X$. The induced map $|X| \to |\mathrm{FF}_X|$ is a continuous section of the projection map $|\mathrm{FF}_X| \cong |\mathrm{FF}_{X'}| \to |X'| \cong |X|$. However, despite there no longer being a mismatch of characteristics, the map $|\mathrm{FF}_X| \to |X|$ is still not induced by a morphism $\mathrm{FF}_X \to X$ of adic spaces.

Definition 8.7.5. — Define the graded ring P_R as in Definition 6.3.1. The natural morphism $P_R \to \tilde{\mathcal{R}}_R^{[s,r]}$ then defines a morphism $U_R \to \operatorname{Proj}(P_R)$ of locally ringed spaces, which factors through a morphism $\mathrm{FF}_R \to \operatorname{Proj}(P_R)$. For $n \in \mathbb{Z}$, we define the line bundle $\mathcal{O}(n)$ on FF_R by pulling back the line bundle $\mathcal{O}(n)$ on $\operatorname{Proj}(P_R)$ defined in Remark 6.3.16.

Remark 8.7.6. — Let L be a perfect analytic field of characteristic p over which R is a Banach algebra. (For example, with notation as in Remark 8.7.2, we may take L to be the completed perfect closure of $\mathbb{F}_p((\overline{z}))$.) By Proposition 6.2.4 there exist nonzero homogeneous elements $f_1, f_2 \in P_{L,+}$ which generate the unit ideal in $\tilde{\mathcal{R}}_L$, and hence also in $\tilde{\mathcal{R}}_R$. (In fact, one can even take $f_1, f_2 \in P_{L,1}$.) This has the following further consequences.

(a) For any positive integers m, n such that f_1^m and f_2^n have the same degree d, if we write $P_{R,(d)} = \oplus_{h=0}^{\infty} P_{R,hd}$, then the scheme $\mathrm{Proj}(P_R/(f_1^m))$ is isomorphic to

$$\mathrm{Spec}(P_{R,(d)}[f_2^{-n}]_0/(f_1^m f_2^{-n})),$$

and hence is affine.

(b) By Lemma 6.3.7, $\mathrm{Proj}(P_R)$ is covered by the two open affine subsets $D_+(f_1), D_+(f_2)$. Since $\mathrm{Proj}(P_R)$ is separated, we may use Čech cohomology for this covering to compute sheaf cohomology for quasicoherent sheaves [**58**, Proposition 1.4.1], so the cohomology of any quasicoherent sheaf on $\mathrm{Proj}(P_R)$ vanishes in degree greater than 1.

See Remark 8.9.5 for a related observation.

The morphism $\mathrm{FF}_R \to \mathrm{Proj}(P_R)$ leads to a GAGA-style extension of Theorem 6.3.12.

Theorem 8.7.7. — *Pullback along the morphism $\mathrm{FF}_R \to \mathrm{Proj}(P_R)$ of locally ringed spaces defines an equivalence of categories between vector bundles on $\mathrm{Proj}(P_R)$ and on FF_R. Consequently, by Theorem 6.3.12, the latter is equivalent to the category of φ^a-modules over $\tilde{\mathcal{R}}_R$ (which is independent of R^+).*

Proof. — By Theorem 8.2.22, pulling back along the functor $U_R \to \mathrm{FF}_R$ defines an equivalence of categories between vector bundles on FF_R and φ-bundles on $\tilde{\mathcal{R}}_R$. Since the latter category is equivalent to the category of vector bundles on $\mathrm{Proj}(P_R)$ by Theorem 6.3.12, we deduce the desired result. □

This result immediately globalizes as follows.

Theorem 8.7.8. — *The category of vector bundles on FF_X is functorially equivalent to the category of φ^a-modules over $\tilde{\mathcal{R}}_X$.*

Corollary 8.7.9. — *For $n \in \mathbb{Z}$, we have*

$$H^0(\mathrm{Proj}(P_R), \mathcal{O}(n)) = H^0(\mathrm{FF}_R, \mathcal{O}(n)) = \begin{cases} P_{R,n} & n \geqslant 0 \\ 0 & n < 0. \end{cases}$$

Corollary 8.7.10. — *Fix a morphism $X \to \mathrm{Spa}(\mathbb{Q}_{p^a}, \mathbb{Z}_{p^a})$. Then the category of étale $\mathbb{Q}_{p^a}$-local systems on X is functorially equivalent to the category of vector bundles on FF_X which are* pointwise semistable *of degree* 0, i.e., *whose restriction to $\mathrm{FF}_{\mathcal{H}(x)}$ is semistable of degree 0 for any $x \in X$.*

Proof. — This is immediate from Remark 4.2.18, Theorem 8.5.12, and Theorem 8.7.8. □

We will also need some higher-rank vector bundles on FF_R.

Definition 8.7.11. — Write $P^{(a)}_{R,n}, P^{(a)}_R, U^{(a)}_R, \mathrm{FF}^{(a)}_R, \mathrm{FF}^{(a)}_X$ instead of $P_{R,n}$, P_R, U_R, FF_R, FF_X to record the dependence on the positive integer a. Then for any positive integer d, we have $P^{(a)}_{R,n} \subseteq P^{(ad)}_{R,nd}$ as subsets of $\tilde{\mathcal{R}}_R$. We thus get a morphism $P^{(a)}_R \to P^{(ad)}_R$ of graded rings and hence a morphism $\mathrm{Proj}(P^{(ad)}_R) \to \mathrm{Proj}(P^{(a)}_R)$ of schemes. This morphism is finite étale of degree d: it is the quotient by the automorphism induced by φ^a. Similarly, the corresponding morphism $U^{(ad)}_R \to U^{(a)}_R$ is an isomorphism, so the induced morphism $\mathrm{FF}^{(ad)}_R \to \mathrm{FF}^{(a)}_R$ is finite étale of degree d and is the quotient by the action of φ^{a*}.

For $n, d \in \mathbb{Z}$ with $\gcd(n,d) = 1$ and $d > 0$, let $\mathcal{O}(n,d)$ be the vector bundle of rank d on $\mathrm{FF}^{(a)}_R$ obtained by pushing forward the bundle $\mathcal{O}(n)$ on $\mathrm{FF}^{(ad)}_R$.

Corollary 8.7.12. — *For $R = L$ an algebraically closed analytic field, every vector bundle on either* $\mathrm{Proj}(P_R)$ *or* FF_R *is (nonuniquely) isomorphic to a direct sum $\oplus_i \mathcal{O}(n_i, s_i)$ for some $n_i, s_i \in \mathbb{Z}$ with* $\gcd(n_i, s_i) = 1$ *and $s_i > 0$.*

Proof. — Immediate from Theorem 8.7.7 and Proposition 4.2.16. □

Continuing in the GAGA vein, we have the following comparison result. We will later add étale cohomology once we make sense of it in §9.

Theorem 8.7.13. — *Let M be a φ^a-module over $\tilde{\mathcal{R}}_R$ corresponding to the vector bundle V on* FF_R via *Theorem 8.7.8. Then there are natural (in M and R) isomorphisms $H^i_{\varphi^a}(M) \cong H^i(\mathrm{Proj}(P_R), V) \cong H^i(\mathrm{FF}_R, V)$ for $i \geqslant 0$.*

Proof. — For Z a scheme which is quasicompact and semiseparated (*i.e.*, Z is covered by finitely many open affine subschemes, any two of which have affine intersection), the category of quasicoherent sheaves on Z has enough injectives, and the resulting derived functors agree with sheaf cohomology as defined using the full category of sheaves of abelian groups [**125**, Proposition B.8]. These results apply to $\mathrm{Proj}(P_R)$ because this scheme is quasicompact (by Lemma 6.3.7) and separated.

By the previous paragraph plus [**75**, Theorem IV.9.1], we may identify the sheaf cohomology groups $H^i(\mathrm{Proj}(P_R), V)$ with the Yoneda extension groups $\mathrm{Ext}^i(\mathcal{O}, V)$ in the category of quasicoherent sheaves. For $i = 0, 1$, this computation involves only quasicoherent finite locally free sheaves, so we may apply Theorem 6.3.12 to obtain the desired isomorphisms. For $i \geqslant 2$, $H^i(\mathrm{Proj}(P_R), V) = 0$ by Remark 8.7.6 (b) while $H^i_{\varphi^a}(M) = 0$ by definition, so we again obtain an isomorphism.

Similarly, thanks to Theorem 5.3.9 we may compute $H^i(\mathrm{FF}_R, V)$ using Čech cohomology for the covering of FF_R by $\mathrm{Spa}(\tilde{\mathcal{R}}^{[rq^{-1/2},r]}_R, \tilde{\mathcal{R}}^{[rq^{-1/2},r],+}_R)$ and $\mathrm{Spa}(\tilde{\mathcal{R}}^{[rq^{-1},rq^{-1/2}]}_R, \tilde{\mathcal{R}}^{[rq^{-1},rq^{-1/2}],+}_R)$ for any fixed $r > 0$. This immediately yields $H^i(\mathrm{FF}_R, V) = 0$ for $i \geqslant 2$, and again the comparison for $i = 0, 1$ follows by comparing Yoneda extension groups using Theorem 6.3.12 and Theorem 8.7.8. □

Corollary 8.7.14. — *Let V be a quasicoherent finite locally free sheaf on* $\mathrm{Proj}(P_R)$. *Then there exists $N \in \mathbb{Z}$ such that for all $n \geqslant N$, $H^0(\mathrm{Proj}(P_R), V(n))$ generates $V(n)$ and $H^1(\mathrm{Proj}(P_R), V(n)) = H^1(\mathrm{FF}_R, V(n)) = 0$.*

Proof. — This follows from Theorem 8.7.13 plus Propositions 6.2.4 and 6.2.2. □

We next consider the compatibility of the functor FF with étale morphisms.

Lemma 8.7.15. — *For $Y \to X$ a morphism of perfectoid adic spaces which is étale (resp. finite étale, faithfully finite étale), the induced morphism $\mathrm{FF}_Y \to \mathrm{FF}_X$ is also étale (resp. finite étale, faithfully finite étale).*

Proof. — By the perfectoid correspondence, the morphism $Y' \to X'$ in characteristic p is also étale (resp. finite étale, faithfully finite étale). By Proposition 5.5.4, if $Y' \to X'$ is finite étale then so is $\mathrm{FF}_Y \to \mathrm{FF}_X$; the other cases follow immediately. □

Remark 8.7.16. — Lemma 8.7.15 provides a promotion of the morphism $\mathrm{FF}_X \to X$ of topological spaces to a morphism of étale topoi. This morphism behaves in many ways like a circle bundle; see Remark 5.4.7.

8.8. Ampleness on relative curves

We now make a more detailed study of positivity of vector bundles on relative Fargues-Fontaine curves; the analogy with vector bundles on projective curves turns out to be rather fruitful. Throughout §8.8, continue to retain Hypothesis 8.7.1.

Definition 8.8.1. — Throughout §8.8, take L, f_1, f_2 as in Remark 8.7.6; we can and will assume that $f_1, f_2 \in P_{L,1}$. Let Z_1 be the zero locus of f_1 and put $U_1 = \mathrm{Proj}(P_R) \setminus Z_1$; by Remark 8.7.6, both Z_1 and U_1 are affine schemes.

Definition 8.8.2. — A vector bundle $\mathcal{F}$ on $\mathrm{Proj}(P_R)$ is *globally ample* if for every quasicoherent sheaf of finite type $\mathcal{G}$ on $\mathrm{Proj}(P_R)$, there exists $n_0 \in \mathbb{Z}$ such that $\mathcal{F}^{\otimes n} \otimes \mathcal{G}$ is generated by global sections for all $n \geqslant n_0$. By Corollary 8.8.5 below, it will suffice to check the condition for $\mathcal{G} = \mathcal{O}(e)$ for all $e \in \mathbb{Z}$.

We say that $\mathcal{F}$ is *ample* if there exists a strong rational covering $\{(R, R^+) \to (R_i, R_i^+)\}$ such that the pullback of $\mathcal{F}$ to $\mathrm{Proj}(P_{R_i})$ is globally ample for each i.

Lemma 8.8.3. — *Let $\mathcal{F}$ be a vector bundle on* $\mathrm{Proj}(P_R)$ *and let m be a positive integer. Then $\mathcal{F}$ is (globally) ample if and only if $\mathcal{F}^{\otimes m}$ is (globally) ample.*

Proof. — It is evident that if $\mathcal{F}$ is globally ample, then so is $\mathcal{F}^{\otimes m}$. On the other hand, if $\mathcal{F}^{\otimes m}$ is globally ample, then for any given $\mathcal{G}$, there exists $n_0 \in \mathbb{Z}$ such that for $n \geqslant n_0$, each of the bundles $(\mathcal{F}^{\otimes m})^{\otimes n} \otimes (\mathcal{F}^{\otimes i} \otimes \mathcal{G})$ for $i = 0, \dots, m-1$ is generated by global sections. Consequently, $\mathcal{F}^{\otimes n} \otimes \mathcal{G}$ is generated by global sections for $n \geqslant mn_0$, so $\mathcal{F}$ is globally ample. The ample case is similar. □

Lemma 8.8.4. — *The vector bundles $\mathcal{O}(e)$ on* $\mathrm{Proj}(P_R)$ *are globally ample for all* $e > 0$.

Proof. — Let $\mathcal{G}$ be a quasicoherent sheaf of finite type on $\mathrm{Proj}(P_R)$. Since $U_1 = \mathrm{Spec}(P_R[f_1^{-1}]_0)$ is affine, $H^0(U_1, \mathcal{G})$ is a finitely generated $P_R[f_1^{-1}]_0$-module (see Remark 1.1.5), and so $\mathcal{G}(md)$ is generated by global sections for any sufficiently large m. By Lemma 8.8.3, this proves the claim. □

Corollary 8.8.5. — *Any quasicoherent sheaf of finite type on* $\mathrm{Proj}(P_R)$ *is a quotient of a vector bundle of the form* $\oplus_{i=1}^m \mathcal{O}(e_i)$ *for some* $m \geqslant 0$ *and* $e_1, \dots, e_m \in \mathbb{Z}$.

We have the following variant of the cohomological criterion for ampleness for projective schemes [**68**, Proposition III.5.3].

Proposition 8.8.6. — *For $\mathcal{F}$ a vector bundle on* $\mathrm{Proj}(P_R)$, *the following conditions are equivalent.*

(a) *The vector bundle $\mathcal{F}$ is globally ample.*

(b) *For every quasicoherent sheaf of finite type $\mathcal{G}$ on* $\mathrm{Proj}(P_R)$, *there exists $n_0 \in \mathbb{Z}$ such that for all $n \geqslant n_0$,* $H^1(\mathrm{Proj}(P_R), \mathcal{F}^{\otimes n} \otimes \mathcal{G}) = 0$.

(c) *For each $e \in \mathbb{Z}$, there exists $n_0 \in \mathbb{Z}$ such that* $H^1(\mathrm{Proj}(P_R), \mathcal{F}^{\otimes n}(e)) = 0$ *for all* $n \geqslant n_0$.

Proof. — We first observe that (b) and (c) are equivalent. Indeed, (b) trivially implies (c), whereas (c) implies (b) by Corollary 8.8.5 and Remark 8.7.6 (b).

We next check that (a) implies (c). By Lemma 8.8.4, there exists $e' > 0$ such that $H^1(\mathrm{Proj}(P_R), \mathcal{O}(e')) = 0$. (In fact we may take $e' = 1$ by Proposition 6.2.2, but this is not crucial here.) Given (a), we may choose n_0 so that for $n \geqslant n_0$, $\mathcal{F}^{\otimes n}(e - e')$ is generated by global sections; then for each such n, there exists a surjective homomorphism $\mathcal{O}(e')^{\oplus m} \to \mathcal{F}^{\otimes n}(e)$ for some m (depending on e). By Remark 8.7.6 (b), the long exact sequence in cohomology yields $H^1(\mathrm{Proj}(P_R), \mathcal{F}^{\otimes n}(e)) = 0$. Hence (a) implies (b).

We finally check that (b) implies (a). Fix $e \in \mathbb{Z}$. For any $n \in \mathbb{Z}$, we have an exact sequence

$$H^0(\mathrm{Proj}(P_R), \mathcal{F}^{\otimes n}(e)) \longrightarrow H^0(Z_1, \mathcal{F}^{\otimes n}(e)) \longrightarrow H^1(\mathrm{Proj}(P_R), \mathcal{F}^{\otimes n}(e-1)).$$

By Remark 1.1.5, $H^0(Z_1, \mathcal{F}^{\otimes n}(e))$ is a finitely generated module over the coordinate ring of Z_1. Choose n sufficiently large so that $H^1(\mathrm{Proj}(P_R), \mathcal{F}^{\otimes n}(e-1)) = 0$; there then exist finitely many sections $s_1, \dots, s_m \in H^0(\mathrm{Proj}(P_R), \mathcal{F}^{\otimes n}(e))$ which generate the restriction of $\mathcal{F}^{\otimes n}(e)$ to Z_1.

Let $\mathcal{G}$ be the subsheaf of $\mathcal{F}^{\otimes n}(e)$ generated by $s_1, \dots, s_m$. Let Z' be the support of $\mathcal{F}^{\otimes n}(e)/\mathcal{G}$; it is disjoint from Z_1 and hence is a closed subscheme of U_1. Since U_1 is affine, so then is Z'. Put $U' = \mathrm{Proj}(P_R) \setminus Z'$, which is open in $\mathrm{Proj}(P_R)$ but not necessarily affine.

Since $\mathcal{F}^{\otimes n}(e)/\mathcal{G}$ is finitely presented, Z' can also be realized as the support of a finitely generated ideal sheaf $\mathcal{I}$ (*e.g.*, using Fitting ideals). For n' sufficiently large, we have

$$H^1(\mathrm{Proj}(P_R), \mathcal{F}^{\otimes(n+n')}(e) \otimes \mathcal{I}) = H^1(\mathrm{Proj}(P_R), \mathcal{F}^{\otimes n'} \otimes (\mathcal{F}^{\otimes n}(e) \otimes \mathcal{I})) = 0$$

and hence another exact sequence

$$H^0(\mathrm{Proj}(P_R), \mathcal{F}^{\otimes(n+n')}(e)) \longrightarrow H^0(Z', \mathcal{F}^{\otimes n+n'}(e) \otimes (\mathcal{O}/\mathcal{I})) \longrightarrow 0.$$

Since Z' is affine, by Remark 1.1.5 $\mathcal{F}^{\otimes(n+n')}(e) \otimes (\mathcal{O}/\mathcal{I})$ corresponds to a finitely generated module over the coordinate ring of Z'. Consequently, for n' sufficiently large, there exist finitely many global sections of $\mathcal{F}^{\otimes(n+n')}(e)$ which generate the restriction of $\mathcal{F}^{\otimes(n+n')}(e)$ to Z'.

In case $e = 0$, we also know that for n' divisible by n, the restriction of $\mathcal{F}^{\otimes(n+n')}$ to U' is also generated by finitely many global sections (namely the $(n'/n+1)$-fold products of $s_1, \dots, s_m$). Consequently, for all n' sufficiently large and divisible by n, $\mathcal{F}^{\otimes(n+n')}$ is generated by finitely many global sections.

For general e, for n' sufficiently large and divisible by n, the restrictions of $\mathcal{F}^{\otimes n}(e)$ and $\mathcal{F}^{\otimes n'}$ to U' are both generated by finitely many global sections (the latter thanks to the previous paragraph). For such n', $\mathcal{F}^{\otimes(n+n')}(e)$ is generated by finitely many global sections, so $\mathcal{F}^{\otimes n}$ is globally ample; by Lemma 8.8.3, $\mathcal{F}$ is also globally ample. □

Corollary 8.8.7. — *Let $\mathcal{F}$ be a globally étale vector bundle on* $\mathrm{Proj}(P_R)$*. Then for any positive integer n, $H^1(\mathrm{Proj}(P_R), \mathcal{F}(n)) = 0$ and $\mathcal{F}(n)$ is globally ample.*

Proof. — The equality $H^1(\mathrm{Proj}(P_R), \mathcal{F}(n)) = 0$ is immediate from Proposition 6.2.2. Given this equality (for all $\mathcal{F}$ and n), we may check criterion (c) of Proposition 8.8.6 to deduce that $\mathcal{F}(n)$ is globally ample. □

Lemma 8.8.8. — *Let $\mathcal{F}$ be a globally ample line bundle on* $\mathrm{Proj}(P_R)$*. Then for any $s \in H^0(\mathrm{Proj}(P_R), \mathcal{F})$, the open subscheme U_s of* $\mathrm{Proj}(P_R)$ *on which s generates $\mathcal{F}$ is affine.*

Proof. — Since $\mathrm{Proj}(P_R)$ is quasicompact and quasiseparated by Remark 8.7.6, so then is U_s. By the cohomological criterion for affinity [**122**, Tag 01XG], it suffices to check that $H^1(U_s, \mathcal{G}) = 0$ where $\mathcal{G}$ is an arbitrary quasicoherent sheaf of finite type on U_s. Let $j : U_s \to \mathrm{Proj}(P_R)$ be the canonical inclusion, so that $H^1(U_s, \mathcal{G}) = H^1(\mathrm{Proj}(P_R), j_*\mathcal{G})$. We may then write $j_*\mathcal{G}$ as the direct limit of $\mathcal{H} \otimes \mathcal{F}^{\otimes d}$ as $d \to \infty$ for some quasicoherent sheaf of finite type $\mathcal{H}$ on $\mathrm{Proj}(P_R)$. Since cohomology commutes with direct limits, we deduce the claim from Proposition 8.8.6. □

Corollary 8.8.9. — *Let $\mathcal{F}$ be a globally ample line bundle on* $\mathrm{Proj}(P_R)$*. There is then a natural isomorphism*

$$\mathrm{Proj}(P_R) \cong \mathrm{Proj}(P_{\mathcal{F}}), \qquad P_{\mathcal{F}} = \bigoplus_{n=0}^{\infty} H^0(\mathrm{Proj}(P_R), \mathcal{F}^{\otimes n}).$$

We next establish a pointwise interpretation of ampleness.

Definition 8.8.10. — Let $\mathcal{F}$ be a vector bundle on $\operatorname{Proj}(P_R)$. Let M be the φ^a-module over $\tilde{\mathcal{R}}_R$ corresponding to $\mathcal{F}$ *via* Theorem 6.3.12. We define the *slope polygon* of $\mathcal{F}$ as a function on $\mathcal{M}(R)$ (and on $\operatorname{Spa}(R, R^+)$ by retraction) as the fiberwise Harder-Narasimhan polygon; this agrees with the slope polygon of M thanks to Remark 4.2.18. We say that $\mathcal{F}$ is *pointwise ample* at $\beta \in \mathcal{M}(R)$ if the slopes of $\mathcal{F}$ at β are everywhere positive; by Theorem 7.4.5, this is an open condition on $\mathcal{M}(R)$. If this condition holds for all β, we say that $\mathcal{F}$ is *pointwise ample.*

Lemma 8.8.11. — *For any pointwise ample vector bundle $\mathcal{F}$ on $\operatorname{Proj}(P_R)$ and any vector bundle $\mathcal{G}$ on $\operatorname{Proj}(P_R)$, there exists $n_0 \in \mathbb{Z}$ such that $\mathcal{F}^{\otimes n} \otimes \mathcal{G}$ is pointwise ample for all $n \geqslant n_0$.*

Proof. — We first observe that this is true if $R = L$ is an analytic field. Namely, if the slopes of $\mathcal{F}$ are $\mu_1, \dots, \mu_m$ and the slopes of $\mathcal{G}$ are $\nu_1, \dots, \nu_l$, then for $n > 0$ each slope of $\mathcal{F}^{\otimes n} \otimes \mathcal{G}$ is equal to some ν_i plus an n-fold sum of the μ_j. If $\mathcal{F}$ is pointwise ample, then $\mu_1, \dots, \mu_m$ are positive, so for n large enough the slopes of $\mathcal{F}^{\otimes n} \otimes \mathcal{G}$ are all positive.

To extend this argument to the general case, it suffices to note that on one hand, by Proposition 7.4.6 the slopes of $\mathcal{F}$ and $\mathcal{G}$ are bounded below; on the other hand, the slopes of $\mathcal{F}$ are limited to a discrete subset of $\mathbb{Q}$, and hence bounded away from 0. □

Although logically unnecessary, the following argument may be helpful on first reading.

Lemma 8.8.12. — *Suppose that $R = L$ is an analytic field. Let $\mathcal{F}$ be a pointwise ample vector bundle on* $\operatorname{Proj}(P_R)$.

(a) *We have* $H^1(\operatorname{Proj}(P_R), \mathcal{F}) = 0$.

(b) *The bundle $\mathcal{F}$ is generated by* $H^0(\operatorname{Proj}(P_R), \mathcal{F})$.

Proof. — Let M be the φ^a-module over $\tilde{\mathcal{R}}_R$ associated to $\mathcal{F}$ *via* Theorem 6.3.12. To prove (a), by Theorem 4.2.13 we may reduce to the case where M is pure of some slope $s > 0$. By Theorem 8.7.13, we must check that $H^1_{\varphi^a}(M) = 0$. There is no harm in enlarging a, so we may assume that $as \in \mathbb{Z}$; this case follows from Corollary 8.8.7.

To prove (b), by (a) and Theorem 4.2.13 we may again reduce to the case where M is pure of some slope $s > 0$. By Theorem 8.7.13, we must check that $H^0_{\varphi^a}(M)$ generates M. There is no harm in enlarging a, so we may assume that $as \in \mathbb{Z}$; this case follows from Proposition 6.2.4. □

To relate ampleness to pointwise ampleness, we use a dévissage argument in the style of [**96**].

Lemma 8.8.13. — *Let $\mathcal{F}$ be a vector bundle on $\operatorname{Proj}(P_R)$ whose slopes at some $\beta \in \mathcal{M}(R)$ are all nonnegative but not all zero. Then there exists a short exact sequence*

$$0 \longrightarrow \mathcal{O}(-1) \longrightarrow \mathcal{G} \longrightarrow \mathcal{F} \longrightarrow 0$$

of vector bundles on $\operatorname{Proj}(P_R)$ such that the slopes of $\mathcal{G}$ at β are also all nonnegative.

Proof. — Let $i : Z_1 \to \mathrm{Proj}(P_R)$ denote the canonical inclusion; we then have an exact sequence

$$0 \longrightarrow \mathcal{O}(-1) \longrightarrow \mathcal{O} \longrightarrow i_* i^* \mathcal{O} \longrightarrow 0$$

of sheaves on $\mathrm{Proj}(P_R)$ in which the map $\mathcal{O}(-1) \to \mathcal{O}$ is multiplication by f_1. Tensoring by $\mathcal{F}^\vee$ yields another exact sequence

$$0 \longrightarrow \mathcal{F}^\vee(-1) \longrightarrow \mathcal{F}^\vee \longrightarrow i_* i^* \mathcal{F}^\vee \longrightarrow 0.$$

Write $Z_{1,\beta}$ for the zero locus of f_1 on $\mathrm{Proj}(P_{\mathcal{H}(\beta)})$; we write i also for the canonical inclusion $Z_{1,\beta} \to \mathrm{Proj}(P_{\mathcal{H}(\beta)})$.

By Theorem 4.2.13, there exists a short exact sequence

$$0 \longrightarrow \mathcal{F}_+ \longrightarrow \mathcal{F}_\beta \longrightarrow \mathcal{F}_0 \longrightarrow 0$$

of vector bundles on $\mathrm{Proj}(P_{\mathcal{H}(\beta)})$ such that $\mathcal{F}_+ \neq 0$ has all positive slopes and $\mathcal{F}_0$ has all zero slopes. In the exact sequence

$$H^0(\mathrm{Proj}(P_{\mathcal{H}(\beta)}), \mathcal{F}_+^\vee) \longrightarrow H^0(\mathrm{Proj}(P_{\mathcal{H}(\beta)}), i_* i^* \mathcal{F}_+^\vee) \longrightarrow H^1(\mathrm{Proj}(P_{\mathcal{H}(\beta)}), \mathcal{F}_+^\vee(-1)),$$

the first term vanishes because $\mathcal{F}_+^\vee$ has all slopes negative. We thus have a commutative diagram

$$\begin{array}{ccccc}
H^0(Z_1, i^*\mathcal{F}^\vee) & \longrightarrow & H^0(Z_{1,\beta}, i^*\mathcal{F}^\vee) & \longrightarrow & H^0(Z_{1,\beta}, i^*\mathcal{F}_+^\vee) \\
\| & & \| & & \| \\
H^0(\mathrm{Proj}(P_R), i_*i^*\mathcal{F}^\vee) & \longrightarrow & H^0(\mathrm{Proj}(P_{\mathcal{H}(\beta)}), i_*i^*\mathcal{F}_\beta^\vee) & \longrightarrow & H^0(\mathrm{Proj}(P_{\mathcal{H}(\beta)}), i_*i^*\mathcal{F}_+^\vee) \\
\downarrow & & \downarrow & & \downarrow \\
H^1(\mathrm{Proj}(P_R), \mathcal{F}^\vee(-1)) & \longrightarrow & H^1(\mathrm{Proj}(P_{\mathcal{H}(\beta)}), \mathcal{F}_\beta^\vee(-1)) & \longrightarrow & H^1(\mathrm{Proj}(P_{\mathcal{H}(\beta)}), \mathcal{F}_+^\vee(-1))
\end{array}$$

Since Z_1 and $Z_{1,\beta}$ are affine, we have

$$H^0(Z_{1,\beta}, i^*\mathcal{F}^\vee) = H^0(Z_1, i^*\mathcal{F}^\vee) \otimes_{\mathcal{O}(Z_1)} \mathcal{O}(Z_{1,\beta}).$$

Since $\mathcal{F}_+ \neq 0$, we can find $x \in H^0(Z_1, i^*\mathcal{F}^\vee)$ whose image in $H^0(Z_{i,\beta}, i^*\mathcal{F}_+^\vee)$ is nonzero. Map x to $y \in H^1(\mathrm{Proj}(P_R), \mathcal{F}^\vee(-1))$; then the image of y in $H^0(\mathrm{Proj}(P_{\mathcal{H}(\beta)}), \mathcal{F}_+^\vee(-1))$ is also nonzero.

We claim that the short exact sequence defined by y has the desired property. To check the claim, we may reduce to the case where $R = \mathcal{H}(\beta)$, so that $\mathcal{F}_\beta = \mathcal{F}$. Let $\mathcal{G}_+$ be the inverse image of $\mathcal{F}_+$ in $\mathcal{G}$; we then have an exact sequence

$$0 \longrightarrow \mathcal{O}(-1) \longrightarrow \mathcal{G}_+ \longrightarrow \mathcal{F}_+ \longrightarrow 0 \qquad (8.8.13.1)$$

which by construction is not split. For any vector bundle quotient $\mathcal{H}$ of $\mathcal{G}_+$, we get an exact sequence

$$0 \longrightarrow \mathcal{H}_1 \longrightarrow \mathcal{H} \longrightarrow \mathcal{H}_2 \longrightarrow 0$$

in which $\mathcal{H}_1$ is the image of $\mathcal{O}(-1)$ in $\mathcal{H}$. Note that on one hand, $\deg(\mathcal{H}_1) \geqslant -1$ with equality only if $\mathcal{H}_1 = \mathcal{O}(-1)$; on the other hand, $\deg(\mathcal{H}_2) \geqslant 0$ with equality only if $\mathcal{H}_2 = 0$. We cannot have both equalities simultaneously because (8.8.13.1) does not split; hence $\deg(\mathcal{H}) = \deg(\mathcal{H}_1) + \deg(\mathcal{H}_2) \geqslant 0$, and by Theorem 4.2.13 it follows that $\mathcal{G}_+$ has all slopes nonnegative. From the exact sequence

$$0 \longrightarrow \mathcal{G}_+ \longrightarrow \mathcal{G} \longrightarrow \mathcal{F}_0 \longrightarrow 0,$$

we see that $\mathcal{G}$ also has all slopes nonnegative. □

Corollary 8.8.14. — *Let $\mathcal{F}$ be a vector bundle on $\mathrm{Proj}(P_R)$ whose slopes at some $\beta \in \mathcal{M}(R)$ are all nonnegative. Then there exists a short exact sequence*

$$0 \longrightarrow \mathcal{H} \longrightarrow \mathcal{G} \longrightarrow \mathcal{F} \longrightarrow 0$$

of vector bundles on $\mathrm{Proj}(P_R)$ such that $\mathcal{G}$ is étale at β.

Proof. — This follows from Lemma 8.8.13 by induction on $\deg(\mathcal{F}_\beta)$, plus Theorem 7.3.7. □

Theorem 8.8.15. — *A vector bundle $\mathcal{F}$ on $\mathrm{Proj}(P_R)$ is ample if and only if it is pointwise ample.*

Proof. — Suppose that $\mathcal{F}$ is ample; to prove that $\mathcal{F}$ is pointwise ample, we may even assume that $\mathcal{F}$ is globally ample. Choose any $\beta \in \mathcal{M}(R)$ and let $\alpha_1 \geqslant \cdots \geqslant \alpha_m$ be the slopes of $\mathcal{F}$ at β listed with multiplicity. For $n > 0$, the slopes of $\mathcal{F}^{\otimes n}$ at β are the n-fold sums of $\alpha_1, \dots, \alpha_m$. For some n, $\mathcal{F}^{\otimes n}(-1)$ is generated by global sections, and so $n\alpha_m - 1 \geqslant 0$ and $\alpha_m > 0$. Hence $\mathcal{F}$ is pointwise ample.

Conversely, suppose that $\mathcal{F}$ is pointwise ample. To prove that $\mathcal{F}$ is ample, it is harmless to first replace R by a rational localization encircling β. By Lemma 8.8.11, there exists $n_0 \in \mathbb{Z}$ such that for $n \geqslant n_0$, $\mathcal{F}^{\otimes n}(-1)$ is pointwise ample at β. By Corollary 8.8.14, there exists a short exact sequence

$$0 \longrightarrow \mathcal{H} \longrightarrow \mathcal{G} \longrightarrow \mathcal{F}^{\otimes n}(-1) \longrightarrow 0$$

of vector bundles on $\mathrm{Proj}(P_R)$ such that $\mathcal{G}$ is étale at β. By replacing R by a suitable rational localization, we may ensure that $\mathcal{G}$ is globally étale; by Corollary 8.8.7, $\mathcal{G}(1)$ is globally ample, as then is its quotient $\mathcal{F}^{\otimes n}$. By Lemma 8.8.3, $\mathcal{F}$ is ample. □

Remark 8.8.16. — We do not know whether pointwise ample (or equivalently ample) implies globally ample. For example, we do not know whether $\mathcal{F} = \mathcal{L}^{\otimes n}(e)$ is necessarily globally ample in case $e, n > 0$ and $\mathcal{L}$ is étale but not globally étale.

We next globalize the construction.

Definition 8.8.17. — Let $\mathcal{F}$ be a vector bundle on FF_X. We say that $\mathcal{F}$ is *ample* if for every choice of A and R and every morphism $f : \mathrm{Spa}(A, A^+) \to X$, the vector bundle $f^*\mathcal{F}$ on FF_R corresponds *via* Theorem 8.7.7 to an ample vector bundle on $\mathrm{Proj}(P_R)$. By Theorem 8.8.15, this is equivalent to requiring that the slopes of $\mathcal{F}$ (as functions on X) be everywhere positive; this equivalence has the following consequences.

- If $X = \mathrm{Spa}(A, A^+)$, then a vector bundle on $\mathrm{Proj}(P_R)$ is ample if and only if the corresponding vector bundle on FF_X is ample.

- Ampleness descends along surjective morphisms on the base. That is, if $f : Y \to X$ is a surjective morphism of perfectoid adic spaces, $\mathcal{F}$ is a vector bundle on FF_X, and $f^*\mathcal{F}$ is ample, then $\mathcal{F}$ is ample. In particular, ampleness is local on the base.
- By Theorem 7.4.5, ampleness is an open condition on the base, and even on the real quotient of the base. That is, if $\mathcal{F}$ is a vector bundle on X and the restriction of $\mathcal{F}$ to $\mathrm{FF}_{\mathcal{H}(x)}$ is ample for some $x \in X$, then there exists a partially proper open neighborhood U of x in X such that the restriction of $\mathcal{F}$ to FF_U is also ample.

We finally introduce a key example of an ample line bundle.

Definition 8.8.18. — For z as in Remark 8.7.2, the inclusion of modules $\tilde{\mathcal{R}}_R^{[q^{-1/2},q^{1/2}]} \to z^{-1}\tilde{\mathcal{R}}_R^{[q^{-1/2},q^{1/2}]}$ induces an inclusion $\mathcal{O} \to \mathcal{L}$ of vector bundles on FF_R. Since this construction is canonically independent of z, it globalizes to define an inclusion $\mathcal{O}_X \to \mathcal{L}_X$ of line bundles on FF_X corresponding to an element $t_X \in H^0(\mathrm{FF}_X, \mathcal{L}_X)$ whose divisor is the image of the canonical section $X \to \mathrm{FF}_X$. Note that $\mathcal{L}_X$ is pure of slope 1 and hence ample by Theorem 8.8.15.

Lemma 8.8.19. — *For $X = \mathrm{Spa}(A, A^+)$, the φ^a-module corresponding to $\mathcal{L}_X$ is globally pure of slope* 1.

Proof. — Write $z = [\overline{z}] + pz_1$. Let M be the φ^a-module over $\tilde{\mathcal{R}}_R$ free on a single generator $\mathbf{v}$ satisfying $\varphi^a(\mathbf{v}) = z_1^{-1}z\mathbf{v}$; it is evidently globally étale. Define the convergent product

$$u = \prod_{n=0}^{\infty} \varphi^{an}(1 + p^{-1}z_1^{-1}[\overline{z}]) \in \tilde{\mathcal{R}}_R^+.$$

In $\tilde{\mathcal{R}}_R$, we have $\varphi^a(u) = pz_1z^{-1}u$; consequently, $u\mathbf{v}$ defines an inclusion $\tilde{\mathcal{R}}_R \to M(1)$ of φ^a-modules. Computing in $\tilde{\mathcal{R}}_R^{[q^{-1/2},q^{1/2}]}$ shows that $M(1)$ must be the φ^a-module corresponding to $\mathcal{L}_X$. This proves the claim. □

In the previous example, when $X = \mathrm{Spa}(A, A^+)$ the zero locus of the section t_X is isomorphic to $\mathrm{Spec}(A)$. This suggests the following conjectures.

Conjecture 8.8.20. — *Let $\mathcal{F}$ be a line bundle on $\mathrm{Proj}(P_R)$ such that $\deg(\mathcal{F}) > 0$ and $\mathcal{F}(-\deg(\mathcal{F}))$ is globally étale. (Note that $\mathcal{F}(-\deg(\mathcal{F}))$ makes sense because $\deg(\mathcal{F}) : X \to \mathbb{Z}$ is continuous for the discrete topology on $\mathbb{Z}$. Note also that $\mathcal{F}$ is globally ample by Corollary 8.8.7.)*

(a) *Choose $t \in H^0(\mathrm{Proj}(P_R), \mathcal{F})$ whose restriction to FF_x is nonzero for all $x \in \mathrm{Spa}(R, R^+)$. Then the closed subscheme Z of $\mathrm{Proj}(P_R)$ cut out by t is affine. (This would follow from Lemma 8.8.8 given the existence of a second section s whose zero locus is disjoint from that of t.)*

(b) *With notation as in (a), equip the coordinate ring of Z with a uniform norm by identifying it with the global sections of the subspace of* FF_R *cut out by* t. *Then this Banach algebra over* $\mathbb{Q}_p$ *is perfectoid.*

Remark 8.8.21. — In the case where $R = L$ is an analytic field, Conjecture 8.8.20 is established in [**45**]. In the case where $\deg(\mathcal{F}) = 1$, it should be possible to argue by reversing the proof of Lemma 8.8.19 and rescaling the norm on R as in Remark 2.3.11 (d). It is less clear what should happen if $\mathcal{F}(-\deg(\mathcal{F}))$, which is necessarily étale, fails to be globally étale; compare Remark 8.8.16.

8.9. B-pairs

We next make contact with another interpretation of the functor FF inspired by Berger's construction of *B-pairs* [**11**].

Hypothesis 8.9.1. — Throughout §8.9, retain Hypothesis 8.7.1, but assume in addition that $X = \mathrm{Spa}(A, A^+)$.

Convention 8.9.2. — We use the notation $\mathcal{L}_X$ to represent not only the line bundle on FF_X described in Definition 8.8.18, but also the line bundle on $\mathrm{Proj}(P_R)$ giving rise to it *via* Theorem 8.7.7.

Lemma 8.9.3. — *Let Z be the image of the canonical section* $\mathrm{Spec}(A) \to \mathrm{Proj}(P_R)$.

(a) *The open subscheme* $\mathrm{Proj}(P_R) - Z$ *of* $\mathrm{Proj}(P_R)$ *is affine.*

(b) *The closed subscheme Z of* $\mathrm{Proj}(P_R)$ *is a Cartier divisor contained in an open affine subscheme of* $\mathrm{Proj}(P_R)$.

Proof. — Thanks to the interpretation of Z as the divisor of the section t_X of the line bundle $\mathcal{L}_X$, we may invoke Lemma 8.8.8 to deduce (a). To prove (b), define L as in Remark 8.7.6; it then suffices to exhibit some $t_L \in P_{L,1}$ whose support in $\mathrm{Proj}(P_R)$ is disjoint from Z. For this, it suffices to follow the proof of Proposition 6.2.4 to force t not to have the slope 1 in its Newton polygon. □

Definition 8.9.4. — By Lemma 8.9.3 (a), the complement of Z in $\mathrm{Proj}(P_R)$ is an affine scheme $\mathrm{Spec}(R_1)$. By Lemma 8.9.3 (b), the completion of $\mathrm{Proj}(P_R)$ along Z is another affine scheme $\mathrm{Spec}(R_2)$, and $\mathrm{Spec}(R_1) \times_{\mathrm{Proj}(P_R)} \mathrm{Spec}(R_2)$ is yet another affine scheme $\mathrm{Spec}(R_3)$. One can also identify R_2 with the $\ker(\theta)$-adic completion of $\tilde{\mathcal{R}}_R^{\mathrm{int},1}$ and R_3 with $R_2[z^{-1}]$ for some z generating $\ker(\theta)$.

Remark 8.9.5. — In case $R = L$, the morphism $\mathrm{Spec}(R_1 \oplus R_2) \to \mathrm{Proj}(P_R)$ is faithfully flat, so it can be used to define quasicoherent sheaves and compute their cohomology. In general, one might expect the same to hold, but one cannot quite prove it using faithfully flat descent because it is unclear whether $\mathrm{Spec}(R_2) \to \mathrm{Proj}(P_R)$ is a flat morphism. Nonetheless, we can salvage something; see Theorem 8.9.6.

Theorem 8.9.6. — *Set notation as in Remark 8.9.5.*

(a) *For any flat quasicoherent sheaf V on $\mathrm{Proj}(P_R)$, the cohomology of the complex*

$$0 \longrightarrow \Gamma(\mathrm{Spec}(R_1), V) \oplus \Gamma(\mathrm{Spec}(R_2), V) \longrightarrow \Gamma(\mathrm{Spec}(R_3), V) \longrightarrow 0$$

where the arrow is given by the difference between the two natural restriction maps, may be naturally identified with $H^i(\mathrm{Proj}(P_R), V)$.

(b) *The morphism $\mathrm{Spec}(R_1 \oplus R_2) \to \mathrm{Proj}(P_R)$ is an effective descent morphism for the category of quasicoherent finite locally free sheaves over schemes (after reversing all arrows).*

(c) *The category of vector bundles on $\mathrm{Proj}(P_R)$ is equivalent to the category of triples (V_1, V_2, ι) where V_1 is a finite projective R_1-module, V_2 is a finite projective R_2-module, and $\iota : V_1 \otimes_{R_1} R_3 \cong V_2 \otimes_{R_2} R_3$ is an isomorphism of R_3-modules.*

Proof. — This follows from Proposition 1.3.6. □

CHAPTER 9

RELATIVE (φ, Γ)-MODULES

To conclude, we indicate how to use the preceding constructions to describe étale local systems on arbitrary adic spaces; this involves certain sheaves for Scholze's *pro-étale topology*. These sheaves generalize the extended Robba ring in ordinary p-adic Hodge theory (as considered in §4.2) but not the Robba ring itself (as considered in §4.1). Generalizing the latter involves passing from the *perfect period rings* that we consider to certain *imperfect period rings* whose construction is somewhat less functorial; we defer discussion of imperfect period rings to a subsequent paper.

9.1. The pro-étale topology for adic spaces

In p-adic Hodge theory, one studies the Galois theory of a p-adic field using certain highly ramified infinite algebraic extensions. To carry out relative p-adic Hodge theory, one needs an analogous geometric construction; one convenient mechanism for this is the *pro-étale topology* introduced by Scholze [**115**, §3].

Definition 9.1.1. — For $\mathcal{C}$ a category, a *pro-object* over $\mathcal{C}$ consists of a pair (I, F) in which I is a directed poset and F is a contravariant functor from I to $\mathcal{C}$. The pro-objects over $\mathcal{C}$ form a category $\widehat{\mathcal{C}}$ in which

$$\operatorname{Hom}((I,F),(I',F')) = \varprojlim_{i' \in I'} \varinjlim_{i \in I} \operatorname{Hom}(F(i), F'(i'))$$

(see [**5**, Exposé I, §8.10]). By design, the category $\widehat{\mathcal{C}}$ (the *pro-category* associated to $\mathcal{C}$) admits inverse limits (modulo set-theoretic difficulties which we gloss over here). There is a natural embedding $\mathcal{C} \to \widehat{\mathcal{C}}$ taking an object $X \in \mathcal{C}$ to the pair $(\{0\}, F)$ in which $\{0\}$ is the singleton poset and F takes 0 to X.

In the cases we are considering, the category $\mathcal{C}$ admits a forgetful functor to topological spaces denoted $X \mapsto |X|$. We extend this to a forgetful functor from $\widehat{\mathcal{C}}$ to topological spaces by setting $\left|\varprojlim_{i \in I} X_i\right| = \varprojlim_{i \in I} |X_i|$.

Definition 9.1.2. — For X a preadic space, a morphism $U \to V$ in $\widehat{X}_{\text{ét}}$ is *étale* (resp. *finite étale, faithfully finite étale*) if it arises by base extension from an étale (resp. finite étale, faithfully finite étale) morphism $Y_0 \to X_0$ in $X_{\text{ét}}$. One checks formally that these properties are stable under composition (as in [**115**, Lemma 3.10 (ii)]) and base change (as in [**115**, Lemma 3.10 (i)]).

A morphism $U \to V$ in $\widehat{X}_{\text{ét}}$ is *pro-étale* if U admits a *pro-étale presentation* as a cofiltered inverse limit $\varprojlim U_i$ of objects which are étale over V, such that $U_i \to U_j$ is faithfully finite étale for sufficiently large j.

Lemma 9.1.3. — *Let X be a preadic space.*

(a) *Let $U \to V$ be a pro-étale morphism in $\widehat{X}_{\text{ét}}$ and let $W \to V$ be an arbitrary morphism in $\widehat{X}_{\text{ét}}$. Then $U \times_V W \to W$ is pro-étale, and the map $|U \times_V W| \to |U| \times_{|V|} |W|$ is surjective.*

(b) *For $U \to V \to W$ pro-étale morphisms in $\widehat{X}_{\text{ét}}$, the composition $U \to W$ is pro-étale.*

Proof. — To prove (a), we follow the proof of [**115**, Lemma 3.10 (i)]. In case $U \to V$ is étale, we may assume that $U, V \in X_{\text{ét}}$ and realize $W \to V$ using a compatible system of morphisms $W_i \to V$ in $X_{\text{ét}}$. We then have maps

$$|U \times_V W| = \varprojlim_i |U \times_V W_i| \longrightarrow \varprojlim_i |U| \times_{|V|} |W_i| = |U| \times_{|V|} |W|.$$

If we put the discrete topology on each fibre over W, then the central arrow is a surjective map of compact spaces by Remark 2.3.15 (c). In the general case, choose a pro-étale presentation $U = \varprojlim_i U_i \to V$; we then have maps

$$|U \times_V W| = \varprojlim_i |U_i \times_V W| \longrightarrow \varprojlim_i |U_i| \times_{|V|} |W| = |U_i| \times_{|V|} |W|.$$

The central arrow is again surjective by Remark 2.3.15 (c).

To prove (b), we follow the proof of [**115**, Lemma 3.10 (vi)]. It suffices to check the case where $U \to V$ is étale. In this case, $U \to V$ is the pullback of some étale morphism $U_0 \to V_0$ in $X_{\text{ét}}$ along some morphism $V \to V_0$ in $\widehat{X}_{\text{ét}}$. Choose a pro-étale presentation $V = \varprojlim_i V_i \to W$; then $V \to V_0$ arises from a compatible family of morphisms $V_i \to V_0$ in $X_{\text{ét}}$ (for i large). Hence $U = \varprojlim_i U_0 \times_{V_0} V_i$ is pro-étale over W. □

Definition 9.1.4. — Let $X_{\text{proét}}$ denote the full subcategory of $\widehat{X}_{\text{ét}}$ consisting of objects which are pro-étale over X. By Lemma 9.1.3, we may view $X_{\text{proét}}$ as a site by taking coverings to be families $\{U_i \to Y\}_i$ such that for some (hence any) pro-étale presentations $\varprojlim_j U_{i,j}$ of U_i over Y and some (hence any) choice of indices $j = j(i)$ such that $U_{j''} \to U_{j'}$ is faithfully finite étale for any $j' \geqslant j(i)$, the family $\{U_{i,j(i)} \to Y\}_i$ is a set-theoretic covering. The resulting site maps to the usual étale site $X_{\text{ét}}$ *via* the embedding $X_{\text{ét}} \to \widehat{X}_{\text{ét}}$.

In case X is stably adic, we may characterize coverings in $X_{\text{proét}}$ more simply: they are the families $\{U_i \to Y\}_i$ such that the maps $\{|U_i| \to |Y|\}$ form a set-theoretic covering.

Remark 9.1.5. — Recall that an inverse limit of spectral spaces is a spectral space by Corollary 8.1.3. Consequently, for any preadic space X and any $Y \in X_{\text{proét}}$ admitting a pro-étale presentation in which the underlying space of each term is qcqs, $|Y|$ is a spectral space. It follows that for any $Y \in X_{\text{proét}}$, $|Y|$ is a locally spectral space.

In case X is stably adic, we can emulate more of [**115**, Lemma 3.10].

Lemma 9.1.6. — *Let X be a stably adic space.*

(a) *For $U \in \widehat{X}_{\text{ét}}$ and $W \subseteq |U|$ a quasicompact open set, there exist $V \in \widehat{X}_{\text{ét}}$ and an étale map $V \to U$ such that $|V| \to |U|$ induces a homeomorphism $|V| \cong W$. Moreover, if $U \in X_{\text{proét}}$, one can take $V \in X_{\text{proét}}$, and then any morphism $V' \to U$ in $X_{\text{proét}}$ with image contained in $|W|$ factors through V.*

(b) *For any pro-étale morphism $U \to V$ in $\widehat{X}_{\text{ét}}$, the map $|U| \to |V|$ is open.*

(c) *Any (finite) étale map $U \to V$ in $\widehat{X}_{\text{ét}}$ with $V \in X_{\text{proét}}$ and $|U| \to |V|$ surjective is the base extension of some surjective (finite) étale morphism in $X_{\text{ét}}$.*

Proof. — Given Lemma 9.1.3 and the fact that étale maps are open (Lemma 8.2.17 (b)), we may deduce (a)–(c) as in [**115**, Lemma 3.10 (iii)–(v)]. □

Remark 9.1.7. — In [**115**], the only preadic spaces considered are adic spaces which are *locally noetherian* (*i.e.*, they are covered by the adic spectra of strongly noetherian adic Banach rings, which are sheafy by Proposition 2.4.16). This hypothesis is used in an essential way in [**115**, Lemma 3.10 (vii)], which asserts that $X_{\text{proét}}$ admits arbitrary finite projective limits (not just fibred products, which exist by virtue of Lemma 9.1.3). Namely, it is necessary to ensure that objects of $X_{\text{ét}}$ locally have only finitely many connected components, so that arbitrary intersections of closed-open subsets are again closed-open. In the absence of a noetherian hypothesis, we may work with the pro-étale site without incident, but we cannot freely apply topos-theoretic machinery as in [**115**, Proposition 3.12].

Remark 9.1.8. — One can similarly define a pro-étale topology for rigid analytic spaces over K, Berkovich analytic spaces, or schemes. However, in the cases of rigid or Berkovich analytic spaces, one must take suitable care with the definition of coverings. In the case of schemes, one can take advantage of the properties of flatness to introduce a simpler variant of the pro-étale topology; see Definition 1.4.10.

Using the pro-étale topology, we may reinterpret the definition of étale local systems.

Lemma 9.1.9. — *Let X be a preadic space. For any topological space T, the functor $\mathcal{F}_T : Y \mapsto \mathrm{Map}_{\mathrm{cont}}(|Y|, T)$ is a sheaf on $X_{\text{proét}}$. In particular, if T is discrete, then $\mathcal{F}_T$ is the usual constant sheaf associated to T.*

Proof. — It is clear that $\mathcal{F}_T$ is a sheaf on X. To verify that $\mathcal{F}_T$ is a sheaf on $X_{\text{ét}}$, by Proposition 8.2.21 it suffices to observe that for $(A, A^+) \to (B, B^+)$ a faithfully finite étale morphism, by Lemma 8.2.17 (b) the morphism $\operatorname{Spa}(B, B^+) \to \operatorname{Spa}(A, A^+)$ of topological spaces is open and hence a quotient map. To complete the proof, it suffices to observe that for $Y \in X_{\text{proét}}$ a tower of faithfully finite étale morphisms, the morphism $|Y| \to |X|$ is an inverse limit of surjective open morphisms, so it is again a quotient map. □

Definition 9.1.10. — By an *étale $\mathbb{Z}_p$-local system* (resp. an *étale $\mathbb{Q}_p$-local system*) on $X_{\text{proét}}$, we mean a sheaf in $\mathbb{Z}_p$-modules (resp. in $\mathbb{Q}_p$-vector spaces) locally of the form $\mathcal{F}_T$ for T a finite free $\mathbb{Z}_p$-module (resp. finite-dimensional $\mathbb{Q}_p$-vector space) carrying its usual p-adic topology.

Lemma 9.1.11. — *For X a preadic space, the categories of étale $\mathbb{Z}_p$-local systems and $\mathbb{Q}_p$-local systems on X are equivalent to the corresponding categories on $X_{\text{proét}}$.*

Proof. — The functors from local systems on X to local systems on $X_{\text{proét}}$ are defined by pullback as in Definition 1.4.10. To check that these are fully faithful, we may reduce to considering local systems on $\widetilde{\operatorname{Spa}}(A, A^+)$ which become constant on some faithfully finite étale tower; however, such towers descend to $\operatorname{Spec}(A)$ by Lemma 8.2.17 (a), so we may appeal to Theorem 1.4.11. By a similar argument, we may also deduce essential surjectivity from Theorem 1.4.11. (See also [**115**, Proposition 8.2].) □

Definition 9.1.12. — For X a preadic space and V an étale $\mathbb{Z}_p$-local system or $\mathbb{Q}_p$-local system on X, following [**115**] we define the *pro-étale cohomology* of V to be the cohomology of the corresponding étale local system on $X_{\text{proét}}$, and denote it by $H^i_{\text{proét}}(X, V)$.

9.2. Perfectoid subdomains

We next identify the *perfectoid subdomains* of the pro-étale site of a preadic space, which will be used to construct period sheaves.

Hypothesis 9.2.1. — Throughout §9.2, let X be a preadic space over an analytic field K of residue characteristic p.

Definition 9.2.2. — We define the *structure presheaf* $\mathcal{O}$ and the sub-presheaves $\mathcal{O}^\circ, \mathcal{O}^+$ on $X_{\text{proét}}$ as follows. For $Y = \varprojlim_i Y_i \in X_{\text{proét}}$, put

$$\mathcal{O}(Y) = \varinjlim_i \Gamma(Y_i, \mathcal{O}_{Y_i})$$

$$\mathcal{O}^\circ(Y) = \varinjlim_i \Gamma(Y_i, \mathcal{O}^\circ_{Y_i})$$

$$\mathcal{O}^+(Y) = \varinjlim_i \Gamma(Y_i, \mathcal{O}^+_{Y_i}).$$

We define the *spectral seminorm* on $\mathcal{O}(Y)$ as follows. Choose j so that the maps $Y_i \to Y_j$ are faithfully finite étale for all $i \geqslant j$. Then for each $f \in \mathcal{O}(Y)$, choose $i \geqslant j$ for which $f \in \mathcal{O}_{Y_i}(Y_i)$ and define the spectral seminorm of f in $\mathcal{O}(Y)$ to be the spectral seminorm of f in $\mathcal{O}_{Y_i}(Y_i)$.

Remark 9.2.3. — Another way to interpret the spectral seminorm introduced in Definition 9.2.2 is to observe that each element of $|Y|$ defines a multiplicative seminorm on $\mathcal{O}(Y)$, and that the spectral seminorm is the supremum of these.

To define the sheaves we are interested in, we use a special neighborhood basis for the pro-étale topology.

Definition 9.2.4. — An element Y of $X_{\text{proét}}$ is a *perfectoid subdomain* if it admits a pro-étale presentation $\varprojlim_i Y_i$ satisfying the following conditions.

(a) There exists an index $j \in I$ such that the maps $Y_i \to Y_j$ are faithfully finite étale for all $i \geqslant j$ and the space Y_j is a preadic affinoid space over K (as then are the spaces Y_i for all $i \geqslant j$).

(b) The completion of $\mathcal{O}(Y)$ for the spectral seminorm is a perfectoid (if K is of characteristic 0) or perfect (if K is of characteristic p) Banach algebra over K.

These satisfy the following properties.

(i) If Y is a perfectoid subdomain and $Z \to Y$ is a morphism in $X_{\text{proét}}$ which is the pullback of a rational subdomain embedding of elements of $X_{\text{ét}}$, then Z is also a perfectoid subdomain (Theorem 3.6.14).

(ii) Any finite étale cover of a perfectoid subdomain is a perfectoid subdomain (Theorem 3.6.21).

(iii) The fibred product of two perfectoid subdomains is a perfectoid subdomain (by (i) and (ii) plus the local factorization of étale morphisms).

Lemma 9.2.5. — *If $X = \operatorname{Spa}(A, A^+)$ for some adic Banach algebra (A, A^+) over K, then there exists $Y = \varprojlim_i Y_i \in X_{\text{proét}}$ which is a pro-finite étale covering of X and is a perfectoid subdomain of $X_{\text{proét}}$. Consequently, for arbitrary X, the perfectoid subdomains of $X_{\text{proét}}$ form a neighborhood basis of $X_{\text{proét}}$.*

Proof. — Apply Lemma 3.6.26 to A^u. □

We have the following analogue of Proposition 8.2.20.

Proposition 9.2.6. — *Let (A, A^+) be a perfectoid adic Banach algebra over $\mathbb{Q}_p$ and put $X = \operatorname{Spa}(A, A^+)$. Let $\mathcal{B}$ be the collection of perfectoid subdomains in $X_{\text{proét}}$ which can written as faithfully finite étale towers over perfectoid subodmains in $X_{\text{ét}}$. (By Lemma 9.2.5 and the local factorization of étale morphisms, these form a stable basis for $X_{\text{proét}}$.) Let $\mathcal{P}$ be a property of coverings in $X_{\text{proét}}$ of and by elements of $\mathcal{B}$, and assume that the following conditions hold.*

(a) *Any covering admitting a refinement having property $\mathcal{P}$ also has property $\mathcal{P}$.*

(b) *Any composition of coverings having property $\mathcal{P}$ also has property $\mathcal{P}$.*

(c) *For any $Y \in \mathcal{B}$, any rational covering of Y has property $\mathcal{P}$.*

(d) *For any $Y \in \mathcal{B}$, any tower $Y' \to Y$ of faithfully finite étale morphisms, viewed as a covering, has property $\mathcal{P}$.*

Then every covering in $X_{\text{proét}}$ of and by elements of $\mathcal{B}$ has property $\mathcal{P}$.

Proof. — By Proposition 8.2.20, any covering in $X_{\text{ét}}$ of and by elements of $\mathcal{B}$ has property $\mathcal{P}$. The claim then follows by (b) and (d). □

Definition 9.2.7. — We define the *completed structure presheaf* $\widehat{\mathcal{O}}$ (or $\widehat{\mathcal{O}}_X$ for clarity) on $X_{\text{proét}}$ as follows: for $Y \in X_{\text{proét}}$, let $\widehat{\mathcal{O}}(Y)$ be the completion of $\mathcal{O}(Y)$ for the spectral seminorm. We similarly define the subpresheaf $\widehat{\mathcal{O}}^+$ of $\widehat{\mathcal{O}}$.

Lemma 9.2.8. — *Let (A, A^+) be a perfectoid adic Banach algebra over $\mathbb{Q}_p$ or a perfect uniform Banach algebra over $\mathbb{F}_p$ and put $X = \operatorname{Spa}(A, A^+)$.*

(a) *We have $H^0(X_{\text{proét}}, \widehat{\mathcal{O}}) = A$.*

(b) *For $i > 0$, $H^i(X_{\text{proét}}, \widehat{\mathcal{O}}) = 0$.*

(c) *For $i > 0$, the group $H^i(X_{\text{proét}}, \widehat{\mathcal{O}}^+)$ is annihilated by $\mathfrak{m}_A$.*

Proof. — Again as in the proof of Proposition 2.4.21, it suffices to check universal (almost) Čech-acyclicity for coverings of and by elements of a suitable basis. Using Proposition 9.2.6 and Theorem 2.4.23, we reduce to checking Čech-acyclicity for a tower of étale surjective morphisms of affinoid perfectoid spaces. Let $\{Y_i\}_i$ be the terms in such a tower and put $B_i = \mathcal{O}(Y_i)$.

Suppose first that we are in the perfect case. To clarify notation, we write R, S_i in place of A, B_i. For each i, the Čech sequence

$$0 \longrightarrow R \longrightarrow S_i \longrightarrow S_i \otimes_R S_i \longrightarrow \cdots$$

is strict exact and hence almost optimal exact (Remark 3.1.6); we may thus take completed direct limits to obtain another strict exact sequence

$$0 \longrightarrow R \longrightarrow S_\infty \longrightarrow S_\infty \widehat{\otimes}_R S_\infty \longrightarrow \cdots, \tag{9.2.8.1}$$

where S_∞ is the completed direct limit of the S_i.

In the perfectoid case, we apply the perfectoid correspondence to each morphism in the tower to obtain a tower of étale surjective morphism of affinoid perfect uniform spaces. By applying Proposition 3.6.9 (b) to (9.2.8.1), we deduce the desired result. □

Corollary 9.2.9. — *The completed structure presheaf on $X_{\text{proét}}$ agrees with its sheafification on perfectoid subdomains.*

Remark 9.2.10. — Lemma 9.2.8 (c) implies that the subsheaf $\widehat{\mathcal{O}}^+$ of $\widehat{\mathcal{O}}$ is *almost acyclic* in the sense of Proposition 8.3.2.

Definition 9.2.11. — For K of characteristic zero, we define the sheaves $\overline{\mathcal{O}}, \overline{\mathcal{O}}^{\circ}, \overline{\mathcal{O}}^{+}$ on $X_{\text{proét}}$ *via* the perfectoid correspondence: for $Y \in X_{\text{proét}}$ a perfectoid subdomain, $(\overline{\mathcal{O}}(Y), \overline{\mathcal{O}}^{+}(Y))$ is the perfect uniform adic Banach algebra of characteristic p corresponding to $(\widehat{\mathcal{O}}(Y), \widehat{\mathcal{O}}^{+}(Y))$ *via* Theorem 3.6.5.

Definition 9.2.12. — Let $\nu_X : X_{\text{proét}} \to X_{\text{ét}}$ be the natural morphism. We say that $X = \operatorname{Spa}(A, A^+)$ is *pro-sheafy* if X is uniform and sheafy and the morphism $\mathcal{O}_X \to \nu_{X*}\widehat{\mathcal{O}}_X$ is an isomorphism. For example, if A is perfectoid, then X is pro-sheafy by Lemma 9.2.8. For another example, if $A = K$, then X is pro-sheafy by the Ax-Sen-Tate theorem [**6**].

Remark 9.2.13. — Suppose that $X = \operatorname{Spa}(A, A^+)$ and that there exists a perfectoid Banach algebra B over K admitting a bounded homomorphism $A \to B$ which splits in the category of Banach modules over A. Then X is pro-sheafy. For instance, this is the case if A is strongly preperfectoid (by choosing a perfectoid field which admits a Schauder basis over $\mathbb{Q}_p$). Also, if A has this property, then so does $A\{T_1, \dots, T_n\}$.

Remark 9.2.14. — In case $X = \operatorname{Spa}(A, A^+)$ for A a reduced normal affinoid algebra over K, one can prove that X is pro-sheafy by using Temkin's resolution of singularities for affinoid algebras [**124**] to reduce to the case where A is smooth over K, then making a construction of imperfect period rings generalizing that of Andreatta-Brinon [**2**]. We will discuss this point in a subsequent paper.

One has a Kiehl glueing property for the pro-étale topology.

Theorem 9.2.15. — *Suppose X is perfectoid or perfect uniform. Then pullback of finite locally free $\mathcal{O}_X$-modules to finite locally free $\widehat{\mathcal{O}}_X$-modules defines an equivalence of categories.*

Proof. — We treat the case where K is of characteristic 0, the characteristic p case being similar but easier. Full faithfulness of the pullback functor is immediate from Lemma 9.2.8, so we need only check essential surjectivity. By Theorem 8.2.22 and Proposition 8.2.20, it suffices to check descent in case $X = \operatorname{Spa}(A, A^+)$ for some perfectoid adic Banach algebra (A, A^+) and $Y \to X$ is a tower of faithfully finite étale morphisms; moreover, we may formally reduce to the case of a countable tower. For $B = \mathcal{O}(Y)$, it then suffices to construct an A-linear splitting $B \to A$ of the inclusion $A \to B$.

Write B as the completed direct limit of an increasing sequence of faithfully finite étale A-subalgebras B_i. By Theorem 5.5.9, one can find a B_i-linear splitting of $B_i \to B_{i+1}$ such that the operator norm of $B_{i+1} \to B_i \to B_{i+1}$ is at most $p^{p^{-i}}$. Chaining these together gives the desired splitting $B \to A$. □

Remark 9.2.16. — For X perfectoid or perfect uniform, Scholze has suggested a variant of the pro-étale topology more in the spirit of the definition for schemes (Definition 1.4.10). In this approach, one says that a morphism $f : Y \to X$ of perfectoid spaces is *pro-étale* if Y is "similar to" an inverse limit of étale spaces over X in the sense of [**116**, §2.4]. More precisely, one posits the existence of a cofiltered inverse system $\{Y_i\}_{i\in I}$ in $X_{\text{ét}}$ and a compatible family of morphisms $Y \to Y_i$ with the following properties.

(a) The transition morphisms $Y_j \to Y_i$ are qcqs (*i.e.*, for any morphism $U \to V_i$ with U a preadic affinoid space, $Y_j \times_{Y_i} U$ is qcqs).

(b) The induced map $|Y| \to \varprojlim_i |Y_i|$ is a homeomorphism.

(c) There exists a covering of Y by perfectoid affinoid subspaces $\mathrm{Spa}(A, A^+)$, each with the following property: for each $i \in I$, consider all perfectoid affinoid subspaces $\mathrm{Spa}(A_{ij}, A_{ij}^+)$ of Y_i through which $\mathrm{Spa}(A, A^+) \to Y_i$ factors. Then the direct limit of the maps $A_{ij} \to A$ has dense image. (Note that Y then admits a basis of such subspaces.)

One then defines a *pro-étale covering* of a perfectoid space X to be a family of pro-étale morphisms $\{U_i \to X\}_{i\in I}$ with the property that for any quasicompact open subspace V of X, there exist a finite subset J of I and some quasicompact open subsets V_j of $U_j \times_X V$ for each $j \in J$ such that $\{V_j \to V\}_{j\in J}$ is a set-theoretic covering. (The auxiliary finiteness condition on coverings is needed because the analogue of Lemma 9.1.6 does not hold; this is typical for "large" topologies such as the fpqc topology on schemes.)

With this definition, it is not difficult to check that all of the acyclicity assertions we make about the pro-étale topology on perfectoid or perfectoid uniform spaces, such as Lemma 9.2.8 and Lemma 9.3.4, remain true for this finer topology. However, it is not immediately clear how to adapt the proof of Theorem 9.2.15 to the finer topology.

9.3. φ-modules and local systems

Using the completed structure sheaf, we proceed to construct perfect period sheaves on preadic spaces over $\mathbb{Q}_p$. We then relate φ-modules over these period sheaves to étale local systems on the spaces. Even over perfectoid spaces, this adds to the discussion in §8 because now we can also say something about the pro-étale cohomology of local systems.

Remark 9.3.1. — The suite of notations introduced below is similar to the "Colmez style" of notations in p-adic Hodge theory (as distinguished from the "Fontaine style" of notations). One key difference is that those notations primarily distinguish between rings with and without integral structure, by basing their notations on the letters $\mathbf{A}$ and $\mathbf{B}$ respectively. We prefer to further emphasize the difference between bounded and unbounded rings without integral structure (*e.g.*, the bounded Robba ring $\mathcal{R}^{\mathrm{bd}}$ versus the full Robba ring $\mathcal{R}$), so we derive our notations for the latter from the letter $\mathbf{C}$.

Hypothesis 9.3.2. — For the remainder of §9, let X be a preadic space over $\mathbb{Q}_{p^d}$ for some positive integer d, and let $Y \in X_{\text{proét}}$ denote an arbitrary perfectoid subdomain. Let Y' be the perfect uniform adic space over $\mathbb{F}_{p^d}$ associated to Y *via* the perfectoid correspondence.

Definition 9.3.3. — Define sheaves on $X_{\text{proét}}$ by the following formulas:

$$\tilde{\mathbf{A}}_X(Y) = W(\overline{\mathcal{O}}(Y)), \quad \tilde{\mathbf{A}}_X^+(Y) = \tilde{\mathcal{R}}^{\text{int},+}_{\overline{\mathcal{O}}(Y)}, \quad \tilde{\mathbf{A}}_X^{\dagger,r}(Y) = \tilde{\mathcal{R}}^{\text{int},r}_{\overline{\mathcal{O}}(Y)}, \quad \tilde{\mathbf{A}}_X^{\dagger}(Y) = \tilde{\mathcal{R}}^{\text{int}}_{\overline{\mathcal{O}}(Y)},$$
$$\tilde{\mathbf{B}}_X(Y) = W(\overline{\mathcal{O}}(Y))[p^{-1}], \quad \tilde{\mathbf{B}}_X^{\dagger,r}(Y) = \tilde{\mathcal{R}}^{\text{bd},r}_{\overline{\mathcal{O}}(Y)}, \quad \tilde{\mathbf{B}}_X^{\dagger}(Y) = \tilde{\mathcal{R}}^{\text{bd}}_{\overline{\mathcal{O}}(Y)},$$
$$\tilde{\mathbf{C}}_X^+(Y) = \tilde{\mathcal{R}}^+_{\overline{\mathcal{O}}(Y)}, \quad \tilde{\mathbf{C}}_X^r(Y) = \tilde{\mathcal{R}}^r_{\overline{\mathcal{O}}(Y)}, \quad \tilde{\mathbf{C}}_X^I(Y) = \tilde{\mathcal{R}}^I_{\overline{\mathcal{O}}(Y)}, \quad \tilde{\mathbf{C}}_X(Y) = \tilde{\mathcal{R}}_{\overline{\mathcal{O}}(Y)}.$$

Lemma 9.3.4. — *The sheaves defined in Definition 9.3.3 are acyclic on Y.*

Proof. — Imitate the proof of Lemma 9.2.8. □

Definition 9.3.5. — The sheaves $\tilde{\mathbf{A}}_X, \tilde{\mathbf{A}}_X^+, \tilde{\mathbf{A}}_X^\dagger, \tilde{\mathbf{B}}_X, \tilde{\mathbf{B}}_X^\dagger, \tilde{\mathbf{C}}_X^+, \tilde{\mathbf{C}}_X$ carry actions of φ; we refer to the sheaves collectively as *perfect period sheaves.* By a *φ^d-module* over one of these sheaves, we will mean a sheaf of finite projective modules over the corresponding sheaf of rings equipped with an isomorphism with its φ^d-pullback.

We say that a φ^d-module M over $\tilde{\mathbf{C}}_X$ is *pure* (resp. *étale*) at a point $x \in X$ if its restriction to some perfectoid subdomain is pure (resp. étale) at some lift of x. The same is then true for any other lift of x to any other perfectoid subdomain, by virtue of the pointwise nature of the pure and étale conditions (Corollary 7.3.10).

Theorem 9.3.6. — *For M a φ^d-module over $\tilde{\mathbf{C}}_X$, the pure locus (resp. the étale locus) of M is a partially proper open subset of X. In particular, by Lemma 8.2.12, if X is taut, then so are the pure locus and the étale locus.*

Proof. — The claim is local, so we may assume $X = \operatorname{Spa}(A, A^+)$. In this case, we deduce the claim by constructing $A \to B$ as in Lemma 3.6.26, applying Corollary 7.3.8 to $\mathcal{M}(B)$, and applying Remark 2.3.15 (b) to the map $\mathcal{M}(B) \to \mathcal{M}(A)$. □

Theorem 9.3.7. — *The following categories are equivalent* via *functors which preserve rank and are natural for pullbacks on X.*

(a) *The category of étale $\mathbb{Z}_{p^d}$-local systems over X.*

(b) *The category of φ^d-modules over $\tilde{\mathbf{A}}_X$.*

(c) *The category of φ^d-modules over $\tilde{\mathbf{A}}_X^\dagger$.*

Proof. — Immediate from Theorem 8.5.5 and Lemma 9.1.11. □

Corollary 9.3.8. — *For $X = Y$, the base extension functor from φ^d-modules over $\tilde{\mathcal{E}}^{\text{int}}_{\overline{\mathcal{O}}(Y)}$ (resp. $\tilde{\mathcal{R}}^{\text{int}}_{\overline{\mathcal{O}}(Y)}$) to φ^d-modules over $\tilde{\mathbf{A}}_X$ (resp. $\tilde{\mathbf{A}}_X^\dagger$) is an equivalence of categories.*

Proof. — Combine Theorem 8.5.3 with Theorem 9.3.7. □

In our description of étale $\mathbb{Q}_p$-local systems, we would like to include "vector bundles over FF_X" but it is not so straightforward to make sense of what FF_X would be. Instead, we settle for some indirect descriptions of the category of vector bundles.

Definition 9.3.9. — Define the sheaf of graded rings $P_X = \oplus_{n=0}^{\infty} P_{X,n}$ by the formula

$$P_{X,n}(Y) = \{x \in \tilde{\mathbf{C}}_X(Y) : \varphi^d(x) = p^n x\}.$$

By a *vector bundle over* FF_X, we will mean a sheaf of graded P_X-modules whose restriction to each Y has the form $\oplus_{n=0}^{\infty}\Gamma(\mathrm{FF}_Y, V^{\otimes n})$ for some vector bundle V on FF_Y. Morphisms of vector bundles are morphisms of graded modules whose restrictions to each Y arise from morphisms of vector bundles.

Remark 9.3.10. — In case $X = Y$, then FF_Y is a well-defined adic space and we have already introduced the category of vector bundles over FF_Y. In order to make sense of Definition 9.3.9, one needs to check that base extension of true vector bundles over FF_Y to vector bundles over FF_X in this sense is an equivalence of categories. Fortunately, this follows from Theorem 9.2.15 by Theorem 5.3.9 and a splitting argument.

Definition 9.3.11. — Define the sheaf $\mathbf{B}^+_{\mathrm{dR},X}$ by setting $\mathbf{B}^+_{\mathrm{dR},X}(Y)$ to be the $\ker(\theta)$-adic completion of $\tilde{\mathbf{B}}^{\dagger,1}_X(Y)$. Let $\mathbf{B}_{\mathrm{dR},X}$ be the localization of $\mathbf{B}^+_{\mathrm{dR},X}$ obtained by inverting a generator of $\ker(\theta)$.

Define the sheaf $\mathbf{B}_{\mathrm{e},X}$ by setting $\mathbf{B}_{\mathrm{e},X}(Y)$ to be the coordinate ring of the open affine subscheme of $\mathrm{Proj}(P_X(Y))$ obtained by removing the zero locus of the canonical section $Y \to \mathrm{FF}_Y$. The rings $\mathbf{B}_{\mathrm{e},X}(Y), \mathbf{B}^+_{\mathrm{dR},X}(Y), \mathbf{B}_{\mathrm{dR},X}(Y)$ correspond to the rings R_1, R_2, R_3 introduced in Definition 8.9.4.

By a *B-pair* over X, we will mean a triple $(M_{\mathrm{e}}, M^+_{\mathrm{dR}}, M_{\mathrm{dR}})$ in which M_{e} is a finite projective $\mathbf{B}_{\mathrm{e},X}$-module, M^+_{dR} is a finite projective $\mathbf{B}^+_{\mathrm{dR},X}$-module, and M_{dR} is a finite projective $\mathbf{B}_{\mathrm{dR},X}$-module equipped with isomorphisms

$$M_{\mathrm{dR}} \cong M_{\mathrm{e}} \otimes_{\mathbf{B}_{\mathrm{e},X}} \mathbf{B}_{\mathrm{dR},X} \cong M^+_{\mathrm{dR}} \otimes_{\mathbf{B}^+_{\mathrm{dR},X}} \mathbf{B}_{\mathrm{dR},X}.$$

Theorem 9.3.12. — *The categories of φ^d-modules over $\tilde{\mathbf{C}}_X$, vector bundles on FF_X, and B-pairs on X are functorially equivalent.*

Proof. — Immediate from Theorems 8.7.8, 8.9.6. □

Theorem 9.3.13. — *The following categories are equivalent* via *functors which preserve rank and are natural for pullbacks on X.*

(a) *The category of étale $\mathbb{Q}_{p^d}$-local systems over X.*

(b) *The category of étale φ^d-modules over $\tilde{\mathbf{B}}_X$.*

(c) *The category of étale φ^d-modules over $\tilde{\mathbf{B}}^{\dagger}_X$.*

(d) *The category of étale φ^d-modules over $\tilde{\mathbf{C}}_X$.*

(e) *The category of vector bundles on* FF_X *which are pointwise semistable of degree* 0.

Proof. — Immediate from Theorem 8.5.12, Theorem 8.7.7 and Lemma 9.1.11. □

Corollary 9.3.14. — *For* $X = Y$*, the base extension functor from* φ^d*-modules over* $\tilde{\mathcal{R}}_{\overline{\mathcal{O}}(Y)}$ *to* φ^d*-modules over* $\tilde{\mathbf{C}}_X$ *is an equivalence of categories.*

Proof. — Immediate from Theorem 9.3.13 and Remark 9.3.10. □

Corollary 9.3.15. — *For* $X = Y$*, the base extension functor from pure* φ^d*-modules over* $\tilde{\mathcal{E}}_{Y'}$ *(resp.* $\tilde{\mathcal{R}}^{\mathrm{bd}}_{Y'}$*) to pure* φ^d*-modules over* $\tilde{\mathbf{B}}_X$ *(resp.* $\tilde{\mathbf{B}}^{\dagger}_X$*) is an equivalence of categories.*

Proof. — Immediate from Theorem 9.3.13 and the fact that étale $\mathbb{Q}_p$-local systems arise adic-locally from isogeny $\mathbb{Z}_p$-local systems. □

Remark 9.3.16. — The analogue of Fontaine's functor D_{dR} in this context is the functor taking a B-pair M to the global sections of M_{dR}; for instance, this functor appears in Scholze's approach to the comparison isomorphism in [**115**]. We will return to this point, and to the analogues of the functors D_{crys} and D_{st}, in a subsequent paper.

Remark 9.3.17. — Thanks to Lemma 9.3.4, any φ^d-module over any of the sheaves described in Definition 9.3.3 may be evaluated at any perfectoid space mapping to X, not just perfectoid subdomains.

9.4. Comparison of cohomology

We next compare the pro-étale cohomology of local systems with φ-cohomology.

Definition 9.4.1. — For M a φ^d-module over a perfect period sheaf, define the cohomology groups $H^i_{\varphi^d}(M)$ as the hypercohomology groups of the complex

$$0 \longrightarrow M \stackrel{\varphi^d-1}{\longrightarrow} M \longrightarrow 0$$

of sheaves on $X_{\mathrm{pro\acute{e}t}}$.

Theorem 9.4.2. — *Let E be an étale* $\mathbb{Z}_{p^d}$*-local system on X. Apply Theorem 9.3.7 to produce a* φ^d*-module M over* $\tilde{\mathbf{A}}_X$ *and a* φ^d*-module* $M^\dagger$ *over* $\tilde{\mathbf{A}}^{\dagger}_X$*. Then for* $i \geqslant 0$*, there are functorial (in E and X) isomorphisms*

$$H^i_{\mathrm{pro\acute{e}t}}(X, E) \longrightarrow H^i_{\varphi^d}(M^\dagger) \longrightarrow H^i_{\varphi^d}(M).$$

Proof. — The morphism $H^i_{\varphi^d}(M^\dagger) \to H^i_{\varphi^d}(M)$ is induced by the inclusion $M^\dagger \to M$. By Theorem 8.6.2, this morphism induces quasi-isomorphisms of the kernel and cokernel sheaves of φ^d-1. By computing the hypercohomology groups using matching spectral sequences, we see that $H^i_{\varphi^d}(M^\dagger)\to H^i_{\varphi^d}(M)$ is itself an isomorphism.

To produce the isomorphism $H^i_{\text{proét}}(X,E)\to H^i_{\varphi^d}(M)$, we reinterpret the proof of Theorem 8.6.2 in the pro-étale context using the interpretation of E as a locally constant sheaf $\mathcal{F}_T$ on $X_{\text{proét}}$ in the sense of Lemma 9.1.11. To begin with, we may view M as the sheafification of the presheaf taking Y to $W(\overline{\mathcal{O}}(Y))\widehat{\otimes}_{\mathbb{Z}_p}E(Y)$. In this interpretation, the arrow $E\to M$ in the sequence

$$0\longrightarrow E\longrightarrow M\stackrel{\varphi^d-1}{\longrightarrow} M\longrightarrow 0 \tag{9.4.2.1}$$

is the natural map $1\otimes\mathrm{id}$, so it is clear that we obtain a complex.

To prove the theorem, it suffices to check that the sequence (9.4.2.1) is exact. It is enough to compare sections over a perfectoid subdomain Y on which $E\cong\mathcal{F}_T$ for some finite free $\mathbb{Z}_p$-module T. By the perfectoid correspondence, we may identify $E(Y)$ not only with $\mathrm{Map}_{\text{cont}}(|Y|,T)$ but also with $\mathrm{Map}_{\text{cont}}(|Y'|,T)$; the injectivity and the exactness at the middle are now immediate from Corollary 3.1.4. For any $\mathbf{v}\in M(Y)$, by Proposition 7.3.6, there exists a perfectoid subdomain Z which is pro-étale over Y (namely, it is a tower of faithfully finite étale morphisms over Y) such that $\mathbf{v}$ can be lifted to $M(Z)$ *via* the map φ^d-1. This proves the surjectivity. □

Remark 9.4.3. — One aspect of Theorem 8.6.2 that is hidden in the statement and proof of Theorem 9.4.2 is that while E is determined by its sections only on sufficiently small perfectoid subdomains, M is determined by its sections on arbitrary perfectoid subdomains by Corollary 9.3.8. The same will happen in Theorem 9.4.5 by virtue of Corollary 9.3.14.

Definition 9.4.4. — For $M=(M_{\mathrm{e}},M^+_{\mathrm{dR}},M_{\mathrm{dR}})$ a B-pair over X, define the cohomology groups $H^i_B(M)$ as the hypercohomology groups of the complex

$$0\longrightarrow M_{\mathrm{e}}\times M^+_{\mathrm{dR}}\longrightarrow M_{\mathrm{dR}}\longrightarrow 0\qquad (x,y)\longmapsto x-y$$

of sheaves on $X_{\text{proét}}$.

Theorem 9.4.5. — *Let V be an étale $\mathbb{Q}_{p^d}$-local system over X. Apply Theorem 9.3.13 to produce an étale φ^d-module M over $\tilde{\mathbf{B}}_X$, an étale φ^d-module $M^\dagger$ over $\tilde{\mathbf{B}}^\dagger_X$, an étale φ^d-module $M_{\mathbf{C}}$ over $\tilde{\mathbf{C}}_X$, and a B-pair $M_B=(M_{\mathrm{e}},M^+_{\mathrm{dR}},M_{\mathrm{dR}})$ over X. Then for $i\geqslant 0$, there are functorial (in E and X) isomorphisms*

$$H^i_{\text{proét}}(X,V)\longrightarrow H^i_{\varphi^d}(M)\longrightarrow H^i_{\varphi^d}(M^\dagger)\longrightarrow H^i_{\varphi^d}(M_{\mathbf{C}})\longrightarrow H^i_B(M_B).$$

Proof. — We reduce at once to the case where V is an isogeny $\mathbb{Z}_{p^d}$-local system. The comparisons among

$$H^i_{\text{proét}}(X,V),H^i_{\varphi^d}(M),H^i_{\varphi^d}(M^\dagger),H^i_{\varphi^d}(M_{\mathbf{C}})$$

are then achieved by constructing exact sequences

$$0 \longrightarrow V \longrightarrow M \stackrel{\varphi^d - 1}{\longrightarrow} M \longrightarrow 0$$

$$0 \longrightarrow V \longrightarrow M^\dagger \stackrel{\varphi^d - 1}{\longrightarrow} M^\dagger \longrightarrow 0$$

$$0 \longrightarrow V \longrightarrow M_{\mathbf{C}} \stackrel{\varphi^d - 1}{\longrightarrow} M_{\mathbf{C}} \longrightarrow 0$$

analogous to (9.4.2.1); for the exactness we depend on Corollary 5.2.4 in addition to Corollary 3.1.4 and Proposition 7.3.6. The comparison between $H^i_{\varphi^d}(M_{\mathbf{C}})$ and $H^i_B(M_B)$ follows from Theorem 8.7.13 and Theorem 8.9.6 (a). □

9.5. Comparison with classical p-adic Hodge theory

To conclude, we compare this construction to the classical theory of (φ, Γ)-modules.

Definition 9.5.1. — With notation as in §4.1, define the rings

$$\mathbf{A}_{\mathbb{Q}_p} = \mathfrak{o}_{\mathcal{E}_{\mathbb{Q}_p}},\ \mathbf{A}^\dagger_{\mathbb{Q}_p} = \mathcal{R}^{\mathrm{int}}_{\mathbb{Q}_p},\ \mathbf{B}_{\mathbb{Q}_p} = \mathcal{E}_{\mathbb{Q}_p},\ \mathbf{B}^\dagger_{\mathbb{Q}_p} = \mathcal{R}^{\mathrm{bd}}_{\mathbb{Q}_p},\ \mathbf{C}_{\mathbb{Q}_p} = \mathcal{R}_{\mathbb{Q}_p},$$
$$\tilde{\mathbf{A}}_{\mathbb{Q}_p} = \mathfrak{o}_{\tilde{\mathcal{E}}_{\mathbb{Q}_p}},\ \tilde{\mathbf{A}}^\dagger_{\mathbb{Q}_p} = \tilde{\mathcal{R}}^{\mathrm{int}}_{\mathbb{Q}_p},\ \tilde{\mathbf{B}}_{\mathbb{Q}_p} = \tilde{\mathcal{E}}_{\mathbb{Q}_p},\ \tilde{\mathbf{B}}^\dagger_{\mathbb{Q}_p} = \tilde{\mathcal{R}}^{\mathrm{bd}}_{\mathbb{Q}_p},\ \tilde{\mathbf{C}}_{\mathbb{Q}_p} = \tilde{\mathcal{R}}_{\mathbb{Q}_p}.$$

In addition to the endomorphism φ, these rings also carry an action of the group $\Gamma = \mathbb{Z}_p^\times$ characterized by the formula

$$\gamma(1+T) = (1+T)^\gamma = \sum_{n=0}^\infty \binom{\gamma}{n} T^n.$$

Definition 9.5.2. — Let F be the completion of $\mathbb{Q}_p(\mu_{p^\infty})$. This analytic field is perfectoid: it is of characteristic 0, not discretely valued, and $\overline{\varphi}$ is surjective on

$$\mathfrak{o}_F/(p) \cong \mathbb{Z}_p[\mu_{p^\infty}]/(p) \cong \mathbb{F}_p[T_0, T_1, \dots]/(T_0 - 1, T_1^p - T_0, \dots).$$

By the henselian property of $\mathbf{A}^\dagger_{\mathbb{Q}_p}$ (see Definition 4.1.2), Lemma 2.2.4 (or Krasner's lemma), and the perfectoid correspondence for analytic fields (Theorem 3.5.3), we have distinguished equivalences of tensor categories

$$\mathbf{F\acute{E}t}(\mathbb{Q}_p(\mu_{p^\infty})) \cong \mathbf{F\acute{E}t}(F) \cong \mathbf{F\acute{E}t}(\tilde{\mathbf{A}}^\dagger_{\mathbb{Q}_p}/(p)) \cong \mathbf{F\acute{E}t}(\mathbf{A}^\dagger_{\mathbb{Q}_p}/(p)) \cong \mathbf{F\acute{E}t}(\mathbf{A}^\dagger_{\mathbb{Q}_p}).$$

Via these equivalences, for any finite extension K of $\mathbb{Q}_p$, the finite extension $K \otimes_{\mathbb{Q}_p} \mathbb{Q}_p(\mu_{p^\infty})$ of $\mathbb{Q}_p(\mu_{p^\infty})$ corresponds to a finite étale extension $\mathbf{A}^\dagger_K$ of $\mathbf{A}^\dagger_{\mathbb{Q}_p}$.

For $* \in \{\mathbf{A}, \mathbf{A}^\dagger, \mathbf{B}, \mathbf{B}^\dagger, \mathbf{C}, \tilde{\mathbf{A}}, \tilde{\mathbf{A}}^\dagger, \tilde{\mathbf{B}}, \tilde{\mathbf{B}}^\dagger, \tilde{\mathbf{C}}\}$, put $*_K = *_{\mathbb{Q}_p} \otimes_{\mathbf{A}^\dagger_{\mathbb{Q}_p}} \mathbf{A}^\dagger_K$. These rings admit extensions of the actions of φ and Γ. A (φ, Γ)-*module* over one of these rings is a φ-module equipped with an action of Γ which is continuous for the appropriate topology (the weak topology on $\mathbf{A}_K, \tilde{\mathbf{A}}_K, \mathbf{B}_K, \tilde{\mathbf{B}}_K$, or the LF-topology on the other rings).

Remark 9.5.3. — Our convention for (φ, Γ)-modules for $K \neq \mathbb{Q}_p$ is not the standard one: it is more common to take $*_K$ to be a connected component of the ring we are considering, in which case one gets an action not of Γ but only its subgroup Γ_K corresponding to $\mathrm{Gal}(K(\mu_{p^\infty})/K)$ *via* the cyclotomic character. However, results formulated using one convention convert easily to the other *via* induction and restriction between Γ_K and Γ.

Theorem 9.5.4. — *For K a finite extension of $\mathbb{Q}_p$, the following categories are canonically equivalent.*

(a) *The category of continuous representations of G_K on finite free $\mathbb{Z}_p$-modules.*

(b) *The category of (φ, Γ)-modules over $\mathbf{A}_K$.*

(c) *The category of (φ, Γ)-modules over $\tilde{\mathbf{A}}_K$.*

(d) *The category of (φ, Γ)-modules over $\mathbf{A}_K^\dagger$.*

(e) *The category of (φ, Γ)-modules over $\tilde{\mathbf{A}}_K^\dagger$.*

(f) *The category of φ-modules over $\tilde{\mathbf{A}}_{\mathrm{Spa}(K, \mathfrak{o}_K)}$.*

(g) *The category of φ-modules over $\tilde{\mathbf{A}}^\dagger_{\mathrm{Spa}(K, \mathfrak{o}_K)}$.*

Proof. — The equivalences among (a) and (b)–(e) include Fontaine's original theory of (φ, Γ)-modules and its refinement by Cherbonnier and Colmez; see [**89**, §2]. The equivalences among (a) and (f)-(g) follow from Theorem 9.3.7. □

Definition 9.5.5. — For $* = \mathbf{B}, \mathbf{B}^\dagger, \mathbf{C}, \tilde{\mathbf{B}}, \tilde{\mathbf{B}}^\dagger, \tilde{\mathbf{C}}$, we say that a (φ, Γ)-module over $*_K$ is *étale* if its underlying φ-module is étale, *i.e.*, it descends to a φ-module over $\mathbf{A}_K, \mathbf{A}_K^\dagger, \mathbf{A}_K^\dagger, \tilde{\mathbf{A}}_K, \tilde{\mathbf{A}}_K^\dagger, \tilde{\mathbf{A}}_K^\dagger$, respectively. Note that the Γ-action does act on some such descent, but not necessarily on all of them.

Theorem 9.5.6. — *For K a finite extension of $\mathbb{Q}_p$, the following categories are canonically equivalent.*

(a) *The category of continuous representations of G_K on finite-dimensional $\mathbb{Q}_p$-vector spaces.*

(b) *The category of étale (φ, Γ)-modules over $\mathbf{B}_K$.*

(c) *The category of étale (φ, Γ)-modules over $\mathbf{B}_K^\dagger$.*

(d) *The category of étale (φ, Γ)-modules over $\mathbf{C}_K$.*

(e) *The category of étale (φ, Γ)-modules over $\tilde{\mathbf{B}}_K$.*

(f) *The category of étale (φ, Γ)-modules over $\tilde{\mathbf{B}}_K^\dagger$.*

(g) *The category of étale (φ, Γ)-modules over $\tilde{\mathbf{C}}_K$.*

(h) *The category of étale φ-modules over $\tilde{\mathbf{B}}_{\mathrm{Spa}(K, \mathfrak{o}_K)}$.*

(i) *The category of étale φ-modules over $\tilde{\mathbf{B}}^\dagger_{\mathrm{Spa}(K, \mathfrak{o}_K)}$.*

(j) *The category of étale φ-modules over $\tilde{\mathbf{C}}_{\mathrm{Spa}(K, \mathfrak{o}_K)}$.*

Proof. — The equivalences among (a), (b)-(c), (e)-(f) is immediate from Theorem 9.5.4. The equivalence between (c) and (d) follows from Proposition 4.1.8. The equivalence between (f) and (g) follows from Theorem 8.5.6. The equivalences among (a) and (h)–(j) follow from Theorem 9.3.13. □

We next introduce Berger's concept of a B-pair from [**11**].

Definition 9.5.7. — Let $\mathbb{C}_p$ be a completed algebraic closure of $\mathbb{Q}_p$. For K a finite extension of $\mathbb{Q}_p$ within $\mathbb{C}_p$, a *B-pair over K* is a B-pair over $\mathrm{Spa}(\mathbb{C}_p, \mathfrak{o}_{\mathbb{C}_p})$ equipped with a continuous semilinear action of G_K.

Theorem 9.5.8. — *Let F be the completion of $\mathbb{Q}_p(\mu_{p^\infty})$. For K a finite extension of $\mathbb{Q}_p$, the following categories are canonically equivalent.*

(a) *The category of (φ, Γ)-modules over $\mathbf{C}_K$.*

(b) *The category of (φ, Γ)-modules over $\tilde{\mathbf{C}}_K$.*

(c) *The category of B-pairs over K.*

(d) *The category of continuous Γ-equivariant vector bundles over*
$$\mathrm{FF}_{\mathrm{Spa}(F\otimes_{\mathbb{Q}_p}K, (F\otimes_{\mathbb{Q}_p}K)^\circ)}.$$

(e) *The category of continuous G_K-equivariant vector bundles over* $\mathrm{FF}_{\mathrm{Spa}(\mathbb{C}_p, \mathfrak{o}_{\mathbb{C}_p})}$.

(f) *The category of φ-modules over* $\tilde{\mathbf{C}}_{\mathrm{Spa}(K, \mathfrak{o}_K)}$.

(g) *The category of B-pairs over* $\mathrm{Spa}(K, \mathfrak{o}_K)$.

Proof. — The equivalence between (a) and (c) is a theorem of Berger [**11**, Théorème 2.2.7]; the same argument gives the equivalence between (b) and (c). The equivalences between (b) and (d) and between (c) and (e) follow from Theorem 8.7.7. The equivalence between (b) and (f) follows from Corollary 9.3.14. The equivalence between (f) and (g) follows from Theorem 8.9.6. □

Remark 9.5.9. — Using Remark 9.4.3, one can similarly check that our computation of pro-étale cohomology in terms of φ-cohomology agrees with the analogous computations in the language of classical (φ, Γ)-modules made by Herr [**72, 73**] and the second author [**96**].

Remark 9.5.10. — Theorems 9.5.4, 9.5.6, and 9.5.8 can be extended to the relative (φ, Γ)-modules of Andreatta-Brinon [**2**]. We will discuss this and related generalizations in a subsequent paper.

BIBLIOGRAPHY

[1] F. Andreatta – "Generalized ring of norms and generalized (ϕ,Γ)-modules", *Ann. Sci. École Norm. Sup. (4)* **39** (2006), p. 599–647.

[2] F. Andreatta & O. Brinon – "Surconvergence des représentations p-adiques : le cas relatif", *Astérisque* **319** (2008), p. 39–116.

[3] A. Arabia – "Relèvements des algèbres lisses et de leurs morphismes", *Comment. Math. Helv.* **76** (2001), p. 607–639.

[4] M. Artin – "Algebraization of formal moduli II. Existence of modifications", *Ann. of Math. (2)* **91** (1970), p. 88–135.

[5] M. Artin, A. Grothendieck & J.-L. Verdier – *Théorie des topos et cohomologie étale des schémas (SGA 4) I. Théorie des topos*, Lecture Notes in Math., vol. 269, Springer-Verlag, 1972.

[6] J. Ax – "Zeros of polynomials over local fields – The Galois action", *J. Algebra* **15** (1970), p. 417–428.

[7] H. Bass – *Algebraic K-theory*, W. A. Benjamin, New York-Amsterdam, 1968.

[8] A. Beauville & Y. Laszlo – "Un lemme de descente", *C. R. Acad. Sci. Paris Sér. I Math.* **320** (1995), p. 335–340.

[9] J. Bellaïche – "Ranks of Selmer groups in an analytic family", *Trans. Amer. Math. Soc.* **364** (2012), p. 4735–4761.

[10] L. Berger – "Représentations p-adiques et équations différentielles", *Invent. Math.* **148** (2002), p. 219–284.

[11] ———, "Construction de (ϕ,Γ)-modules : représentations p-adiques et B-paires", *Algebra Number Theory* **2** (2008), p. 91–120.

[12] ———, "Équations différentielles p-adiques et (ϕ, N)-modules filtrés", *Astérisque* **319** (2008), p. 13–38.

[13] L. Berger & P. Colmez – "Familles de représentations de de Rham et monodromie p-adique", *Astérisque* **319** (2008), p. 303–337.

[14] V. G. Berkovich – *Spectral theory and analytic geometry over non-Archimedean fields*, Math. Surveys Monogr., vol. 33, Amer. Math. Soc., Providence, 1990.

[15] ———, "Étale cohomology for non-Archimedean analytic spaces", *Publ. Math. Inst. Hautes Études Sci.* **78** (1993), p. 5–161.

[16] B. Bhatt – "Almost direct summands", *Nagoya Math. J.* **214** (2014), p. 195–204.

[17] B. Bhatt & P. Scholze – "The pro-étale topology for schemes", arXiv: 1309.1198v2, 2014.

[18] S. Bosch, U. Güntzer & R. Remmert – *Non-Archimedean analysis*, Grundlehren Math. Wiss., vol. 261, Springer-Verlag, Berlin, 1984.

[19] N. Bourbaki – *Topologie générale. Chapitres 1 à 4*, Hermann, 1971.

[20] ———, *Algèbre commutative. Chapitres 1 à 4*, Springer, Berlin, 2006, reprint of the 1985 original.

[21] ———, *Espaces Vectoriels Topologiques*, Springer, Berlin, 2007, reprint of the 1981 original.

[22] O. Brinon & B. Conrad – lecture notes on p-adic Hodge theory, available at `http://math.stanford.edu/~conrad/`.

[23] K. Buzzard & A. Verberkmoes – "Stably uniform affinoids are sheafy", arXiv: 1404.7020v1, 2014.

[24] F. Cherbonnier & P. Colmez – "Représentations p-adiques surconvergentes", *Invent. Math.* **133** (1998), p. 581–611.

[25] R. Coleman & B. Mazur – "The eigencurve", in *Galois representations in arithmetic algebraic geometry (Durham, 1996)*, London Math. Soc. Lecture Note Ser., vol. 254, Cambridge Univ. Press, Cambridge, 1998, p. 1–113.

[26] P. Colmez – "Représentations triangulines de dimension 2", *Astérisque* **319** (2008), p. 213–258.

[27] ———, "Représentations de $\mathrm{GL}_2(\mathbf{Q}_p)$ et (ϕ, Γ)-modules", *Astérisque* **330** (2010), p. 281–509.

[28] R. Crew – "F-isocrystals and p-adic representations", in *Algebraic geometry, Bowdoin, 1985 II*, Proc. Sympos. Pure Math., vol. 46, Amer. Math. Soc., Providence, 1987, p. 111–138.

[29] C. Davis & K. S. Kedlaya – "On the Witt vector Frobenius", *Proc. Amer. Math. Soc.* **142** (2014), p. 2211–2226.

[30] C. Davis, A. Langer & T. Zink – "Overconvergent de Rham-Witt cohomology", *Ann. Sci. Éc. Norm. Supér. (4)* **44** (2011), p. 197–262.

[31] ———, "Overconvergent Witt vectors", *J. Reine Angew. Math.* **668** (2012), p. 1–34.

[32] A. J. de Jong – "Étale fundamental groups of non-Archimedean analytic spaces", *Compositio Math.* **97** (1995), p. 89–118.

[33] A. J. de Jong & M. van der Put – "Étale cohomology of rigid analytic spaces", *Doc. Math.* **1** (1996), p. 1–56.

[34] J. Dee – "Φ-Γ modules for families of Galois representations", *J. Algebra* **235** (2001), p. 636–664.

[35] P. Deligne – "Théorie de Hodge II", *Publ. Math. Inst. Hautes Études Sci.* **40** (1971), p. 5–57.

[36] ———, "La conjecture de Weil II", *Publ. Math. Inst. Hautes Études Sci.* **52** (1980), p. 137–252.

[37] P. Deligne & N. Katz – *Groupes de monodromie en géométrie algébrique (sga 7) II*, Lecture Notes in Math., vol. 340, Springer-Verlag, Berlin-New York, 1973.

[38] A. Ducros – "Les espaces de Berkovich sont excellents", *Ann. Inst. Fourier (Grenoble)* **59** (2009), p. 1443–1552.

[39] D. Eisenbud – *Commutative algebra*, Grad. Texts in Math., vol. 150, Springer-Verlag, New York, 1995.

[40] R. Elkik – "Solutions d'équations à coefficients dans un anneau hensélien", *Ann. Sci. École Norm. Sup. (4)* **6** (1973), p. 553–603 (1974).

[41] M. Emerton – "A local-global compatibility conjecture in the p-adic Langlands programme for $\mathrm{GL}_{2/\mathbb{Q}}$", *Pure Appl. Math. Q.* **2** (2006), p. 279–393.

[42] G. Faltings – "p-adic Hodge theory", *J. Amer. Math. Soc.* **1** (1988), p. 255–299.

[43] ______, "Almost étale extensions", *Astérisque* **279** (2002), p. 185–270.

[44] L. Fargues – "Quelques résultats et conjectures concernant la courbe", 2015, to appear in *Astérisque*.

[45] L. Fargues & J.-M. Fontaine – "Courbes et fibrés vectoriels en théorie de Hodge p-adique", in preparation; draft (July 2013) available at `http://webusers.imj-prg.fr/~laurent.fargues/`.

[46] ______, "Vector bundles and p-adic Galois representations", in *Automorphic Forms and Galois Representations I*, London. Math. Soc. Lecture Note Ser., vol. 415, Cambridge Univ. Press, Cambridge, 2015, p. 17–104.

[47] J.-M. Fontaine – "Représentations p-adiques des corps locaux I", in *The Grothendieck Festschrift II*, Progr. Math., vol. 87, Birkhäuser, Boston, 1990, p. 249–309.

[48] ______, "Perfectoïdes, presque pureté et monodromie-poids (d'après Peter Scholze)", *Astérisque* **352** (2013), p. 509–534, Séminaire Bourbaki, Vol. 2011/2012.

[49] J.-M. Fontaine & J.-P. Wintenberger – "Le 'corps des normes' de certaines extensions algébriques de corps locaux", *C. R. Acad. Sci. Paris Sér. A-B* **288** (1979), p. A367–A370.

[50] E. Freitag & R. Kiehl – *Étale cohomology and the Weil conjecture*, Ergeb. Math. Grenzgeb. (3), vol. 13, Springer-Verlag, Berlin, 1988.

[51] J. Fresnel & M. van der Put – *Rigid analytic geometry and its applications*, Progr. Math., vol. 218, Birkhäuser, Boston, 2004.

[52] O. Gabber & L. Ramero – *Almost ring theory*, Lecture Notes in Math., vol. 1800, Springer-Verlag, Berlin, 2003.

[53] ______, "Foundations for almost ring theory theory – release 6.8", 2014.

[54] S. Greco – "Henselization of a ring with respect to an ideal", *Trans. Amer. Math. Soc.* **144** (1969), p. 43–65.

[55] A. Grothendieck – "Torsions homologiques et sections rationnelles", in *Séminaire C. Chevalley, 1958 : Anneaux de Chow et Applications*, vol. 16, Secretariat mathématique, Paris, 1958, exp. 5.

[56] ———, "Éléments de géométrie algébrique I. Le langage des schémas", *Publ. Math. Inst. Hautes Études Sci.* **4** (1960), p. 5–228.

[57] ———, "Éléments de géométrie algébrique. II Étude globale élémentaire de quelques classes de morphismes", *Publ. Math. Inst. Hautes Études Sci.* **8** (1961), p. 5–222.

[58] ———, "Éléments de géométrie algébrique III. Étude cohomologique des faisceaux cohérents I", *Publ. Math. Inst. Hautes Études Sci.* **11** (1961), p. 5–167.

[59] ———, "Éléments de géométrie algébrique. IV. Étude locale des schémas et des morphismes de schémas I", *Publ. Math. Inst. Hautes Études Sci.* **20** (1964), p. 5–259.

[60] ———, "Éléments de géométrie algébrique IV. Étude locale des schémas et des morphismes de schémas IV", *Publ. Math. Inst. Hautes Études Sci.* **32** (1967), p. 5–361.

[61] A. GROTHENDIECK et al. – *Cohomologie ℓ-adique et Fonctions L*, Lecture Notes in Math., vol. 589, Springer-Verlag, Berlin, 1977.

[62] ———, *Revêtements étales et groupe fondamental*, Doc. Math. (Paris), vol. 3, Soc. Math. France, Paris, 2003.

[63] L. GRUSON – "Une propriété des couples henséliens", in *Colloque d'Algèbre Commutative (Rennes, 1972)*, Publ. Univ. Rennes, 1972, exp. no. 10, p. 13.

[64] U. HARTL – "On period spaces for p-divisible groups", *C. R. Math. Acad. Sci. Paris* **346** (2008), p. 1123–1128.

[65] ———, "Period spaces for Hodge structures in equal characteristic", *Ann. of Math. (2)* **173** (2011), p. 1241–1358.

[66] ———, "On a conjecture of Rapoport and Zink", *Invent. Math.* **193** (2013), p. 627–696.

[67] U. HARTL & R. PINK – "Vector bundles with a Frobenius structure on the punctured unit disc", *Compos. Math.* **140** (2004), p. 689–716.

[68] R. HARTSHORNE – *Algebraic geometry*, Grad. Texts in Math., vol. 52, Springer-Verlag, New York, 1977.

[69] E. HELLMANN – "On families of weakly admissible filtered ϕ-modules and the adjoint quotient of GL_d", *Doc. Math.* **16** (2011), p. 969–991.

[70] ______, "On arithmetic families of filtered ϕ-modules and crystalline representations", *J. Inst. Math. Jussieu* **12** (2013), p. 677–726.

[71] T. Henkel – "An Open Mapping Theorem for rings with a zero sequence of units", arXiv: 1407.5647v2, 2014.

[72] L. Herr – "Sur la cohomologie galoisienne des corps p-adiques", *Bull. Soc. Math. France* **126** (1998), p. 563–600.

[73] ______, "Une approche nouvelle de la dualité locale de Tate", *Math. Ann.* **320** (2001), p. 307–337.

[74] H. Hida – "Galois representations into $\mathrm{GL}_2(\mathbf{Z}_p[[X]])$ attached to ordinary cusp forms", *Invent. Math.* **85** (1986), p. 545–613.

[75] P. J. Hilton & U. Stammbach – *A course in homological algebra*, Grad. Texts in Math., vol. 4, Springer-Verlag, 1970.

[76] M. Hochster – "Prime ideal structure in commutative rings", *Trans. Amer. Math. Soc.* **142** (1969), p. 43–60.

[77] R. Huber – "Continuous valuations", *Math. Z.* **212** (1993), p. 455–477.

[78] ______, "A generalization of formal schemes and rigid analytic varieties", *Math. Z.* **217** (1994), p. 513–551.

[79] ______, *Étale cohomology of rigid analytic varieties and adic spaces*, Aspects Math., E30, Friedr. Vieweg & Sohn, Braunschweig, 1996.

[80] L. Illusie – "Complexe de de Rham-Witt et cohomologie cristalline", *Ann. Sci. École Norm. Sup. (4)* **12** (1979), p. 501–661.

[81] N. M. Katz – "p-adic properties of modular schemes and modular forms", in *Modular functions of one variable, III*, Lecture Notes in Math., vol. 350, Springer, Berlin, 1973, p. 69–190.

[82] K. S. Kedlaya – "A p-adic local monodromy theorem", *Ann. of Math. (2)* **160** (2004), p. 93–184.

[83] ______, "Slope filtrations revisited", *Doc. Math.* **10** (2005), p. 447–525, *errata, ibid.* **12** (2007) p. 361-362.

[84] ______, "Slope filtrations for relative Frobenius", *Astérisque* (2008), p. 259–301.

[85] ______, *p-adic differential equations*, Cambridge Stud. Adv. Math., vol. 125, Cambridge Univ. Press, Cambridge, 2010.

[86] ———, "Relative p-adic Hodge theory and Rapoport-Zink period domains", in *Proceedings of the International Congress of Mathematicians II*, Hindustan Book Agency, New Delhi, 2010, p. 258–279.

[87] ———, "Nonarchimedean geometry of Witt vectors", *Nagoya Math. J.* **209** (2013), p. 111–165.

[88] ———, "Some slope theory for multivariate Robba rings", arXiv: 1311.7468v1, 2013.

[89] ———, "New methods for (φ, Γ)-modules", arXiv: 1307.2937v2, 2015.

[90] K. S. Kedlaya & R. Liu – "On families of ϕ, Γ-modules", *Algebra Number Theory* **4** (2010), p. 943–967.

[91] R. Kiehl – "Theorem A und Theorem B in der nichtarchimedischen Funktionentheorie", *Invent. Math.* **2** (1967), p. 256–273.

[92] M. Kisin – "The Fontaine-Mazur conjecture for GL_2", *J. Amer. Math. Soc.* **22** (2009), p. 641–690.

[93] H. Kurke – "Grundlagen der Theorie der Henselschen Ringe und Schemata, und ihrer Anwendungen", Habilitation thesis, Humboldt-Universitat zu Berlin, 1969.

[94] H. Kurke, G. Pfister & M. Roczen – *Henselsche Ringe und algebraische Geometrie*, Math. Monographien, Band II, VEB Deutscher Verlag der Wissenschaften, Berlin, 1975.

[95] M. Lazard – *Commutative formal groups*, Lecture Notes in Math., vol. 443, Springer-Verlag, Berlin-New York, 1975.

[96] R. Liu – "Cohomology and duality for (ϕ, Γ)-modules over the Robba ring", *Int. Math. Res. Not. IMRN* (2008), Art. ID rnm150 (32 p.).

[97] ———, "Slope filtrations in families", *J. Inst. Math. Jussieu* **12** (2013), p. 249–296.

[98] ———, "Triangulations of refined families", arXiv: 1202.2188v4, 2014.

[99] T. Mihara – "On Tate acyclicity and uniformity of Berkovich spectra and adic spectra", arXiv: 1403.7856v1, 2014.

[100] J. S. Milne – *Étale Cohomology*, Princeton Math. Ser., vol. 33, Princeton Univ. Press, Princeton, 1980.

[101] D. Mumford, J. Fogarty & F. Kirwan – *Geometric Invariant Theory*, third enlarged ed., Ergeb. Math. Grenzgeb. (3), vol. 34, Springer-Verlag, 1994.

[102] M. Narasimhan & C. S. Seshadri – "Stable and unitary vector bundles on a compact Riemann surface", *Ann. of Math. (2)* **82** (1965), p. 540–567.

[103] M. C. Olsson – "On Faltings' method of almost étale extensions", in *Algebraic Geometry – Seattle 2005 II*, Proc. Sympos. Pure Math., vol. 80, Amer. Math. Soc., Providence, 2009, p. 811–936.

[104] G. Pappas & M. Rapoport – "φ-modules and coefficient spaces", *Mosc. Math. J.* **9** (2009), p. 625–663.

[105] J. Pottharst – "Analytic families of finite-slope Selmer groups", *Algebra Number Theory* **7** (2013), p. 1571–1612.

[106] M. van der Put & P. Schneider – "Points and topologies in rigid geometry", *Math. Ann.* **302** (1995), p. 81–103.

[107] M. Rapoport & T. Zink – *Period Spaces for p-divisible Groups*, Ann. of Math. Stud., vol. 141, Princeton Univ. Press, Princeton, 1996.

[108] M. Raynaud – *Anneaux Locaux Henséliens*, Lecture Notes in Math., vol. 169, Springer-Verlag, Berlin, 1970.

[109] R. Rodriguez – "Preperfectoid algebras", Ph.D. Thesis, University of California, San Diego, 2014.

[110] P. Schneider – *Nonarchimedean Functional Analysis*, Springer Monogr. Math., Springer-Verlag, Berlin, 2002.

[111] P. Schneider & J. Teitelbaum – "Algebras of p-adic distributions and admissible representations", *Invent. Math.* **153** (2003), p. 145–196.

[112] A. Scholl – "Higher fields of norms and (φ, Γ)-modules", *Doc. Math.* (2006), p. 685–709, extra vol.

[113] P. Scholze – "Perfectoid spaces and their applications", in *Proceedings of the International Congress of Mathematicians, Seoul, 2014*.

[114] ———, "Perfectoid spaces", *Publ. Math. Inst. Hautes Études Sci.* **116** (2012), p. 245–313.

[115] ———, "p-adic Hodge theory for rigid analytic varieties", *Forum Math. Pi* **1** (2013).

[116] P. Scholze & J. Weinstein – "Moduli of p-divisible groups", *Camb. J. Math.* **1** (2013), p. 145–237.

[117] S. Sen – "Ramification in p-adic Lie extensions", *Invent. Math.* **17** (1972), p. 44–50.

[118] J.-P. Serre – "Les espaces fibrés algébriques", in *Séminaire C. Chevalley, 1958 : Anneaux de Chow et Applications*, Secrétariat mathématique, Paris, 1958, exp. 1.

[119] ———, *Local Fields*, Grad. Texts in Math., vol. 67, Springer-Verlag, New York, 1979.

[120] ———, *Algebraic Groups and Class Fields*, Grad. Texts in Math., vol. 117, Springer-Verlag, New York, 1988, translation of the French edition.

[121] J. Silverman – *Advanced Topics in the Arithmetic of Elliptic Curves*, Grad. Texts in Math., vol. 151, Springer-Verlag, New York, 1994.

[122] *The Stacks Project* – `http://stacks.math.columbia.edu`, 2014.

[123] M. Temkin – "A new proof of the Gerritzen-Grauert theorem", *Math. Ann.* **333** (2005), p. 261–269.

[124] ———, "Functorial desingularization of quasi-excellent schemes in characteristic zero: the non-embedded case", *Duke Math. J.* **161** (2012), p. 2207–2254.

[125] R. W. Thomason & T. Trobaugh – "Higher algebraic K-theory of schemes and of derived categories", in *The Grothendieck Festschrift III*, Progr. Math., vol. 88, Birkhäuser, Boston, 1990, p. 247–435.

[126] J. Weinstein – "Gal($\overline{\mathbf{Q}}_p/\mathbf{Q}_p$) as a geometric fundamental group", arXiv: 1404.7192v1, 2014.

[127] J.-P. Wintenberger – "Le corps des normes de certaines extensions infinies des corps locaux ; applications", *Ann. Sci. École Norm. Sup. (4)* **16** (1983), p. 59–89.

ASTÉRISQUE

2015

370. *De la géométrie algébrique aux formes automorphes (II)*, J.-B. BOST, P. BOYER, A. GENESTIER, L. LAFFORGUE, S. LYSENKO, S. MOREL, B. C. NGÔ, éditeurs
369. *De la géométrie algébrique aux formes automorphes (I)*, J.-B. BOST, P. BOYER, A. GENESTIER, L. LAFFORGUE, S. LYSENKO, S. MOREL, B. C. NGÔ, éditeurs
367-368. SÉMINAIRE BOURBAKI, volume 2013/2014, exposés 1074–1088

2014

366. J. MARTÍN, M. MILMAN – *Fractional Sobolev Inequalities: Symmetrization, Isoperimetry and Interpolation*
365. J. LOTT, B. KLEINER – *Local collapsing, Orbifolds, and Geometrization*
363-364. L. ILLUSIE, Y. LASZLO, F. ORGOGOZO – *Travaux de Gabber sur l'uniformisation locale et la cohomologie étale des schémas quasi-excellents, Séminaire à l'École polytechnique 2006-2008*
362. M. JUNGE, M. PERRIN – *Théorie des espaces $\mathcal{H}_p$ pour des filtrations continues dans des algèbres de von Neumann*
361. SÉMINAIRE BOURBAKI, volume 2012/2013, exposés 1059–1073
360. J.I. BURGOS GIL, P. PHILIPPON, M. SOMBRA – *Géométrie arithmétique des variétés toriques. Métriques, mesures et hauteurs*
359. M. BROUÉ, G. MALLE, J. MICHEL – *Split Spetses for primitive reflection group*

2013

358. A. GETMANENKO, D. TAMARKIN – *Microlocal Properties of Sheaves and Complex WKB*
357. A. AVILA, J. SANTAMARIA, M. VIANA, A. WILKINSON – *Cocycles over partially hyperbolic maps*
356. D. SCHÄPPI – *The Formal Theory of Tannaka Duality*
355. J.-P. RAMIS, J. SAULOY, C. ZHANG – *Local analytic classification of q-difference equations*
354. S. CROVISIER – *Perturbation de la dynamique de difféomorphismes en topologie C^1*
353. N.-G. KANG, N. G. MAKAROV – *Gaussian free field and conformal field theory*
352. SÉMINAIRE BOURBAKI, volume 2011/2012, exposés 1043–1058
351. R. MELROSE, A. VASY, J. WUNSCH – *Diffraction of Singularities for the Wave Equation on Manifolds with Corners*
350. F. LE ROUX – *L'ensemble de rotation autour d'un point fixe*
349. J. T. COX, R. DURRETT, E. A. PERKINS – *Voter Model Perturbations and Reaction Diffusion Equations*

2012

348. SÉMINAIRE BOURBAKI, volume 2010/2011, exposés 1027–1042
347. C. MOEGLIN, J.-L. WALDSPURGER – *Sur les conjectures de Gross et Prasad (II)*
346. W. T. GAN, B. H. GROSS, D. PRASAD, J.-L. WALDSPURGER – *Sur les conjectures de Gross et Prasad (I)*
345. M. KASHIWARA, P. SCHAPIRA – *Deformation quantization modules*
344. M. MITREA, M. WRIGHT – *Boundary value problems for the Stokes system in arbitrary Lipschitz domains*
343. K. BEHREND, G. GINOT, B. NOOHI, P. XU – *String topology for stacks*
342. H. BAHOURI, C. FERMANIAN-KAMMERER, I. GALLAGHER – *Phase-space analysis and pseudodifferential calculus on the Heisenberg group*
341. J.-M. DELORT – *A quasi-linear Birkhoff normal forms method. Application to the quasi-linear Klein-Gordon equation on $\mathbb{S}^1$*

2011

340. T. MOCHIZUKI – *Wild harmonic bundles and wild pure twistor D-modules*
339. SÉMINAIRE BOURBAKI, volume 2009/2010, exposés 1012-1026
338. G. ARONE, M. CHING – *Operads and chain rules for the calculus of functors*
337. U. BUNKE, T. SCHICK, M. SPITZWECK – *Periodic twisted cohomology and T-duality*
336. P. GYRYA, L. SALOFF-COSTE – *Neumann and Dirichlet Heat Kernels in Inner Uniform Domains*
335. P. PELAEZ – *Multiplicative Properties of the Slice Filtration*

2010

334. J. POINEAU – *La droite de Berkovich sur* $\mathbf{Z}$
333. K. PONTO – *Fixed point theory and trace for bicategories*
332. SÉMINAIRE BOURBAKI, volume 2008/2009, exposés 997-1011
331. *Représentations p-adiques de groupes p-adiques III : méthodes globales et géométriques*, L. BERGER, C. BREUIL, P. COLMEZ, éditeurs
330. *Représentations p-adiques de groupes p-adiques II : représentations de* $\mathbf{GL}_2(\mathbf{Q}_p)$ *et* (φ, Γ)*-modules*, L. BERGER, C. BREUIL, P. COLMEZ, éditeurs
329. T. LÉVY – *Two-dimensional Markovian holonomy fields*

2009

328. *From probability to geometry (II), Volume in honor of the 60th birthday of Jean-Michel Bismut*, X. DAI, R. LÉANDRE, X. MA, W. ZHANG, editors
327. *From probability to geometry (I), Volume in honor of the 60th birthday of Jean-Michel Bismut*, X. DAI, R. LÉANDRE, X. MA, W. ZHANG, editors
326. SÉMINAIRE BOURBAKI, volume 2007/2008, exposés 982–996
325. P. HAÏSSINSKY, K. M. PILGRIM – *Coarse expanding conformal dynamics*
324. J. BELLAÏCHE, G. CHENEVIER – *Families of Galois representations and Selmer groups*
323. *Équations différentielles et singularités en l'honneur de J. M. Aroca*, F. CANO, F. LORAY, J. J. MORALES-RUIZ, P. SAD, M. SPIVAKOVSKY, éditeurs

2008

322. *Géométrie différentielle, Physique mathématique, Mathématiques et société (II). Volume en l'honneur de Jean Pierre Bourguignon*, O. HIJAZI, éditeur
321. *Géométrie différentielle, Physique mathématique, Mathématiques et société (I). Volume en l'honneur de Jean Pierre Bourguignon*, O. HIJAZI, éditeur
320. J.-L. LODAY – *Generalized bialgebras and triples of operads*
319. *Représentations p-adiques de groupes p-adiques I : représentations galoisiennes et* (φ, Γ)*-modules*, L. BERGER, C. BREUIL, P. COLMEZ, éditeurs
318. X. MA, W. ZHANG – *Bergman kernels and symplectic reduction*
317. SÉMINAIRE BOURBAKI, volume 2006/2007, exposés 967-981

2007

316. M. C. OLSSON – *Crystalline cohomology of algebraic stacks and Hyodo-Kato cohomology*
315. J. AYOUB – *Les six opérations de Grothendieck et le formalisme des cycles évanescents dans le monde motivique (II)*
314. J. AYOUB – *Les six opérations de Grothendieck et le formalisme des cycles évanescents dans le monde motivique (I)*
313. T. NGO DAC – *Compactification des champs de chtoucas et théorie géométrique des invariants*
312. ARGOS seminar on intersections of modular correspondences
311. SÉMINAIRE BOURBAKI, volume 2005/2006, exposés 952–966